DEATH FACTORY USA

"Terrorism at Home" "Our Environment" "Death Factory's" "In Our Midst"

By

**"Gerald and Marilyn Pierce (Authors)
Larry D. Land (CoAuthor)"**

This book is a work of fiction. Places, events, and situations in this story are purely fictional. Any resemblance to actual persons, living or dead, is coincidental.

ISBN: 1-4033-1031-9 (e-book)
ISBN: 1-4033-1032-7 (Paperback)
ISBN: 1-4107-1085-8 (Dust Jacket)

Library of Congress Control Number: 2002096478

This book is printed on acid free paper.

Printed in the United States of America
Bloomington, IN

1stBooks - rev. 12/31/02

TERRORISM AT HOME
DEATH

FACTORY

U.S.A.

"OUR ENVIRONMENT"

A true story of Intrigue, Suspense, Horror, Love, Honor, and Patriotism, For the people and our Beloved Environment.

A story about the Rocky Mountain Arsenal "The Most polluted piece of Land on Earth.

"NOT SINCE RACHEL CARSON'S SILENT SPRING, NOR LEWIS REGENSTEIN'S AMERICA THE POISONED, HAS A BOOK SO DYNAMICALLY DOCUMENTED THE HAZARDOUS BIO-CHEMICAL WARFARE THAT THIS NATIONS GOVERNMENT, THE DEPARTMENT OF DEFENSE THE UNITED STATES ARMY, THE PENTAGON, AND THE INDUSTRY GIANTS SUCH AS SHELL OIL COMPANY HAVE WAGED ON IT'S OWN PEOPLE.

LARRY D. LAND

A TRUE STORY, AS SEEN, HEARD, AND LIVED BY ONE MAN. CO AUTHOR OF OUR ENVIRONMENT WITH DEATH FACTORYS; IN OUR MIDST.

A SALUTE TO THE VETERANS OF THE PERSIAN GULF WAR AND THEIR FAMILIES, SEE-10.

"INTRODUCTION"

"YOU MUST SEE IT, AND READ IT, TO BELIEVE IT!

We the people of the world, are not here to predict the future but to change it for the good. We are not here as helpless creatures, but as son's and daughter's of Adam, and Eve, capable of affecting our own fate, our own destiny.

We are not here to avoid decisions but to make hard, decisive choices between good and evil by using an ethical system not invented by man, but by our own creator. A framework of truth and moral guidelines through which we can find deliverance from despair.

We are not here to glorify in ourselves, but to glorify in that which he gave us, and in he who made us all, and who will eventually judge each of us on how well we did on the journey we all take but once.

There can be no excuse for what happened at the Rocky Mountain Arsenal, and to the people of the United States Army, and Shell Oil Company who let this happen, 'may God have mercy'.

BY Larry Land: "I believe the Government once told the Army," "NEVER TELL A LIE...UNLESS LYING IS ONE OF YOUR STRONG POINTS."

George Washington Plunkett Lloyds of London Insurance Representative stated at trial, "There is a fundamental principle of liability, Insurance Companies pay for mistakes. What Shell Oil Company did over the period of thirty years at the Rocky Mountain Arsenal Denver Colorado, was no mistake. It was deliberate and intentional pollution to save money. Shell knew how to do it right but deliberately chose to do it wrong."

Larry says, "Shell Oil Company either deceived the Army or there were massive pay offs (under the table), and together the Army and Shell Oil Company deceived Congress. The pentagon was aware of and advised the Army that Basin F was leaking into the underground auquifers off the Arsenal… by 1969 had already polluted 35,000 acres off the arsenal, yet the pentagon, along with the Army deceived Congress."

Larry says, "BLESSED ARE THE YOUNG, FOR THEY SHALL INHERIT, A TOXIC AND CHEMICAL POLLUTED ENVIRONMENT HERE ON EARTH."

Larry says lets change it, “BLESSED ARE THE YOUNG FOR THEY SHALL INHERIT, A TOXIC AND POLLUTION FREE ENVIRONMENT.

Larry says, “The Army and Shell, operate under the same Government theory,” “IF YOU CAN’T CONVINCE THEM, CONFUSE THEM.”

HARRY S. TRUMAN

AUTHORS: GERALD & MARILYN PIERCE, CO AUTHOR LARRY D. LAND

There is a fundamental principal of responsibility, mistakes can happen, can be accepted but what has happened with the Government, the Army, the pentagon, Shell chemical company, the Rocky Mountain arsenal Denver Colorado, and Industry giants every where, what you have done was no accident.

AUTHORS NOTE

The story you are about to read is a story of pain and suffering. It is also one of love. Love for ones family and love for humans everywhere.

It is the story of one man and a woman, and the Rocky Mountain Arsenal, United States Army.

The entire history of the Rocky Mountain Arsenal has been one of Malfeasance, Reckless, and Malicious Lies to purposefully cover up Negligent, Wrong doing. One of Misinformation, or no Truthful information at all. They seem to hope our concerns will simply go away, or things just happen to disappear. They have many P R people who are well trained in not telling the truth.

When we decided to write this story, our own personal histories stirred our curiosity. We were warned repeatedly against this endeavor, but felt that its importance was worth the risk. There have been many attempts on Mr. Land's life. Attempts that the government would like to refer to as accidents when they were confronted. Fortunately, in most of these attempts, Mr. Land survived only because of his training, and decisive discipline. Twice the real miracles of life came through, and saved Mr. Land, as if it were known he had much more work to complete.

What I am saying is that some people in high places do not want this story told. If there is nothing to hide, then I ask why not.

As research began we found ourselves buried in mountains of documentation as to the history of this facility. While we thought we knew what the arsenal was all about, we discovered it was not simple, and honest. It was very complex, covered by Government Bureaucracy, dishonesty, lies, and untruthfulness. Our eyes were opened as we learned the hidden story of this issue. We felt it was important that the reading public realize what was really going on behind the cover up of this facility, and others across this country. Even more important, was that people realize that this is not unique to the State of Colorado, Denver or Adam's County. Facilities such as this exist all over the country.

To us, even more important is the realization that toxic wastes do no just go away. It's like the Ostrich burying its head, the

Government did as well, while they buried billions of tons of deadly chemicals. While various agencies speak of "Clean Ups" and "Super Fund Sites," the chemicals that were created over the years do not cease to exist. They may be placed in containers, placed in underground storage sites, or moved from one open basin to another, or try force pumping it under ground causing earthquakes, but they all still exist to be dealt with at a later date. All we have done is to put off the inevitable. As was stated at the Rocky Flats Nuclear Weapons Plant, and the Rocky Mountain Arsenal, a chemical weapons manufacturing center. Clean up in most cases even carries a multiplier with it. To clean up hazardous wastes, we in fact, create even more waste.

As a result, we feel it is important that the people of this country understand that we have created a monster over the years that theoretically can not die.

We freely admit that there was a time when those who experimented with various forms of toxic agents, did not realize the ramification of their actions, not knowing how to take these toxic chemicals apart, and back to their non toxic state. There was not the environmental awareness that we have today. However, all of that has changed dramatically today. With much emphasis placed on the disposal of toxic chemicals, and wastes.

Government and corporate officials can not plead ignorance any longer, but by the same measure the American public can no longer ignore their actions by playing dumb and blind.

The actions of certain Federal, State and corporate leaders can no longer be tolerated when the lives of our citizens are at risk. It has been proven many times in the past that they simply do not care and are only interested in profitability.

Our children have the God given right to live their lives unfettered by the poisons left over from the greed of our trusted leaders.

It is our hope that the eyes of the public will be opened permanently and that we will have served a purpose in telling this story as we continue our part in helping to clean up a poisoned nation.

AMERICA IS POISONED, HOW CAN WE SURVIVE

HOW DEADLY
CHEMICALS HAVE
BEEN DESTROYING
OUR ENVIRONMENT,
OUR WILDLIFE,
OUR SELVES,
OUR CHILDREN, OUR
FUTURE.

The arsenal produced sarin in the 1940s and 1950s, Hueneseld said. It built 1.6 million nerve gas bombs.

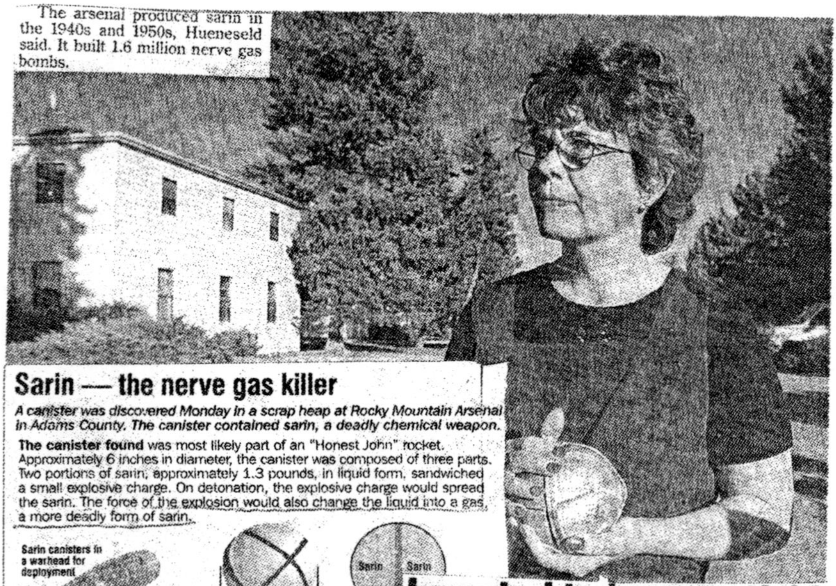

Rocky Mountain Arsenal spokeswoman Ruth Mecham holds a sample nerve-gas container during a news conference Friday. Deadly sarin nerve gas was found by cleanup workers Monday, and officials are debating how best to dispose of it.

Ahmed Terry

Sarin — the nerve gas killer

A canister was discovered Monday in a scrap heap at Rocky Mountain Arsenal in Adams County. The canister contained sarin, a deadly chemical weapon.

The canister found was most likely part of an "Honest John" rocket. Approximately 6 inches in diameter, the canister was composed of three parts. Two portions of sarin, approximately 1.3 pounds, in liquid form, sandwiched a small explosive charge. On detonation, the explosive charge would spread the sarin. The force of the explosion would also change the liquid into a gas, a more deadly form of sarin.

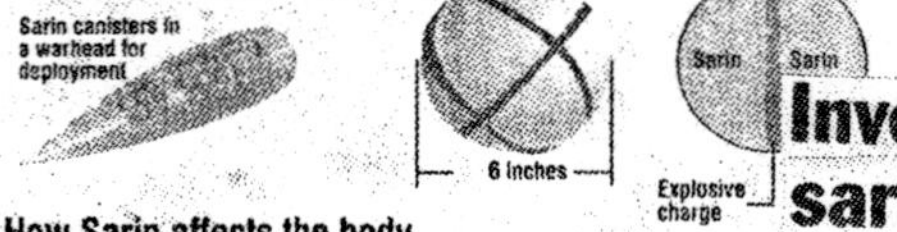

How Sarin affects the body

Developed in Germany in 1938, Sarin is a colorless liquid with almost no odor that disrupts the normal workings of nerve cells in the body and can cause death.

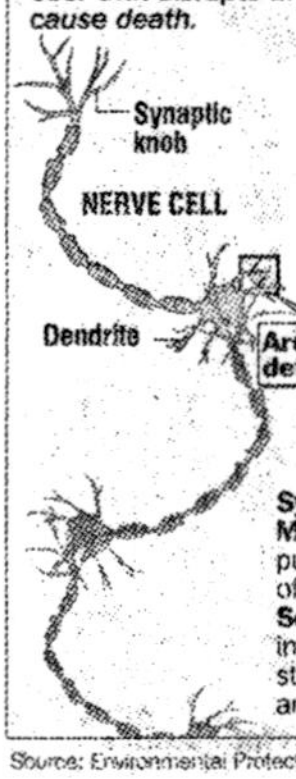

Normal nerve cells pass signals from the synaptic knob of one nerve cell to the dendrite of another. These electrical impulses are passed between cells by neurotransmitter molecules.

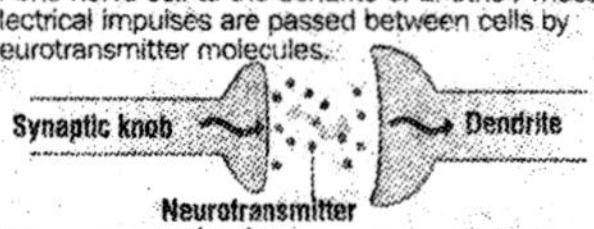

Sarin gas blocks the enzyme that controls when the emission of the molecules is halted. The effect is that the nerve cells continue to send signals when they should have stopped.

Symptoms:

Minor exposure — sweating, muscular twitching, pinpoint pupils, running nose, tightness of chest with shortness of breath and dim vision

Severe exposure — headache, cramps, nausea, vomiting, involuntary defecation and urination, twitching, jerking, staggering, convulsions, drowsiness, coma, respiratory arrest and death

Source: Environmental Protection Agency, Denver Rocky Mountain News research

News staff

Invented to be a pesticide, sarin is 'a horrible way to die'

By Ann Imse

News Staff Writer

Sarin nerve gas was invented as a pesticide in 1930s Germany, and acts like a massive dose of Raid on humans.

Victims stagger and jerk as all muscles convulse, and they die of respiratory arrest, said Jason Pate, a senior research associate at the Center for Non-Proliferation Studies at the Monterey Institute.

"It's a really horrible way to die," Pate added.

But it dissipates quickly, and victims can be treated. Atropine relaxes the muscle contractions, and diazepam — better known as Valium — eases the convulsions, Pate said.

5th nerve gas bomb destroyed

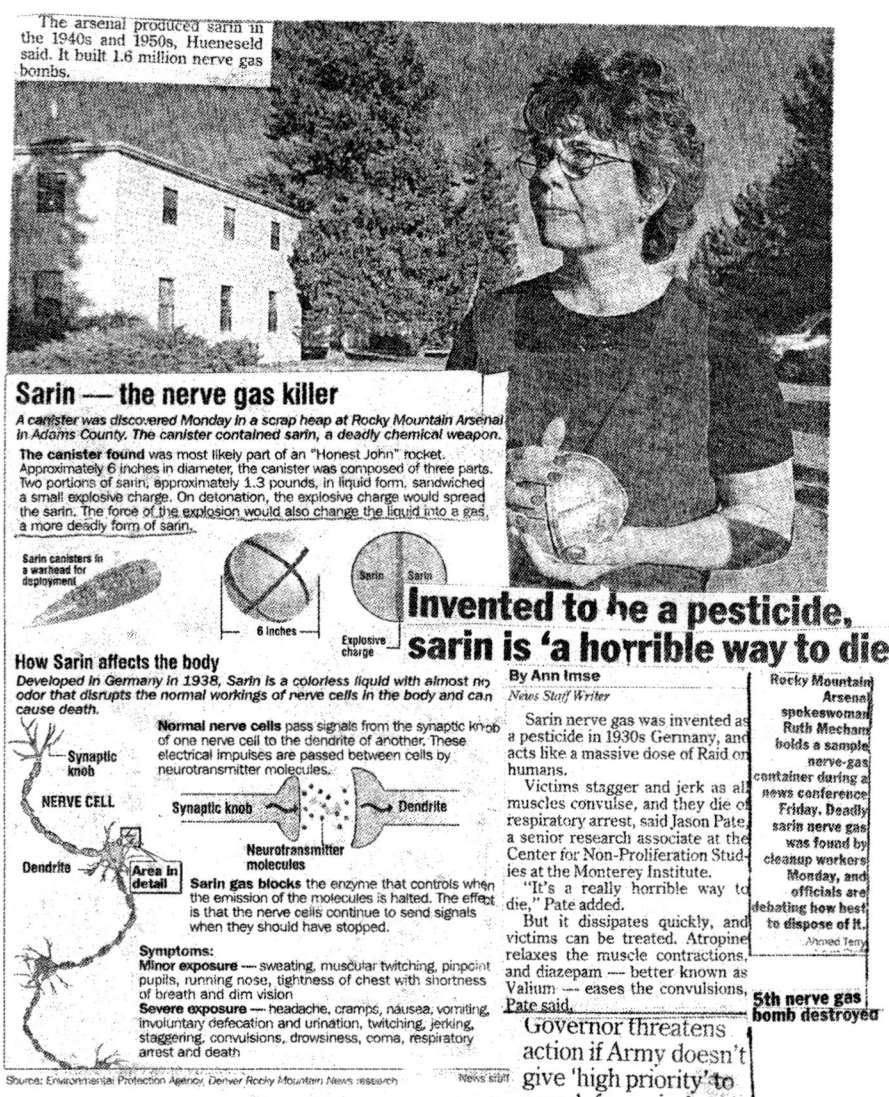

Governor threatens action if Army doesn't give 'high priority' to search for more bombs

Poisoning Our Environment

WHERE HAVE THE TREES GONE
WHERE HAS ALL THE BIRDS GONE

AMERICA'S NATIONAL SYMBOL ARE NEARLY EXTINCT
BECAUSE OF TOXIC CHEMICAL POLLUTION

CHAPTER ONE

"LOOKING BACK AT THE MOST HORRIBLE TRAGEDY KNOWN"

Man has mounted science and it is now running away. I firmly believe that before many centuries more, science will be the master of man. The engines he will have invented, the chemicals he will have made, will be beyond his strength to control. Some day science shall have the existence of mankind in its power, and the human race may commit suicide, by blowing up the world, and/or poisoning it, and all of us to death.

Henry Adams, April 11, 1862.

The old house stood on the crest of a small knoll two hundred feet from the road. No shrubbery or trees obscured the view, nor was there any growth for more than a mile beyond in all directions.

The house was old, weather-beaten, wind torn and long unpainted. By night, lights no longer showed from any window.

Behind the house, a few miles away, the mountains lifted abruptly with steep, ragged slopes broken by ledges and covered by beautiful mountain grasses and timber, and wild flowers.

It was cold in the old house now, and the rooms were dark and shadowed. A little of the light of evening filtered through the old windows as he stood, in what, once was his families' living room. The house was dusty and the air was stale and old.

He thought of his parents, and how they suffered before dying from the poisons of the Rocky Mountain Arsenal, and Shell Oil Company. Not only was the soil poisoned, also was the air and water. A feeling of despair filled his heart as he thought of their memory.

They were gone now, their love reduced to many painful memories and ashes. Tears clouded over his eyes as he looked back, and remembered. His mother and father would never grow old together. They would never see the changing of the seasons that they both had loved. There would be no more springs or summers. They would never again see the gently swelling of the flower buds out on the prairie just before their colors burst forth, or the wild grasses blowing freely in the wind. They would never again see the gradual turning of the leaves as the chill of winter began its deadly grip on the landscape. Now, there were only the memories to a son that loved them dearly.

Walking outside he stood on the porch for one last look around. Off to his right was where a huge old tree had once stood. The tree supported a tree house and tire swing that his father, brothers, and he had built. His fathers two strong hands gripping boards alongside his own young hands, and that of his brothers. We liked that old tree house built in the tree so high, we could look out over the prairie, and see the city of Denver in the horizon to the south, and the mountains to the west.

Walking through what use to be their front yard, he walked across the road down past his old country school "Cactus Ridge". Larry used to ride his horse to and from school nearly every day in the 1940's, and early 50's there were no school buses. The school provided a stable that we all had to maintain to keep our horses safe while we attended school. Rain, shine, snow, or sleet, we made that half mile to school every day. He walked that extra half-mile to the Army's new barrier. A feeling of sadness and desolation seized his insides like a powerful clenched fist, he was trembling and shaking with in.

A slight breeze cooled his already wet cheeks as the tears continued to roll down his face. He remembered his mother and father's pain and suffering and later, that him self, his wife and children, friends and neighbors, and that of other family members.

As a child of the 1940's and 1950's. It was 1953, 1954 before we finally got electricity we experienced television for the very first time. It was like a new world had just opened up for us we put the candles and kerosene lamps up we had electrical lighting. it was real soon we finally got in door plumbing, no more going out in the middle of the night to the out door toilet. There was no more going out side to pump water by hand several times a day especially when it was bath day we used to pump until we were blue in the face, taking buckets of water in so our Mom could heat the water on the old wood stove, and fill the big round bathtubs. At night it was like going to the movies if we had all our chores done we could stay up till ten o'clock watching westerns of Gene Autry and Roy Rogers, this alone left us believing that good always prevailed over evil, that the good guy always won. In the movies of the period the good guy always wore white and the bad guy wore black. It was like the good guy with his ten gallon hat, his country stood for truth and justice for all. He strongly believed in the red, white and blue. In his eyes, the eyes of a small boy, the United States of America could do no wrong. We were always up before sun up to do our chores, milking the cows, feeding the chickens, feeding the pigs, chopping the wood, when that was done we went to school, but on weekends as long as our chores were done we watched the Saturday kids programs, but Sunday we went to Sunday School and to church.

His country stood for decency and had a high moral character that he thought existed no where else in the world. As a boy, he believed America would always do the right thing because his government was made up of good, honorable people, who were like his father. He believed that a country such as this had a responsibility to defend the rights of people no matter where they came from or what color their skin. He was raised with the teachings that all people were the same and entitled to the same rights, happiness and piece of mind that he enjoyed, all because we had a constitution that guaranteed that for all.

Now years later, he was face to face with the error of his ignorant beliefs. People were not treated the same and the United States did not always do what was right. They did not always follow the constitution, they just changed it. They lied and covered up the mistakes they made, it was backed by Congress, the Senate, and the president of the United Sates of America. Even yet today in the millennium, Congress began it's enforcement of a clean up, by the 1970's. Since Mr. Land was able to take this problem to the Congress of the United States Washington D. C. Mr. Land expounded this in D.C., and across the world spurring a global recognition of a very serious and disastrous problem.

Even fearing for his life, Mr. Land felt as if he had reached a point of no return. He tenaciously, and without fear pursued the Government, and targeted large uncaring corporations. He issued the first cease and desist order of the time against the United States Army/United States of America, and to Shell Oil company. When Shell Oil Company knew their days were numbered, and in court where they tried their best to attain permits from the State of Colorado to pump hazardous waste in the Lakota, and the Dakota sands in Brush, and Morgan Counties over 12,000 feet under ground, knowing that this type of disposal had already caused massive earthquakes near Denver Colorado. Yes Mr. Land had done his homework, and during the trial acting as his own attorney, along with seventeen attorneys sitting in to back Mr. Land if necessary. Mr. Land cross examined all the so called expert witnesses of the time for Shell oil company.. actually stultifying the experts. They could not supply the answers that would satisfy the court. For the record after demanding the witness answer the question, when he could not Mr. Land answered the question for the court. At the end of the hearing, before a decision was made by the court. There were several news cameras in court, when a man believed to be an attorney for Shell Oil Company stepped up to Mr. Land with a TV camera in his face stated "You are the most radical little S___ of a B____ that we ever had anything to do with. Mr. Land stated "Well thank you". When Mr. Land stepped out into the corridor of the court room, he had at least one hundred and fifty people actually line up so they could just shake his hand and let him know what a great job he had done. Certain members of the Colorado State Health Department let Mr. Land know how proud they were, and stated you are a hero to us. In 1975 the courts of Colorado rejected any permits to Shell oil for pumping anything under the ground. By 1976, the state of Colorado had cited both the United States Army, and Shell Oil Company with immediate cease and desist orders, upheld by the Governor, and the State Congress, and the Senate. It took another seven years of court battles, while Shell Oil Company not only doubled, but tripled and quadrupled their output of product the most deadly chemicals in the world, and the United States of America supporting this and allowing them to dispose of these chemicals in a basin that the United States Department of Health had banned it's use for any thing in 1965, because it had already polluted 30 square miles of Land to the north, northwest and northeast of the arsenal. finally the doors for both the United Sates Army, and Shell chemical Company were closed for good In 1983. That same year Mr. Land stood with our President of the United States of America Mr. Ronald Reagan as he signed a document disallowing the manufacturing of any nerve agents in the United States, ever again.

Mr. Land was back in Colorado again, staring at part of the great American prairie just outside of Denver, where in the 1970's he had began his very successful cattle operation, that is until they were killed by the malicious, reckless negligent acts of the United States Army at the Rocky Mountain Arsenal. He was witnessing first hand the end result of his country's inability to do what was right.

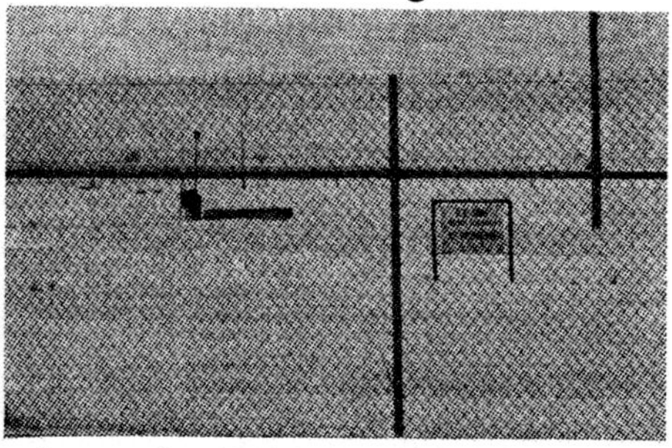

Over the centuries the prairie had remained relatively unchanged. A great elm or cottonwood grew here, or bush died there. An animal would sometimes leave its sign to show of its passing. The constants of this land were the sun, the stars and the wind which moved endlessly over the prairie.

Over the years the rains pounded at the land, yet, it was the wind that always troubled it. Some how mother nature had always fought to keep her land clean and pure. She had written her laws millenniums ago, which all living creatures including man, must obey or cease to exist.

All things living on this planet would die eventually. It was the way of nature. Take the young tree before him, growing along the fence. As it has happened many times in the past, it would soon be old enough to drop its seeds and with nature permitting, give new life to the prairie. Maybe with the aid of a gentle breeze, a seed would fall into a crack like the one by his foot, find nourishment and begin to grow.

Water from a late fall rain would find its way into the crack, and the cold winter would freeze it, expanding the crack still wider, giving it more room to grow. The growing tree, over many years would thicken its roots, growing stronger like the mature one just a short distance away.

Some lived and grew strong, while others would wither and die, to be eventually reclaimed by mother nature.

Suddenly, out of the corner of his eye something captured his attention. Lifting his eyes to the tree again, he watched as a fox, apparently more dead than alive, dragged itself partially under the branches of the old tree.

Even at that distance he could tell it was in pain. As he continued to watch he began to hear the animal whimper as it lowered itself to the ground, resting in an unnatural position.

Unlike humans, the fox had no awareness of history beyond the memory of where his food had been obtained in the past. Nor had he any realization of the events that had taken place on this piece of land he now called home.

This one, like all the other animals the man had seen, was sick. In the time since coming here the fox's health had steadily grown worse,, something was happening to him and he knew not what. The fox obviously thought nothing of this, rather his instincts warned him that he would not live much longer. He had grown weak. Hunting was no longer easy. Yet, now he rarely thought of food. What filled his mind now was the pain deep inside his body and the thought of being able to rest for maybe just for a short while.

Crawling still further under the branches which protected it from the weather, it was apparent that the fox would soon be dead. Like all living things, it would die alone and afraid, then in due time, it would be reclaimed by nature's land, still cowering under the sprawling branches of the tree.

The age old laws of mother nature were not to be taken lightly, but somehow, something had changed. Before his eyesight began to fail the fox had seen many animals come to this land, a land that seemed to offer so much, but in the end, offered only death. They had all made this same mistake. Now, like the rest, the fox would soon die, paying the supreme price for his innocence.

From where he now stood, he knew this land to be the most devastating environmental disaster in American history. While there are other Governmental Agencies that have shown the Army, had purposefully, and with malicious negligence, created the most contaminated piece of land on earth. There is over 18,000 acres and it sits just north of Denver Colorado less than a mile. It can never be used for anything. The Government has been trying to clean it up now for over twenty (20) years. They have spent billions and billions of tax dollars cleaning up their negligent ways. The people of Denver are really lucky, most of the winds are from the south and east, to the north and west keeping the air pollution away. Then all the various water tables begin on the south and west sides of Denver and flow away from Denver to the North and east keeping their water untouched. When construction began in the 1940's I'm sure the Army had thought of this. Since we only have a few farmers living to the north and north west, they were surely thought of as being expendable exactly like they were in later years.

He also felt that unless he told someone about what he knew of this place, few Americans would know the true story of what was happening in their country.

What started as wartime expediency had now become bureaucratic carelessness in peacetime, and a persistent nightmare for thousands of American Citizens.

On this once productive piece of the high plains, nestled comfortably in the shadows of the eastern slope of the Rocky Mountains, the United States Army, along with Shell Oil company, deliberately, for the sake of efficiency and profit, had turned this area into a toxic wasteland.

This area became a twenty-seven square mile federal installation and has been called by the Environmental Protection Agency within recent years, the most contaminated land in this nation, and the world. It has also become the largest and most expensive Government cover up in our nation's history. By none other than the United States Army, and a lot of yes people. The first facilities at the new Rocky Mountain Arsenal were completed in1943. The sole means of waste disposal was a low spot in the natural drainage of the land which was modified into basin A by the construction of dike on the lower portion of the drainage area. Although the soil in the Arsenal is wind laid sandy and loamy soils, the record fails to reflect that there were No percolation testing conducted by the Army prior to the construction and use of this basin. The record also fails to disclose any hydrological studies prior to or along with the initial use of the basin.

The facilities constructed, now known as the South plant area, were used during World War II for the manufacture of chlorine, Mustard and lewisite agents. Materials placed into Basin A during World War II consisted of chlorine, Mustard, Lewisite, sulfur, sulfur mono chlorine, Ethyl alcohol, and Mercury, Sodium Hydroxide, water and other associated materials. At that time there was also work being done on many other nerve agents such as G-B Nerve agent, Sarin Nerve agent. Secrets brought over by captured German Scientists, was the exact nerve agent Germany had already manufactured, and was using to gas the Jewish People.

Welcome to the Rocky Mountain Arsenal, sprawled across this once productive piece of the high plains, nestled comfortably in the shadows of the eastern slope of the Rocky Mountains. The United States Army, along with Shell Oil Company, deliberately, for the sake of efficiency and profit, turned this entire area into a toxic wasteland. People getting sick, and dying they for years didn't listen, didn't hear. "Did not want to hear or listen. Only one Congressman even bothered to listen, the others all turned a deaf ear while this was happening. I can remember trying to see then senator Gary Hart, who not only turned a deaf ear, but he also turned blind.

This became a twenty-seven square mile, federal installation and has been called within recent years the most badly poisoned land on earth. With hundreds of enormous waste pits where drums are stacked ten (10) to (15) high. Then the disastrous waste ponds as they were called, covering hundreds, and hundreds of acres, holding hundreds of millions of gallons of the most toxic chemicals known to man. These chemicals evaporating into our air thousands upon thousands of gallons of this liquid daily.

What happened here is a story of human ignorance, military and corporate indifference, and plain, old fashioned American greed and ignorance. A story that will not end in our lifetime, or our children's, or our children's children. Welcome to the Rocky Mountain Arsenal, just like a powder keg sitting just north of the sprawling city of Denver, Colorado, and just south of Brighton Colorado. The Army has so many unregistered pits, they don't even know what is in them, they themselves are afraid to find out, or even open the pits.

My name is Larry Land and I grew up one mile north of the arsenal. I would like to tell you a story, a story so terrible, so unbelievable, that the few people who know the truth, that is the whole truth, still cringe in fear when word is spoken of it.

To some people the Arsenal meant vast profits, to others it meant financial freedom every week in the form of a paycheck. Yet to others, it meant sickness, lost savings that had taken a lifetime to build, a slow brutally painful death.

With the ever present threat of chemical warfare staring the United States in the face, the question kept being asked by Army leaders was not “IF” the enemy would use chemical gasses, but when, since they were already being used and experimented with. Mustard agents, and Lucite were already being used, very powerful burning, and suffocating agents, used by both the United States, and Germany.

An answer had to be found to combat this threat and give the United States a decisive edge in the World war II.

On May 23, 1942, when the United States Army was battling the nazis in Europe, they announced plans to build a ten-million dollar chemical warfare plant just northeast of Denver. Local residents were given just sixty-days to remove themselves, their machinery and their livestock off of the newly seized Federal Property. As soon as the Army moved in, construction was started at a feverish pace, 24 hours a day, seven days a week non stop. Promising many civilian army jobs to those in Denver, and Brighton.

On January 1, 1943 the first of more than one-thousand army workers filed into the new plant and produced the first stockpiles of Levinstein, Mustard gas and Lewisite. These were first manufactured and used by the Germans, then the United States obtained the recipes from German defectors. These three chemicals were designed to inflict horrible pain by blistering the skin and searing the eyes of enemy troops permanently. They have said that none of these three chemicals were used during World War II. Other arsenal facilities manufactured more than three-million white phosphorus shells, M-47 napalm bombs, and M-74 incendiary bombs that triggered massive firestorms over Tokyo. In addition, millions of munitions were manufactured and filled with phosgene and mustard distillation. Phosgene another chemical never used. Classified records were located that indicated, the Army at the Rocky Mountain Arsenal had in fact manufactured near 500,000 tons of phosgene and put in 55 gallon containers for storage and shipping. Phosgene was never used and records indicate a huge pit was dug near Brighton Colorado, large enough to hold all the phosgene that was made, There are no records now as to where this huge pit is, or what it may do.

From 1945 to 1950, the Arsenal's principle operation involved the reconditioning of one ton containers and the demilitarization of approximately one and a half million Mustard-filled shells. They also allowed other chemical companies to use Arsenal property to dispose of chemicals, later proved to be some of the deadliest in the world.

As a boy, Larry can remember the continuous testing that was done with bombs and artillery shells that were detonated on the Arsenal property. He, and his family would see large pieces of shrapnel, fire, dust, and smoke rising into the air. In his bedroom at night, he lay trembling under the blankets, as small earthquakes shook his home apart. During the day he could remember the airplanes as they rumbled across the sky, sometimes from horizon to horizon. In large v shaped formations the planes dark forms covering the sky. The Rocky Mountain Arsenal, this American death factory, was virtually in his backyard. If it weren't for the bombs, and the real earthquakes, the planes made it feel as if their were a constant earthquake.

At that time in America's history, no thought was given to the environment, we were at war. Disposal of the chemical waste derived from these death gases, was simply piped through clay pipes into huge unlined basins or put into tanker trucks or fifty-five gallon drums and dumped into trenches. These trenches were made with huge earth moving equipment. The clay pipes were eaten away almost immediately, and they continued the use for nearly 30 years until it was brought to their attention in the 1970's as Mr. Land is pointing out so many negligent wrong doings. This equipment cut severe wounds into the earth that exposed the gravel that made the upper stratum of the aquifer. These trenches were 100's of yards long, and 50 to 100 yards wide and 20 to 60 feet deep. Inside were placed thousands of 55 gallon drums containing poisonous waste and finished products that were either over produced or did not meet specifications. The drums were stacked from 10 to 15 high then buried. In a lot of cases no records were kept, of what or where. The tanker trucks would dump their waste in large unlined basins. with the clay pipes not transporting the waste to the basins, allowing 250,000 gallons a day of the most hazardous poisons known to man just leak into the ground, and the water

Pesticides and Herbicides were also placed into trenches. Neither the environment nor the local populace were ever considered. Even during the Vietnam war when it was rumored that Agent Orange was being produced at the Arsenal, and into the early 1980's, was there concern for the environment or the local citizens of the area. "NO" It was totally nonexistent.

In the mid 1940's his father was still following through with the government recommendation that they plant trees. This was and effort to block the wind, which was destroying the top soil as it moved with the harsh winds, much in the same way that it did during the "dust bowl day's of the 1920's and 1930's.

They planted trees up fence lines, down fence lines, dual lines of trees. they were planted around water holes, along the roads, and across the center of fields, wherever they could. Larry's Dad was a great believer in this because his government was supplying many of the trees and had asked him to do this, so it had to be right. All the various pine trees, elm trees and cottonwood trees were supplied by the government.

They planted apple trees, Peach trees, cherry trees, plum trees, both red and purple and they planted blackberry, raspberry and gooseberry bushes, it was like a forest where we lived.

This was a fantastic place for a young person to grow up. We were all very close, and close to our parents, what a life here in the country. We grew what seemed like everything we needed, and we raised our own, chickens, ducks, turkeys, pigs, and live stock too, several cows for milk, whole cream, buttermilk, butter, buttermilk, and ice cream. The wonderful thing about this is we grew everything and we made everything. We grew carrots, horseradish, potatoes, tomatoes, onions, cucumbers, pickles, lettuce, cabbage, corn, squash, pumpkins, peppers, and we also had an old dirt vegetable cave to store enough for the winter.

The hardest part of this whole thing was hoeing the weeds, they seem to try and outgrow what you had planted or even the trees, row after row, if the weeds got to big to hoe we then had to pull them by hand. Good thing I had a lot of brothers and sisters, we always wore gloves, but sometimes they would wear and the next thing to wear was our hands. we worked everyday till our hands bled, bandage them up then keep going. the weeds had to be hoed, away from the young trees as they were growing to give them the best possible chance for survival. His father would say, "These trees cost a lot of money, we are not going to waste them by not taking care of them."

By the mid to late 1950's the trees began dying. Each year a few more trees died. Each year it became harder to grow even their garden. The plants would start good but before long they would wilt and die, the fruit trees began to die not producing any fruit. Even the animals looked very scruffy, and did not produce.

Now, he looked on with saddened eyes and a lump in his throat. Where the trees had once stood, there was nothing, just empty prairie. They were all dead now. Killed from the poisoned water, the toxic air they breathed and the contaminated soil they sat in.

In 1954 under a lease agreement with the Army, Shell Oil Company, a company owned by the Royal Dutch-Shell Group of Companies based in London and the Hague, the Netherlands, leased part of the Federal Land and started producing pesticides. Aldrin and Dieldrin, and Endrin. Aldrin and Dieldrin were so lethal that their use was banned in the United States. However the manufacturing of these chemicals is still allowed and sold to South American Countries where it is used on their crops, which in turn are then exported back to the United States in the form of citrus, fruits, vegetables, and other edible products that we are again unwittingly consuming. These two chemicals follow the entire food chain, from plant to fruit and back to humans.

Soon after Shell moved in, an odor periodically filled the air. It was so bad that it would actually make their eyes water, and be crusted shut by morning. It also it very difficult to breathe. It seemed it would always begin at night. They had no idea what was causing their discomfort. Larry did not think that anyone had any concept as to the toxicity of their neighbor's products. All they knew was that something was wrong.

Being a child, he was not aware as his parents were of the subtle changes taking place around them. His world seem to be much as it had always been, just different. However, he does remember one evening when they sat down to eat. His father questioned his mother about how she had washed the dishes. Could there be soap left on them were they rinsed properly? He said that the food tasted funny. He had to admit that there was something odd about the flavor. His mother was a good cook. He never heard complaints at their dinner table before. Looking back, Larry thought that his father slowly came to the realization that something was wrong, and he believed that his father suspected something was happening to their well. Although he doubts that his father had any idea as to the scope of the problems they were, actually about to face. During the 60's and into the early 70's as Larry could see his father's physical and mental health deteriorate.

This was not the case however and their problems became worse with time. Finally his father lodged a formal complaint against the Army. He felt they had something to do with the problem though he couldn't put his finger on it. His sparring with the army went on through the late 50's, and all through the 60's,. His father would only see a short part of the 70's, but the fight and the problems would only intensify as the evidence piled up against the arsenal, the United States Army, as well as Shell Oil Company. Larry kept his father's fight going all the way to Washington D. C.. Larry lived by the saying, "It's not over till it's over".

Little did his mother and father know, that the water they were drinking, would literally be the cause of their deaths. The same water that has flowed beneath the western prairie since time began. Endless aquifers that had given life to this land.

On the prairies of Colorado you see the twisted and gnarled trunks of the cottonwood trees, and the huge Chinese elm trees. They seem to typify the tough resolve necessary to survive in this country. Some of them stand with hollow trunks. Their insides hollowed out by disease and drought, Yet they continue to stand and each spring they bring forth the soft green growth that will provide shade for yet another hot Colorado summer. Even the Cottonwoods would be hard pressed to endure the introduction of the toxic invaders into their life's blood, the water.

Larry's parents stood like the cottonwoods, and elms, proud and fiercely independent. Determined to make it on their own. They too drank from the water which came to them through the well which supplied their home.

He applied to the State of Colorado and the Tri-County Health department for help and they did test the water for bacteria. The found none. However, no one knew that there would be no bacteria where pesticides were contaminating the water. The Health Department could not test for pesticides and herbicides because it was classified by Shell Oil Company, and the United States Army, because from an inside source Shell Oil Company was under contract and was believed in making seven different types of agent orange for the Government and what ever experimenting they wanted to do was also classified by the Army. Corporate business became the Army's partner in crime. The nerve agents, the mustard agents, was also classified top secret by the military.

His father was the first to show signs of an unknown but debilitating illness. He under went exploratory surgery. His diagnosis was not good. The doctors did not understand why he had such a serious problem with colitis, and inflammation of the colon. His liver, kidneys, spleen, small intestine, stomach area and lower lungs were all inflamed and no one knew why. They began by putting him on many different diets, thinking it was something he was eating. Most of his food was cleaned, soaked, boiled, or stewed in water taken from the tap.

Eventually he was placed in the hospital but there was nothing anyone could do, he was to far gone. His entire body was racked continually with pain. He now suffered from Ataxia, and Aphasia, both critical injuries to the central nervous system. The toxins that were mixed in with the water he drank, had destroyed his brain and his internal organs.

It struck and before they knew it, he was suddenly gone. He left an empty spot in their lives, a void in time. Luckily, he had been spared only some of the agony of a prolonged departure. His mother would not be as lucky.

Her symptoms came on gradually. Whenever she ate she became very nauseated and her blood pressure instantly began to rise as if it were out of control. The doctors prescribed blood pressure medication in an attempt to control it but many of the doctors passed much of it off as the ravages of age. She began getting strange bright red bruises any time she rubbed or bumped into something. Larry's mother developed a terrible rash on her hands, arms, underarms, and groin area that soon turned into hideous looking sores. She experienced severe diarrhea just like his father. She was given antibiotics and pain killers. Still, as the months passed she only became worse. Pain became her constant companion rather than an occasional visitor. Her eyesight was failing too quickly to be a matter of nature's process. Then this once vital woman became confused and out of touch. Hideous sores appeared all over her body and the pain increased. Finally her doctor agreed that something beyond the normal was happening to her. His diagnosis was the of gall bladder problems requiring surgery.

She entered Brighton Community hospital. They all prayed that the surgeons would finally give her the relief that they had all hoped would come for so long. However, that was not to be. When they operated, they found her internal organs were enlarged. Her gall bladder, along with many gall stones, were removed.

A letter from her attorney was sent to the doctors prior to surgery asking them to keep everything for an outside examination with an independent laboratory.

Larry's mother died, but not until she had endured excruciating suffering and pain. They stood by her side, helpless, watching her fade into oblivion, more each day. Her eyes seemed to plead for the end to come, as the woman who was his mother, disappeared behind glazed eyes filled with agony. Mercifully, the end came and they laid her to rest beneath the Colorado soil next to the man she had loved for so many years. At least, for their sake, this sad chapter of their lives was over and his parents were at rest.

This was a land that could only be described as beautiful. On the open prairie you could almost feel a gentle roughness about the land. Looking to the west, only a short distance away, was the inspiring eastern slope of the Rocky Mountains.

This was the playground that Larry grew up in, a playground with the now famous spirits of explorer's like Zebulon Pike, Jerimiah Johnson and Jedediah Smith.

In the spring of 1971 two important events took place. Marie gave birth to their second child, and Marie and Larry bought eighty acres of land. He dreamed of being able to build a house and raise children in a country atmosphere and to make a profitable business on the land they both loved. He had intentions of raising dairy cattle for breeding and resale.

Larry and Marie were still living on their other ranch where they began when they were first married known as Box Elder Creek.

They had nearly a thousand head of Santa Gertrudis, White Face, Black Baldies, Black Angus, Bramas and of course the cattle he started with, Pure bred Holstein heifers. He had eight full blooded Holstein Bulls that would be registered as the best of the milk producing breed.

Larry had invented and begun an entirely new breeding style and program that only he and his veterinarian were knowledgeable of and about. This unique new style in genetic engineering had given the word cattle a new and different meaning. This style of breeding had produced what would be called the widest of outcrosses known in the livestock industry. This technique of hybridization, would have given the dairy industry a much smaller calf at birth. This would stop injury too so many, in the industry today especially first calf heifers. Even though the calf would be and was smaller at birth, they would grow to be as large or larger than their predecessors. Baby to adulthood was a little slower, about four months over the two years it takes for one to grow up They would be much more resistant to the many diseases That plague the livestock industry of today. The animal would be much leaner, healthier and would be able to thrive on a lesser amount of feed.

Larry worked hard at his new ranch as he constructed corrals, pens and buildings to hold up to one thousand head of cattle.

That Autumn, Larry was busy getting his new ranch ready to go and at the same time running his other ranch. He hired another man to work with him as they began to harvest the sugar beets and corn that Larry also raised. This was to be Larry's last year of farming.

From this day forward, he would do well with the livestock. He could see a bright new future on what had begun as only a dream and his new ranch.

Larry was getting caught up in the excitement. He soon realized that if he were to get out of farming he would need to set up a fall auction and sell all his farm machinery and to sell a large part of his livestock, the Santa Gertrudas, Black Baldies, White Face, and the Black Angus, since most were now at a marketable size weighing between 1200 to 1400 lbs. He would keep his bred heifers, and cows. Since his new endeavor would be dealing mostly with the raising and breeding of dairy heifers.

Larry and Marie had little time to themselves and soon realized that the dreams of a new house by fall, would not happen this year. Still they were like two kids who were in love for the first time. Their heads were filled with the dreams of a life time, dreams of the perfect life that they would build for themselves. Theirs was a time of hard work, joy an happiness. A time for turning dreams into reality. Yet they wanted to move as soon as possible.

After much consideration Larry and Marie purchased a mobile home so they could at least move onto the property. This would have to do until they could build a new home in the spring.

The mobile home was moved onto the land and set up. A new sewage tank and drain field was installed. Then the electricity was hooked up and the telephone followed.

The wells were completed and the water lines ran to all the various pens. Larry remembers asking the well man why the water was so red. His reply was, "It appears to be iron. Pump it for awhile then have the State Health Department check it out.", They would be able to tell you everything there is to know.

In a few days they called the State Health Department to test the water and also to inspect the new sewage system.

The State Health Department agreed with the well man. The redness was only iron, it passed the test as good clean potable drinking water containing no bacteria so it was hooked up to the house with an additional line running to the cattle tanks.

Soon after, they settled into their new home.

In the meantime Larry continued to work on the pens, lean too's, and the barns for the additional livestock.

All this was Larry's secret. The excitement was growing. The time was rapidly approaching for Larry to share with the world his new breed of cattle.

The cattle began suffering terribly. They were unable to eat, they vomited, tongues hanging out of the mouth "drooling" eyes bulging out of their heads, staggering, they had real serious diarrhea. It was an awful sight to see. To bring some of them through Larry would mix a milk supplement in large baby bottles and feed them, trying desperately to keep their strength up until he could find an answer, but nothing worked, absolutely nothing. No one, then or now, had ever seen cattle vomit, actual projectile vomit, it was unheard of, cattle just did not vomit, but Larry's did.

Cattle have four stomachs. From the first stomach they regurgitate what they have swallowed and rechew it. When you see a cow standing or lying around chewing continuously they are rechewing or chewing their cud. This is then swallowed and it goes second stomach, then the third and fourth for digestion.

Cattle have in their stomachs a bacteria called Rumen that is used in their digestive tract. With out rumen they cannot digest what they eat.

The water that the cattle drank contained among other things, mustard agents, nerve agents, pesticides, and herbicides, and Hydrazine (Rocket fuel). These chemicals destroyed the Rumen bacteria thus making it impossible for them to digest what they ate, and consequently immediately fell victim to the chemicals, and then secondary infections.

Larry brought one hundred-fifty head of the bred heifers to his new ranch. The bulls which now weighed over a thousand pounds, and other heifers that were nearing the breeding age and size were also brought over. Ten of his exciting new breed of cattle would also make the trip. Also at this time his new livestock began arriving, thirty to fifty head at a time. These would be mostly Holstein heifer's, some were purchased locally at the Greely, Brush, Ft. Collins and North Heights livestock markets, some from markets in Wisconsin, Nebraska, and Kansas.

When the trucks began to arrive the cattle came off the trucks running, jumping and kicking as cattle will do when they are happy and healthy.

Within a week of moving them onto their land and using the new water system, their hopes began to shatter. The cattle became sick. Many died and many more had to be destroyed.

Larry and Marie had no knowledge as to the ominous threat that now hung over their heads.

The veterinarian worked feverishly around the clock trying to figure out what was causing the problem. They did autopsies, (Necropsies) by the 100's on the dead animals, finding that all their major organs, lungs, heart, kidneys, and liver had been destroyed.

In the beginning stages their eyes turned completely red and watered constantly. Until the final stages just before death, their eyes turned white and were by this time so infected and matted shut that some just disintegrated.

Their nose and mouth would have large red blister type sores that continued to get worse. They walked as if all their joints were stiff and rigid. Their ears would begin to droop, a little more each day, down along side their faces. Their teeth turned black, black as a piece of coal, and would actually fall out of their mouths. As bad as he felt, and as bad as he hurt for them, he could only imagine what those poor animals were going through. However Larry and his family were beginning to feel some of what the cattle were going through. Crusted red infected eyes in the morning, red splotches on the lips and tongue, feeling a little nauseated, a little down and dizzy.

When death was near they would lie on their sides in convulsions their legs, and eyes, would go into a rythmatic spasm. He didn't know what to do with them at this point.

Larry could hear their labored breathing. Many of them seemed to stare at him with their clouded eyes, as if pleading for help. Tears falling down his cheek, he knew what he had to do.

Finally he turned and walked back to the house, sobbing Larry went into the bedroom, crossed the room to the dresser and opened the drawer. He removed his old .38 caliber service revolver from the holster. Walking over to the closet he took down a box of cartridges from the shelf. Still sobbing over what he must do, Larry sat heavily upon the edge of the bed and slipped the cartridges into the chamber. How had their dream turned into this nightmare? His heart ached for his family and for his cattle.

Slowly as if in a death march, he left the house to do what he must do. Entering the pens he walked up to one of the calves. There was no sense in her suffering any longer. She was not going to get better. He knelt down and patted her head. The tears came so easily, they ran down both cheeks and dropped to the ground. Larry raised the gun to her head. Compassion was all that drove him now. rubbing her head as if to say he was sorry for bringing her to this point, he pulled the trigger, the gun kicked, and with an awful explosion, it was over for her. It should have gotten easier but it never did. Larry ended up shooting hundreds so it seemed, just to stop those final throws of death, those rythmatic spasms, for which sarin nerve agent is noted for just a day or so before death, or as death comes. By this time the cows and the heifers began aborting their calves, one after another their was nothing that could be done. Most of them was between seven and eight months along, gestation for livestock is nine months. It was absolutely awful there was dead animals every where, Larry knelt down crying and saying why me? Lord, why me?, Please help end this terrible nightmare. There were so many they began stacking them in large piles until the rendering works could get there to pick up a few at a time on each truckload. It got so bad that Larry had to take the tractor out and dig two large pits, got some sacks of lime to spread over the dead animals, they began filling up just like mass graves. It was not only the calves being born dead but the cows were also dying as they gave birth, or shortly there after. Some went to the rendering works where they became dog food, many were buried in mass graves. Larry shot what seemed like hundreds of animals as they went into the final throws of death by nerve agent.

Then if that wasn't enough the very next day, Marie called me into the house, and back to the bedroom where she had gotten some clothes ready for the hospital. Looking up at him, tears running down her cheeks she said I know its several months early, but I think the baby is coming. Larry said oh my God, picked her up carried her out to the car, got her personal items and off to the hospital. Where Marie did have a miscarriage, the baby was lost. We had told them very clearly what was happening, with the livestock and how we were all getting sick also, contaminants from the Rocky Mountain Arsenal had already been found in the water, and in blood tests. They promised they would get to the bottom of it all. That it did appear to be a chemical induced abortion, much like that of the over 165 head of livestock.

On April 25, 1972, Doctor Scott wrote a letter describing the situation to the Colorado Department of Health.

"On March 25, 1972, I vaccinated thirty calves with Nasalgen. This was to prevent respiratory infection. We did not get good results. On April 15, 1972, I performed the first of many autopsies. We would do so many autopsies that I did not even keep track of the number, but they were in the hundreds.

We medicated and treated the calves. The death losses were terrible. I began to suspect some sort of poison. I don't know what the toxic poison is, but I do know that it will kill."

Then one day Doctor Scott raced up the driveway, got out and began yelling, "we've got to move these animals! Larry, it's in the water."

That afternoon, August 12, 1972, they started moving all the remaining livestock to another farm that Doctor Scott had arranged. Many of the cattle were so weak they had to be carried on and off the truck, or lifted on and off with a hydraulic bucket loader.

Larry believed Doctor Scott had a few ideas as to where the problem came from, but knowing him as he did, he knew that the Doctor wanted facts. His idea was simple. They would move Larry's cattle to another parcel of land about six miles away. The animals would be separated from another heard already there by an open but strong barbed wire fence. (Animal contact possible).

Jerry, the owner of the other herd was taking a big risk, but like all the ranchers in the area, he trusted Doctor Scott's judgment, and at the same time he wanted to help Larry. Neighbors are like that out here.

More days passed and Doctor Scott spent an anxious week fearing that he may have spread some new type of disease to a healthy herd. Several more cattle died from the poisons. For the most part they were the weak ones, but absolutely no symptoms were to be found in Jerry's livestock. From that point Larry's cattle began to improve The symptoms and the health maladies that affected them was or had been eradicated. The moving of the cattle and their improvement was passed on to State, and Federal Officials, and also the Colorado Health Department, who had insisted it was shipping fever or bronchopneumonia, a very contagious disease. Needless to say they were not happy to be proved wrong or that Larry and Doctor Scott had acted on their own, without their so-called professional help.

Out of the original six-hundred and thirty-five head, they managed to save two-hundred and fifty head, but it was only a temporary reprieve. Death was still patiently waiting for the rest.

Weeks turned to months,. Many of the cattle should have been weighing in between six and eight hundred pounds. Larry would have bet the title to his pickup truck that none of the cattle would have hit the three to four hundred pound mark on the scale. They were alive, but that was all. Many of them were showing the signs of dwarfism.

To further prove their point to the State Health Department that something was in the water, Larry brought two of the animals back to his farm. They were the strongest of the animals and both were fed exactly the same feed The only difference was in the water. One of the animals was given water out of the newly dug well, while the other was given water brought over from Jerry's farm. Within days the animal that drank the water from the farm water became sick, and several days later the animal died. No symptoms of sickness were observed in the other animal. Again this proved that Doctor Scott had indeed been right: the water was contaminated, but with what, and from where? they already felt they knew exactly where, and what, because the chemists outside Shell Oil Company, and the United States Army could not test for them since the Army and Shell Oil Company had all data secret, "Classified". Not even the Colorado Department of Health could test directly for what was there. Any tests done only proved something was there, and the cattle died. Bacteria could not live in the water, that alone should have rang a bell for someone like the Department of Health. Since they were aware no bacteria could live in that water.

By this time his family and himself were showing the deathly signs of the sickness, the same as the cattle had died with, and the same as the parents had died from. Although they had quit drinking or cooking with the water, they still washed and bathed in it. No one had told them not to, and they were unaware that the toxic chemicals in the water could actually be absorbed through the skin. Although they quit drinking and cooking with the water the very fact they were bathing in it just extended the time when it was their turn.

They now experienced nausea, difficulty in breathing, severe headaches almost to the point of madness. Their hair began falling out. Their eyes constantly burned, their teeth slowly turning black. They suffered from extreme stiffness in their joints. A severe rash covered most of their body and was transformed into huge burn like blisters.

Night after night the scenario was the same. A child would cry out for them. They would jump to their feet and run to the bathroom where Larry and Marie would find him bent over the toilet, violently heaving and vomiting. Then the tears would start to flow as they always did. They didn't know why this was happening, all they had were guesses. How could they be expected to understand their continual nausea? Their children were frail and hard pressed to keep any body weight. fear was beginning to overtake them. After each episode ended, they would return to their bedroom and lie in the dark wondering what to do. Their feelings of guilt were almost overwhelming. Larry and Marie had repeatedly taken the children to the doctor and yet there still was no diagnosis.

Marie would wash the kid hair during their evening bath. As she lathered their heads, she would look at her hands in horror as she found clumps of their hair clinging to her fingers. Small round bald spots would form and Marie would painstakingly try to brush their hair each day to cover the damage. The last thing their children needed was to be teased by the other children.

As adults their health was no better, Larry remembered one day in particular. His cousin was in town and he had a truck load of feed to move. He asked his cousin to come along with him. He had been suffering from one of the Splitting headaches that had become a regular part of his day to day existence. It had begun to let up but Larry knew he would feel better if his cousin was with him just in case something unexpected occurred. As they turned onto the highway, blood began to pour from his nose. It seemed to be unstoppable. His cousin grabbed a rag from the seat of the truck and tried to help him while he struggled to get the vehicle out of traffic and on to the side of the road His cousin was shocked at the intensity of the bleeding. It took time but it finally stopped. This was just another in a long series of symptoms which he would experience.

They were now going to the family doctor, who had called in the Tri County Doctors, for extensive blood tests, but the doctor's were at a complete loss, however they had some good ideas of what they needed to test for, that wasn't classified. In all eight different Doctor's who work in toxicology and blood work. They all examined Larry and his family. Their unanimous conclusion was they had all been poisoned with some very corrosive, and deadly poisons. One toxic chemical had been identified in all blood samples, by 1976 Those records were no where to be found, totally disappeared. The Doctor's had all recanted what they had reported and what we were told was denied.

On the way home from the Doctors Larry thought of his father. Like the trees on his property who had long since died, his father died and then his mother soon followed. Now is was their turn to die.

Looking at his wife, a cold, icy chill swept through him. Were they, each in their own way, witnessing the closing chapter in the life of one another?

God, what had he done? Unknowingly, by coming here, he had brought his wife and children to their own rendezvous with death. In his mind he saw her gentle face and could feel her heartaches. He could see the traces of the tears she had cried and the silent sounds of hurt that were left unspoken, hurt that she was strong enough to hide. Throughout her young life, no one had ever showed her what it was like to have someone hold her. That all changed when they were married.

We search this small planet of ours for a little compassion or just one more look through the innocent eyes of a child. We search for a change to be believer in something, while there's still a chance in time. Larry rewarded his wife by bringing her and his children to this place of death.

He was suppose to be her light in the darkness, her shield from the storm,, her shelter from the rain. In his arms of love, he was going to lift her from the sadness and away from pain. It was more than love that he felt inside for this woman. She came to him like the dawn through the light. Out of his dreams she came into his life. After seeing her the first time, he knew she was the one. He had failed miserably at protecting what he cherished most. A slow, gently moving tear rolled down his cheek as he remembered.

There was only one place around here that could cause this kind of pain and death, the Rocky Mountain Arsenal. They were no different from their cattle. His father had guessed it was the source and Dr. Scott had proven to them that the water had caused the death of his livestock. The thought sent chills through his body. They too had ingested the by-products of our neighbor's deadly concoctions.

Months went by, and Larry and his wife had started buying bottled water from Deep Rock, as many others in town did. It seemed bottled water was their only means answer of relief.

Later they would realize that the severity of their sickness was in direct proportion to the amount of well water they had drank. Within a very few days, of not drinking the well water, the symptoms began to ease, and decrease in severity.

They were only average people out in the world, trying to make a living. They were not Doctors, Chemists, or Toxicologists. Larry and Marie had no knowledge or understanding that would allow them to deal with any of this. They were literally dying on their feet, living in pure hell.

By not drinking the well water it helped a great deal, but it was only much later that they learned they should not have been taking showers, or washing clothes with their well water. As it turned out, they were continually being poisoned through the skin. From the United States Army GB Nerve Agent (Sarin Nerve Gas), Mustard and lewisite agents, then from Shell Oil Company we had Dieldrin, Aldrin, Endrin, and nemagon, several types heavy metals. Yet all these deadly Chemicals were found in the Land wells, blood tests, and fatty tests.

They begged for help. They begged the Colorado Health Department, Their Senators (one by the name of Gary Hart), Their Congressmen, Their Governor, it all seemed to fall on deaf ears, and it seemed as if they were getting nowhere. Their family Doctors kept telling them that with the amount of the poisons in their systems they should all be dead. It seems as if we had built an immunity to all this little at a time, as we were slowly being poisoned each day. Doctors had no other explanation. Larry's hired hand became sick, his teeth falling from his mouth after slowly turning black. everyone was experiencing blackening of their teeth, and became very tender and ached constantly. Keith at the time was just a baby beginning to get his first teeth, and it was noticed they were blackened and deformed. When he was about four years old we had to take him in for oral surgery, because all his teeth under the gums were terribly deformed, and began causing him to have severe pain caused from the teeth. All his teeth were removed except for a few that had came through in the front, even though they were deformed, the orthodontist's, and the surgical dentist were able to do a lot with them teeth so he could at least keep his front teeth.

Everyone had sores in their noses, and in their mouths, just like the livestock.

Diarrhea was almost constant. Everyone had a terrible vomiting problem and they were constantly Nauseous, blurred vision, ringing in the ears, Aching and hurting all over, all joints and radiated between joints, dizziness, staggering, Loss of vertigo (would lose balance and fall, tingling in all extremities, Tightness of the chest, difficulty breathing, or taking a deep breath, headaches to the extreme. We lived with all the above and more day in and day out.

Larry had an upper G.I. ordered at the Colorado General Hospital, since he was spitting a lot of blood. The X-rays showed an enormous amount of Ulcerations, the Doctor said this is the most I've seen in any one person's stomach. They immediately started treatment for the stomach and ordered him to stay away from that water, stating if it gets any worse we will have to operate and possibly remove part of the stomach.

My God, he thought, what the hell are these idiots doing to us? They are purposely destroying the people and the environment and they don't really give a damn. Only the Army and Shell Oil Company knew what was truly happening, so Larry started to complain to the Army, and to Shell Oil Company and again to the Colorado Department of Health. They kept denying everything, it was as if the Colorado Department of Health officials were only covering up for the Army, and Shell Oil Company, or seriously trying to cover up the serious blunder they had made by not seeing this sooner, as they were responsible to the people not to let something like this happen. By one Geologic study Mr. Land had paid for began to open a lot of eyes and doors when it was discovered that 1000's and 1000's of acres outside the Arsenal had already been contaminated, not only to the Northwest, but to the North. The Colorado Department of Health was suppose to be protecting the people so nothing like this would ever happen. Now that it has begun it will do nothing but grow.

The top chemist for the Colorado Department of Health Bill Dunn, said it was physically impossible for any contamination to be leaving the Arsenal. He said Larry must be some kind of a nut. While all the time he was the cover-up man for the United States Army they had conveniently placed in the State Health Department, The watch dog within. Well he knew exactly what was happening every step of the way, and he kept watching, and watching, and telling everyone there is no problem here all the while he knew exactly what was really going on but seemed to keep a lid on it for the Army. It was later learned all the time Mr. Dunn was saying it's all ok here. A letter that had been written to the Land's explaining we have a real serious problem here, You're wells are contaminated with toxic materials from the arsenal this was in 1994. It is believed Mr. Dunn kept that letter from ever being mailed to the Lands. It was then at the same time he had made the statement stating it is Physically impossible for any contamination to be leaving the Arsenal. It would be late in 1996 this letter was retrieved, and became a part of the file on contamination, In Washington D.C. Court of Claims.

The years 1972, 1973, 1974, and 1975, sick and living hell on earth, since the Doctors didn't know how to treat the problem, and could not treat the problem, since all the materials we had been poisoned with were classified material, to include that of Shell Oil Company, since it is now understood, Shell was in fact under contract to the Army, producing seven types of agent orange for the Army, (A 2-4-D type herbicide used in Viet Nam). They wondered what the reddish colored clouds were that hung over the arsenal continuously, during the early evening and on moon lit night it would be easy to see the reddish black cloud stretching from the arsenal to the north and north east. Then in the early morning hours each day, the wind would blow to the north dispersing the whole nights waste by about 7-a.m. when the air monitors would be turned back on. Larry was told that is how they were doing it, shut the scrubbers off at night and go like hell, then the next morning they would shut down, making use of the day time, and compliance time to be cleaning scrubbers, so they would stay in compliance with EPA. They feel they now knew it was agent orange they were making up to each morning, this helped explain the feeling of death. A certain person within air quality control whom Larry got to know, came out in the middle of the nights, and did check the air quality. Even though he was a Government Employee he could not tell his boss what he had been doing. He reported many times, and they tried to force, their administration to do, night time testing but they refused over, and over to ever allow this. This makes it a real nice story on how big business can actually also run parts of our Government. Especially so,

Shell Oil Company is a foreign owned corporation, who has no responsibility to the United States people, they are Dutch owned, supposedly by the Monarch **FACT** Shell Oil Companies own Country would not let them do in their Country, what they did in the United States, that is why they came over here and had no problem. Took control of part of the Government, and began polluting the air, the water, and the ground, making billions while they raped our lands. The place where Shell Oil Company once stood just North of Denver, Colorado Rocky Mountain Arsenal, remains today the most polluted piece of Land on Earth. The United States Government does a much nicer job of hiding wrong doing, "THEY JUST CLASSIFY IT", then if something happens, they have people who are trained to explain to you only what they want you to hear. It is called Bureaucratic Mumbo Jumbo. The United States Government, spends billions telling the people only what they want them to know, whether it is the truth or not.

Finally to keep their heads above water Larry took a job as an Auctioneer, in Brush Colorado, while still working for Erlich and Sears Auctioneers, he sold livestock, farm machinery, and actually worked at selling some farms at Auction, with Mr. Fritz Erlich and J. Lee Sears, Two of the greatest in the World of Auctioneering. These two people are the very ones Larry as a child when he would go to farm sales or dairy sales with his Father he would often say to his Dad, boy I wish I could do that. there were many a time his Dad would see Larry out behind the barn self tutoring himself to be an Auctioneer, He spent more time out behind the barn that anywhere else. His parents would come outside to find Larry, sure nuff all they had to do is listen and they could hear him a whalen. Larry growed up, still wanting to be an Auctioneer took the course of Auctioneering from the Kansas City Missouri Auction School, spent many weeks in Missouri, by gosh he graduated, it hit the hometown newspaper. Then in 1969 Larry began working for Erlich and Sears Auction Company. Larry was one of the best, he was in several world Auctioneer contests, never won 1st place, but always placed. He could whip out the numbers as fast as a car doin 90. He had also learned to be and excellent tobacco Auctioneer. He was a pure Country Auctioneer, a real Country Gentleman.

Larry and his family spent many day, that went into years, a physical, and psychological struggle every day, day after day. When you are sick all the time it only compounds the problems of everyday life. A simple thing like running up to the local convenience store was a major task.

Then early in 1974, Marie told Larry that she was pregnant again. They still wanted another child, so the news was like a ray of sunlight in an otherwise dismal existence. They had been using bottled water for quite some time now, Consequently they thought that she and the baby would be all right.

They went to bed early that evening, laying quietly in each others arms. While they had rejoiced earlier, there was a strange silence that surrounded them as they had more time to think. The past and everything they feared began to close in on them. Memories of the agonizing deaths which occurred in his family and all of the unanswered questions loomed large in their minds. Were they tempting fate? What kind of future had Larry and Marie set in motion for their yet unborn child? They wanted this baby, yet,, they wondered if they had the right to subject anyone to the possibilities that the Arsenal created for them. The pregnancy was difficult from the beginning. Marie was not gaining weight as she should have and her obstetrician was clearly concerned. Marie never seemed to look well anymore. Getting from day to day was difficult for her. She could never seem to get enough rest. She began to notice spotty bleeding that would start and stop with no apparent reason. At first it didn't seem like much more than the normal discomfort that she had experienced before as her body was transforming itself during pregnancy. Then the subtle difference began to appear. Something was not right.

The day soon came when all their dreams of a healthy baby were extinguished. Larry had gone into Brighton to pick up some groceries for her. When he returned he pulled into the driveway and loaded the grocery bags into his arms. He crossed the lawn and entered the house through the back door. The first thing he heard was sobbing. Something in those sobs scared the hell out of him. He dropped the bags and ran toward the sound. He found Marie doubled up on the bedroom floor. Blood was everywhere. She had obviously tried to reach the bed but was unable to walk that far. She was holding her stomach and crying his name. Scooping her up in h is arms he ran for the truck, seat belted her in using some pillows to make her comfortable and raced to the hospital. She seemed so vulnerable.

Pulling up to the emergency room entrance he gave a long blast on the horn as he jumped from the truck and raced to the other side. He pulled open the door as Marie grabbed for him with tears streaming down her cheeks. He could see the terror in her eyes. She wanted this child so much, yet, deep down she already knew this was the end.

The hospital staff came running with a gurney and Marie was rushed inside with Larry running helplessly behind. Never before had he felt so alone and so helpless.

A nurse asked him to stay in the waiting room while they assessed her condition. He could hear her crying and it broke his heart. It seemed like an eternity before the Doctor appeared.

"Mr. Land, your wife has had a miscarriage. There was nothing we could do, to save the child, but your wife has stabilized". He was in shock but was grateful that Marie was alive.

The days passed slowly as Marie recovered. Many times he came to the door of her hospital room only to see her staring into space, lost in her own private thoughts. They shed many tears during those awful days, always wondering why this was happening. It still felt as though they must have done something wrong. Their sense of guilt, while not logical, was still very real and very overwhelming, It brought a feeling of sadness to their lives that was difficult for them to deal with. The Arsenal was stealing their sense of well being. This was, but one more loss, but it was one too many.

Where had the honor and trust gone in America?

It was then that he decided to go all the way to Washington D.C.

At first he decided to sue Shell Oil Company because they didn't have the Army's protection of sovereign immunity.

Larry sought out an another attorney, since his first attorney, and practicing physician by the name of Doctor Donald Tyler, had disappeared. He took with him all the videos, photos, tests, and legal papers, all events that had been chronicled in 1972, and 1973. This man abandoned him after the third meeting with Shell Oil Company Officials, it seemed only maybe three days after a deposition.

In mid 1974, Tyler called, and announced he was going into Federal Court and demand the case be returned to State Court. Tyler told Larry and Marie the motion was a routine matter and that they need not attend the proceedings.

A short time later they learned that Tyler had gone in to Federal Court and asked that the case be dismissed. He didn't care if it was dismissed with prejudice or not, which meant Larry could never bring suit against Shell Oil Company again, if it were brought with prejudice. In an instant, all of their documentation, photos, videos, and hard work was reduced to nothing.

It was at that time Larry began to realize what was taking place. If Shell Oil Company was capable of this, what is the Government capable of, He learned very soon into this never say never, because they will.

He was a Doctor and an Attorney both, He quit his position on the board of directors at the Brighton Hospital, sold his Doctor's office, and his Attorney's office, He owned them plus a large parcel of Land they sat on, and he sold his home, and flat disappeared. He had walked into Federal Court where he had filed a case for us against Shell Oil Company, and asked it to be dismissed with or without prejudice, he didn't care.

Larry feels the day of the deposition, Shell Oil seen and heard all they needed to, and it was his opinion, he knew they had to get rid of his attorney, pay him off, and get him out of town. They did exactly that and quick.

Larry's whole family was in shock when they found out what Tyler had done. They could only guess what caused his abrupt about face in position. He had been so determined when he first took their case. Now, with the case dismissed, Tyler avoided all contact, and refused to explain. Larry asked him to turn over the court documents, the photographs and the videotapes that he had, so he could pursue the case with another attorney. Tyler refused, saying that any evidence in his possession was his property. Then the other Attorney traced Tyler up to Oregon, He refused to turn over any material to anyone involving that case.

At this point and as a way of holding down the expenses Larry did much of the leg work involving the investigation. Many times he would drive out to the Arsenal's perimeter with a pair of binoculars and camera. He took pictures of anything that he thought would help. Once, on the eastern side of the facility, he took pictures of a creek with pinkish water. Another time, that same creek was running a pale yellow water, minutes later it had changed to a brilliant yellow. This same water soon empty into the South Platte River.

One time as Larry was flying over the Arsenal, mainly near the Shell Plant. It had been raining very heavy, and of course Larry had his good camera with some good sized Zoom lenses, he quick prepared it. The runway extends now out into the Arsenal, and as the plane took off it made a leaning right hand turn and gave me some photos that almost looked as if one were taking photos of a solid rainbow beneath and just North and East of the Shell Oil Plant at least five (5) 640 acre tracts, or sections of property were so contaminated so badly it turned every color in the rainbow plus.

One time he saw men rolling 55 gallon drums into a huge pit, some of which were leaking you could see the waste on the truck, and these drums were taking at least a thirty (30) foot drop, They were all dressed in white rubber, or plastic cover all suits.

Months later, Larry's Lawyer tried to settle the case in Colorado State Court, but Shells Attorneys were able to get the case moved back into Federal Court, where it had been dismissed. It had not been done with prejudice so it gave him some room to move in. They were talking about a new tactic on getting it moved back into State Court, but at that point decided to go to Congress asking for a right to sue the United States Army, and to put all our effort and resources into it, since the United States Army had taken the responsibility for Shell Oil Companies Hazardous waste.

It was about this time Larry received a letter from the U.S. Marshall's office, Department of Justice, Washington D.C., ordering him to stop buying and selling cattle. It stated that his Federal Bond for buying and selling any livestock coming into or leaving the State Of Colorado, had been denied. Earlier when he went to apply for the Federal Bond to License, they suddenly doubled the amount of the bond with no explanation. When Larry was able to put the amount together the Government wrote to his Insurance Company stating they would not accept any bond on him, because of the unusual circumstances surrounding the loss of livestock in 1972. That any attempt on his part to buy or sell livestock for profit in the United

States would be punishable by a five (5) thousand dollar fine, and Six (6) month jail term for each animal bought or sold by him from this day forward.

This was nothing more than a way for the Government to take away his livelihood away, his means of making a living. They knew he was making enough money to resist them and to fight back for what they had done. Remember earlier He said Never Say Never, because when you do they will be there to get ya. They found a way to financially break him, and that is exactly what they did. He was just now getting the State of Colorado to back his cause against the Army. Well they almost broke him, and his spirit, but then he just got spittin mad, “Watch Out”.

Sam, Larry’s Attorney tried to locate Dr. Donald Tyler, but seemed to run into dead ends wherever he went. However he was finally located in Oregon where he was registered with the bar to practice law. Sam was able to get hold of him, and again requested that he return the legal file for Larry. Dr. Tyler flat refused to relinquish any part of the file stating it was his, it is not for sale as he hung up the phone.

It took many more months for Larry and Sam to assemble material similar to that which Tyler Had collected, and absconded with. We would never be able to get the videos, pictures, and the documentation as this was unfolding. In February 1975 his attorney filed claims against the Army on Larry’s behalf. They were administratively denied. He was now without legal recourse.

Acting on advice from his Attorney Larry went to his Congressman to get his backing for private Congressional approval that would allow him to sue the Army and thus have his day in court. In the United States a private citizen can not take the Government to Court unless it is approved by Congress, and that is what Larry wanted.

Larry had originally thought he was going to get that support from Congress Woman Patricia Schroeder, but soon found out she couldn’t help because Larry was just outside her Congressional district. Her and her entire office felt Larry should get his day in court. She then wrote to Congressman Jim Johnson, and informed him of what she knew, and had seen first hand. Congressman Johnson’s office took a real keen interest in the problems Larry was having and finding out first hand that the Army was causing a massive amount of contaminated soil and water, well off the Arsenal property, and was in fact contaminating the Land Wells, along with an additional 25 square miles off the Arsenal property.

On April 3, 1975, the Colorado Department of Health concluded that the toxic waste that came from the Arsenal had contaminated an area of thirty five square miles, of which twenty-five square miles was off the Arsenal property. Contamination was found in the Land Wells that could be, and was fingerprinted directly to the Army at the Rocky Mountain Arsenal.

Following a series of meetings in 1975, the Army graciously agreed to conduct a water testing program on his property.

In June of 1975, Diisopropylmethylphosphonate, or simply Dimp, a hydrolysis by-product of G-B nerve gas, was found in extremely large quantities in Larry's well water samples. There were many other chemicals found in the Land wells at extremely high concentrations. Aldrin, and Deildrin, two pesticides manufactured by Shell Chemical Company, that had been banned in the United States of America, because of the nature of it's toxicity, it's staying in the food chain from the first time it is used all the way to the fruits of those plants. Then from the fruits, or vegetables to the human where it stayed and done the damage, such as cancer (highly carcinogenic), and causing mutenigenic problems, many forms of birth defects that may have been blamed on certain medications. Shell Chemical Company not being able to sell this product in the United States, continued to manufacture it, selling it in other parts of the world., knowing full well the fruits and vegetables were coming right back into the United States, probably something sitting on your table right now, or will be served for lunch, or dinner. Endrin, and Nemagon were two other chemicals manufactured by Shell Chemical Company that were found in the Land wells at toxic levels. Then again another by the United States Army called IMPA, another related product of Disopropylmethylphosphonate, two types Mustard agents were present, along with Hydrozine, and Arizine, two types of Rocket Fuel manufactured by the Army, at the Rocky Mountain Arsenal found at toxic levels. The Colorado State Health Department, could only say get as far away from here as you can as quick as possible you have been terribly poisoned. The Army was so shocked at what they found they refused to do any further testing of his water.

At the same time, his attorney was bringing to the attention of the Army, the information contained in the new Dr. Konikow report concerning underground water flow. Dr. Konikow who is a Hydrologist and Geologist, and a leading expert on water flow with the EPA. The Army denied the results and effects of this report.

Recently discovered information revealed that Shell Oil Company and the Army met with Dr. Konikow in April of 1975, and reached a conclusion that the claims of Nerve Gas or similar contamination in the Larry Land water was valid. Strangely enough the information was intentionally withheld from Larry, the Media, and the Members of Congress.

Larry's patience had come to an end. He became disillusioned and no longer could trust anyone to do what was right. He was angry. Enough is enough he said. The bottom line was: it was him against the a (Giant), the United States Army. What a lonely place to be for someone who had been raised with such a deep belief in the integrity of this Government. Right here at this point is where one man literally took on the United States army, and began better planning his strategy, and using what was available to him. At times it didn't seem much. However Larry was beginning to get the State of Colorado behind him, along with some members of Congress, Except for "Senator Gary Hart" who just did not want to stir the pot. Larry ask him to at least paddle the boat, and for some reason mentioned ripple, and Senator Hart just damn near crawled under his desk.

Well again Larry was absolutely, again just spittin mad about all the do nothing in the Government, the County, the State, and the Federal Government., everyone talked, but failed to act on what was in their power. Larry went on talking on radio, and television, hitting the newspaper. Larry done his work and within two weeks issued to Shell Chemical Company., and the United States Army their very first cease and desist order, as mandated by the State of Colorado. Of course the Army and Shell Chemical Company very simply laughed about it.

Now it seemed that the people who were in charge, were breaking the laws of Government and nature by destroying the environment, thus the lives of people, and of the animals. What was worse was the feeling they were destroying his country or at least its credibility. How could he sit idly by and do nothing to stop it?

Like many of the people who live out in the western part of the United States, he grew up with a fierce sense of independence and loyalty. He like many Americans, there are two things you just do not do, mess with someone's family, or mess with his country, Larry rationalized that the United States Government did not always have the best interests of the American people first, especially when they are doing something wrong.

There was a strong feeling of sorrow in his heart that he could not explain. He had a feeling for this land and it's people for them personally, and for what they represent. Maybe the politicians in Washington should have read the beautiful quotation that his mother had hanging in her kitchen by Stephen Grellet: "I expect to pass through this World but once; any good thing, therefore, that I can do, or any kindness that I can show to any fellow creature, let me do it now. Let me not defer or neglect it, for I shall not pass this way again".

Larry wanted a nation with a Government that was clean and honest, decent and compassionate, and was filled with love like the American People.

The men and women who make up this Land, what kind of people are they? He comes from many different Lands or he is the descendant of several, hence the strange mixture of blood, which you will find in no other Country. Maybe the grandfather was an Englishman, whose wife was Scottish, whose son married a French woman, and whose present four sons have married four wives of different nationalities. They are all Americans, who leave behind all their ancient prejudices and manners and receive new ones from the new life they have embraced.

Here individuals of all nations have melted into a race of American people, whose labors and posterity have caused many great and wonderful changes in the world. Was this not worth fighting for? Lets all work together so we can make it even a better America today.

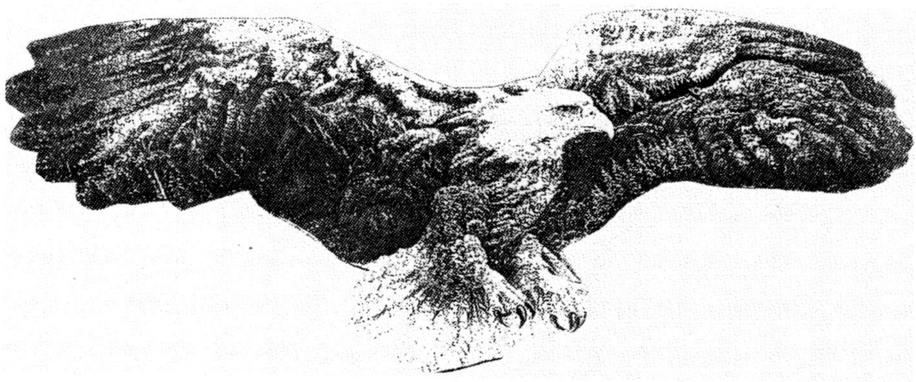

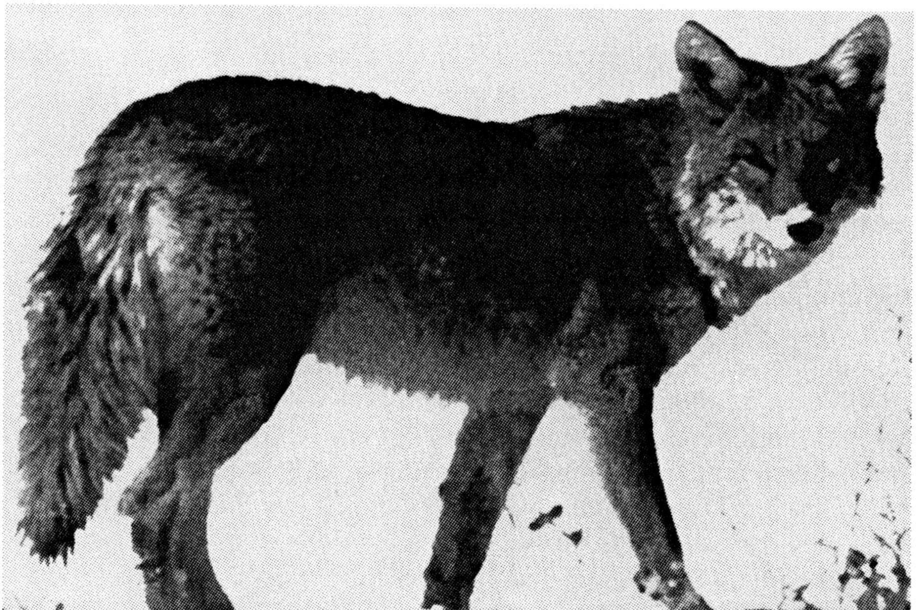

CHAPTER TWO

Science has radically changed the conditions of human life on earth. It has expanded our knowledge, and our power, but not our capacity to use them with wisdom.

J. William Fulbright, 1964.

During February of 1975, Larry's Attorney Sam, filed claims with the Army for administrative relief. Substantial delays were encountered in acting upon the claim, which were somehow lost, then denied, then reinstated and finally denied altogether.

Even though he was married to a wonderful woman and had two beautiful children, he was depressed most of the time. Larry felt naked, exposed, afraid and and even ashamed. His Country, the Country that he loved so dearly, had allowed this tragedy to happen. Then one day he had a strange new thought, something that had never occurred to him before. He realized that his thinking was all wrong. It wasn't the Country who had allowed this to happen. It was the fault of our political leaders, and large corporations protected by our own politicians. Starting in the 1940's and even before, all the way up to the present, the decision makers. He realized that as the years passed these problems were never corrected to begin with, they were side stepped or completely ignored, hidden, or covered up, which in turn only complicated matters allowing things to get worse. The Country, standing by herself, was pure, untouched in what she represented. It was not until greed and financial rewards entered into the picture that our Political and Industrial leaders became corrupt and began not to care what happened to the people, the wild life or the land. The only thing they really cared for was the almighty dollar.

The same Government leaders who had pledged their honor to defend this land and safeguard its citizens under her laws, could not even protect the people from the forces within our own borders. Then they have the audacity to ask why the people do not believe in them. How can we when were consistently lied to? What has happened to our Government, our Corporations, and to our Politicians? Surely not all of them are being paid off or dealt cards under the table to protect corporate decisions? For instance the protection the Government gave to Shell Chemical Company, protecting them from the law itself, and let wrongdoing he ignored by the Government. It was as if the Army had just closed their eyes to this horrible, negligent, wrong doing.

This is the America that Larry has always loved, the land of the free, home of the taxes, where the rich get richer, and the poor pay more, yet something has happened.

He felt as if he had been stabbed in the back by a now unknown and superficial lover.

Once again he approached the officials at the Arsenal. Their medical problems and the bills associated with them were escalating daily. They needed help. Again he was told that it was not their problem. They refused to take any responsibility for the illnesses that were plaguing his family. Larry could no longer contain his anger and, without thinking, he verbally attacked the officer in charge. In a final assault born of desperation, Larry told the Officer in charge that he would release the information to the media, telling the public the exact reasons for what was happening to his family, and where it all began.

The Officer stated to "Mr. Land, you do not strike me as a stupid man, why act like one".

At that point, Larry asked to use the telephone and called one of the local television stations. He graphically told them only part of a long and in-depth story about the United States Army, and Shell Chemical Company, with more to come. The station agreed to meet with him immediately. Larry had been creating Pandora's box for quite some time, so he could begin putting the squeeze on both the Army, and Shell Chemical Company. Larry called this his tool box, and it was full of tools, but now Larry felt a real fear for his life. There had always been a veil of fear cast over this subject and what might happen to someone who dared to talk, every one automatically feared and assumed the worst. No one knew who had started this fear but it was not to hard to figure out that the Army and Shell Chemical Company would be the only ones to benefit from such a story, so logically they had to be the ones behind it.

Larry was about to open pandora's box and tell the secrets about both the Army, and Shell Chemical Company. As Larry hung up the phone and without saying a word, he turned and started to walk out, the sounds of his shoes echoed loudly in the now silent room.

Then suddenly a voice, "Mr. Land, I was wrong. You are a stupid man. Things could get very rough if you follow through with your plan. I would strongly advise against such action, why not just move away. The Army is very powerful and takes a dim view of outsiders meddling in it's business. This is not a threat Mr. Land, just a friendly warning".

Move, where was he suppose to move to, he had no money left, everything had been spent on Doctors and hospital bills.

Larry stated, "Mr., I can tell that you're not from around here so I expect you to be ignorant of our ways out here, but you just made a big mistake. You see out here it doesn't make any difference who you are, there is a kind of ruggedness, and independence in the people who live out here on the front range that you won't find anywhere else in the country. The mistake you made was giving me that so-called friendly warning. You should never have done that. You see, you have let the so-called cat out of the bag and told me that I'm on the right trail. If the Army didn't have anything to hide it wouldn't give a tinkers damn, or a hoot what some broken down rancher had to say. You think that because you're in the Army and have the backing of the Federal Government that you can't be stopped. That you can go right ahead and destroy everything. Things that have taken nature millions of years to build and the life times of men and women to establish, because you think the Army's in the right. A lot of people have recently told me that one person can not make the difference. Well they're wrong, and so is the Army. If you stay out here long enough, you might learn something about the people who live out here and how we feel about our land and just how wrong you are. Then again I doubt you will, You're not the type".

Little did Larry know that he had his lessons to learn as well. Anytime that you take on a powerful entity, You will pay the price. He had been brought up with such a strong belief in the ability of the average citizen to stand up for their rights, and principles, that the reality of where the real power in this country lie, was clouded over in his mind. He realized that most people feared the power structure that represented the Rocky Mountain Arsenal and that of Shell Chemical Company, and other facilities like them around the country, that scare the hell out of people. As of yet he did not understand why. Eventually, it would be that fear, that would bring Larry information, but would frequently bring it under a veil of anonymity. He too had a sense of anxiety as he confronted the bureaucrats, however he did not believe that it ever occurred to him that it would become as intense and frightening as it eventually did. Larry could not imagine the total invasion of their privacy that would ensue, the attempts on his life, nor in their efforts to control his ability in caring for his family.

Just a short time back around June 1974, as Larry Had begun appearing on radio and television News with a few small incidental environmental stories about the Arsenal, and what contamination has done to this county, this kept getting him the information he needed in building pandora's box. He had got a call from Oklahoma, Karen Silkwood just to share a few thoughts as to how scared she was, and whether she was doing the right thing, he assured her she was, she told Larry after seeing him on television, she felt it gave her the same unwavering strength that he had shown to move forward. Soon after that call she was killed by Kerr Magee, and or Government Agents. They had killed her for trying to do the right thing by tailgating her car then drove her straight into a cement bridge. For a little bit it sure took the wind out of his sails, but he picked up the pieces and continued on, cause he knew he had to, there was no stoppin now.

The terror of future attempts on his life were almost pale in comparison to the subtler assault on his everyday well being. Frequent but shrewd entry into his home, which had so long sheltered his family, would have the most profound effect on all of them. It would soon become apparent to Larry and Marie that nothing would be sacred on this battleground, or for them any more. Larry looked at it this way "It's like one man taking on the United States Army, along with one giant world corporation Shell Oil Company. Larry thought this is like David and Goliath, and just stop and think for a moment. Who won! Then there is the war box, the intruders failed to breech. (Pandora's Box)

On July 3, 1975, one day before the scheduled interview, while driving down interstate 76, Larry was lost to his own thoughts, which seemed to be the norm at this time in his life. He was thinking about the interview tomorrow, wondering what they would ask him, what should he offer them. Suddenly he noticed a light colored car in the lane next to him. He realized that another vehicle had been pacing him for some distance, neither speeding up or slowing down. As he glanced over at the driver his head moved to the left and, the car began to pull forward until it was several feet ahead of Larry. Then the man in the other vehicle began slowing down. So Larry slowed down also. Larry was wondering what to do. As the driver in front began pulling ahead several car lengths. He noticed that the darkened window on the passenger side of the Car was being lowered. Larry watched with interest and concern as he prepared his 44 mag. And his twelve gauge Ithica shot gun loaded with double 00 buck shot. He then saw the barrel of a rifle starting to protrude out of the window, it was being aimed at him. This could not be happening. Things like this didn't really happen. "Only in the movies" He thought.

Fear gripped his insides, a knot instantly formed in his throat. Part of him wanted to duck, while his sensible side knew he had to watch where he was going. Larry slammed on the brakes, then suddenly, he heard a sharp crack and steam began to billow from the front of his truck. He swerved, all of his defense mechanisms were in place and he was at full adrenaline high, everything he was doing went into slow motion, Larry said it is one of those life threatening moments he calls "Just like making a movie". Every movement, and things that are happening, and every move you make is in complete slow motion as if you could make no mistakes. when once again two sharp cracks rocked his consciousness, as it struck the windshield on the driver side Larry was leaning over in the seat coming to a much faster stop. The temperature gauge was climbing, as the pick up truck Larry was driving came to a screeching halt, with the back wheels locked up, Larry had pressed the emergency brake on hard when he leaned over in the seat, and turning his wheel slightly to the right, leaving him somewhat in a protective position off the edge of the highway as he bailed out the passenger side. Larry was fully armed, and prepared for whatever may be waiting. As Larry peered over the hood of his truck he could see both vehicles hightailing out of the area, he believes they thought for a while they had completed their contract, and took him out. He thanked God for his police training, and how well it kept him together during this as Larry says "one scary incident". Larry had went to the front of his pick-up only to find the radiator steaming, and the sweet, almost nauseating smell of hot antifreeze drifted through the air. Looking up the entire windshield had been shattered, still in tact. Walking back to the drivers door, he noticed a bullet hole in the shattered windshield, it was at a level straight through the opening of the steering wheel, he opened the door looking in shock at a hole, chest level in the seat, and right out through the back of the cab. Larry sat down in the seat, and with a long sigh, he rested his head buried in his hands, thinking back as he set the brake and rolled over in the seat as his truck came to a long squealing stop. Was it really the hand of God that pushed him over in the seat and saved him. Wondering just how far this whole thing would go. It seemed like a bad novel and yet he was living it day by day, and could not stop, his convictions grew even stronger, his tenacity unending. Now he had to go home and explain this to his loving wife without scaring her to death, and he was thinking how does one do that.

After getting all the reports made, and his truck towed, he finally returned home, the house was quiet. He sat down in his chair and succumbed to a mood of disbelief. His mind would still not completely accept the fact that things like this could really happen. This was straight out of a Ian Fleming novel. He almost expected James Bond to appear. Someone had actually fired shots at him on a road that he traveled daily. Someone or a bunch of someone's felt so threatened by this Adams County rancher, that they found it necessary to launch an assault on him. Were they afraid that if he spoke up, that others who had suffered the same fate from their carelessness might stand up as well?

By the end of July, Larry had appeared on many television and radio stations throughout many of the western states, many of the Midwestern states, and as far east as New York, Chicago, Washington, D. C., and all the way to Miami. Last count he had been on sixty nine (69) radio stations, and over 30 television stations, and so far 16 newspapers and going. Larry was only giving them a little at a time so he could keep them continuously coming back for more. He was getting calls from around the Country. He was called from Germany, and England with two people actually flying over to see him, and see for them selves the real problem. He was flat airing the dirty laundry of Shell Oil Company, and the United States Armies Top Secret Chemical Weapons base, The Rocky Mountain Arsenal. Each time the news media would contact him, he would give them what was ultimately the key to pandora's box which Larry was able to keep full. He would let them only have a peak at a time, and he would give them each a time to come back for more. What the news had to say was the truth and by God the Government was forced to answer. By this time he was also getting some strong political backing from both the State and Federal Governments.

During the last couple of years it seemed as if every news paper, television and radio station out there was calling to set up an interview, as far away as England, France, and Germany. The Stars and Stripes Newspaper based out of England. They done a real big story and Larry was just the one to give it to them. It covered the entire front page, the second page, and the entire back page of the newspaper. Larry is starting to really like the attention, and what was happening. The Army and Shell Chemical Company had to answer what they were being asked, and why? Talk shows were showing some interest. Larry was given the opportunity to tell his story, and in so doing the listening audience became more aware of the issues. As he listened to the questions that began coming in, Larry sensed that many people in the audience had their own suspicions about facilities in their home areas. He wondered how many of them were finding similarities between incidents in his life and those in their own. A lot of calls were coming out of the Broomfield area regarding Rocky Flats to which Larry was able to build a war chest for them also, but could only talk of one at a time. He felt a sort of kinship with many of the people to whom he talked with. Larry also felt the under current of fear. It was as if they were urging him on to do what they were very much afraid to do.

The Rocky Mountain Arsenal was news and the United States Army, and Shell Chemical Company were getting their noses rubbed in something they didn't like. It was the beginning of their own undoing.

The United States Army, and Shell Chemical Company tried many times over the years to find out where in the World was Larry getting all his information, and when he turned it over to the news media it was always invariably on track. They were ransacking his home, garage many times over. Larry finally wrote them a big note, left it on the kitchen table stating very simply it's not here. It was like catching them with their pants down, and being able to spank their bare tusch, over and over and over again. Every time they would open their mouth, or have one of their public meetings, where they could lie to the people trying to make it look as if Larry was a radical. He was and they pushed him right into it, they were wrong and he was right. Larry soon became the hero of many around the World, as he continued to leak information to the news media. Larry became a hero to many of the Colorado state agencies. The Colorado Department of Health thinks of him as their hero, and the hero of millions world wide, for what he done by opening the eyes of the World to toxic and hazardous waste. This opened up a toxic box in order to let the scum out. Like a wound in America's side that was festering, immediate treatment was needed, and Larry found the only way to get that treatment was through the news media. It woke up the politicians around the World, Rattled the doors in Washington D. C., things began to happen, and more and more new discoveries were made. Larry was able to disclose this wrong doing, and waking up the World more each time he released material to the news media, other's began doing the same. Larry feared every day for his life as many attempts were made, but he had God with him as he rooted out the scum. Larry is proud of what he was able to do, its much like getting a movement started, and by God it got started, and it is working. Larry has devoted himself from 1972 on to make sure it never gets out of hand again, and he felt the world should know some of what happened, and that our World is at least a safer place to live. Stringent safety measures have been put in place for the Government and Industry alike so our future may be a little safer than it was.

Then right out of the blue the Colorado Department of Health and the Governor of the State of Colorado issued to the United States Army, and Shell Chemical Company, a final cease and desist order, This ordered them to stop polluting the land, air and water in Colorado immediately. Well Shell Chemical Company knew it was coming, since they had already purchased several thousand acre's of land in Morgan County near the city of Brush Colorado, straight north east of Denver on interstate 76. The Army of course knew what was coming also, they ran out of chances to hide, reminding me of the story of an ostrich who ran away, then suddenly poked his head under the sand, just hoping it would all go away. The Army's demise will soon shock the nation, in fact it reached out and shocked Shell Chemical Company also.

Well Larry had found that Shell Chemical Company had drilled, and cased of several wells on their new property. They had applied to the state for permits to use these new wells, that had been drilled twelve (12) thousand to fourteen (14) thousand feet deep into what is called the lakota and dakota sands, or salt basins, some being completely dry. They had these basins completely cased off, and the huge pumping stations in place. where they maintained they would be able to pump under the ground to these areas, by forcing the water down the pipe at thousands of pounds per square inch (psi) pressure, forcing this waste water into the salt and rock formations below.

In 1966 unbeknownced to any one the Army had drilled three wells on the arsenal, and tried injecting the waste chemicals down into some rock formations nearly thirteen (13) thousand feet using the same method of pumping it into the formations at thousands of pounds per square inch (psi) pressure. well it didn't work, with in a matter of days it was causing earthquakes in the Denver area. Usually an earthquake in the Denver area is very rare. They began having two and three quakes a day any where from 3.4 to 4.2 on the World scale at Boulder Colorado. The Government was very quiet about it as if they knew nothing of the quakes, yet they had been ordered to stop pumping. Well they waited a few days and then decided to start the pumping again. When they were down there were no earthquakes, within eight (8) hours after they started force pumping the waste the earth began to quake, they continued to pump, but by the third quake they again were ordered to stop. They started caused quakes each time then shut down. More Government got involved, they started and stopped each time causing quakes. They finally woke up and realized it was really them and they had to quit pumping in order to quit causing the quakes. When they quit pumping the quakes stopped, and they have gotten no quakes since.

Larry had exposed so many of the blunders out and out negligent blunders, and reckless wrong doing to the biggest at the Rocky Mountain Arsenal, and he knew he just had to stop them or Shell Chemical Company from spreading out using other methods to pollute the earth further. They were all very good at covering things up, and not doing what they were suppose to.

He got the date for the hearings set for Shell Chemical Company to obtain their permits to pump the hazardous materials into the ground near Brush Colorado. Larry met with some of the Commissioners from there, Mayors, and City and County Officials to hear all their thoughts and objections. Larry then began spending a lot of time in the libraries searching history, since he was aware of a lot of drilling for oil in that area in the early 1900's. They were all dry holes, to close a hole it was nearly mudded shut. He was able to find out where darn near every hole was drilled, and to what depth. Many of the holes had been drilled through and to the Lakota, and Dakota stratas then just mudded shut. Some of the wells had been drilled within a quarter mile of where Shell was going to begin force pumping the toxic waste. They would have been forcing the toxic waste down into the ground then back up each tube that was left there. Forcing it up these holes until it would be able to wash out into the well water, the irrigation water. If this were to have happened people would have been poisoned and never know until it was too late. It would be like taking a rubber ball, hollow out the center, then drill about one hundred small holes all around, no problem, but then drill one large hole put a large tube in the hole then hook it up to the kitchen faucet, turn it on slow and water begins to leak from the holes turn up the pressure and you've soon got them all reacting squirting water up with force. this is exactly what would have happened if the State were to allow Shell the permits. Larry went to State Officials, and to City and County Officials, and explained their dilemma. Shell Chemical Company would be the cause of a catastrophe, and no one could possibly live with the effects. They were all shocked, and not sure what to do. Larry registered with the Court so he could present his side of the problem, and was also able to question all Shells Expert witnesses. He was getting a lot of help, legal advice, some ok, and some rather discouraging. He was talking to city hall then to hear some of them say your fighting City Hall, well as far as Larry is concerned a big City Hall is no different than a little City Hall. Boy He found out he was right, the only thing he noticed they sure get a lot madder.

Well the four day hearing began, there was a lot of people present, He recognized many from Brush, Fort Morgan, and Morgan County. Shell Chemical Company did their opening statement, some others, EPA, and the Colorado Health Department spoke. It seemed very long and somewhat dry. as the experts were testifying as to their expertise.

He felt many times I just wanted to get up and object to anything that sounded like it might work for them. what he had to ask the State Experts, and EPA Experts was pretty much straight forward answers he would get regarding the well and the pressures at which the toxic waste would be forced into the underground formations. Good answers regarding the Armies pumping in 1966, and would there be any comparisons, it all sounded like good common sense for all the reasons they did not think it should happen, and there were reasons to fear earthquakes. By the third and fourth days Larry got more of his turns to question the Shell experts. Larry had some large graphs, and a couple of over lays that he presented to the Court, and as he questioned and ask for answers of Shells Experts. He was feeling pretty good when he could lock up an expert on certain questions, he would strike the podium, and demand the man is an expert in this field and demand the Court have him answer the question. When he could not answer the question Larry had ask, he would then ask the Court for the record may I answer that question, the Court would agree and Larry was able to read that answer into the record. Larry in a great finale turned to the people of Morgan County and City of Brush and Fort Morgan and stated to them "wake up people and stand your ground", what Shell Chemical Company is contemplating here will have catastrophic Effects that will last for centuries. Let us all work to gather and Nip it in the Bud. The Court adjourned to until a decision could be made whether Shell gets their permits or not to get permits, this was going to be a long and agonizing wait not knowing what the answer may be.

As Larry was packing up his papers and getting ready to leave. There were TV and Newspapers present asking questions when a short well dressed man walked straight up to the front of Larry, he was just red, his little bald head was wet with sweat. He wasn't sure who this man was but he stated "I want to tell you, that you are the most radical little son of a _____ we've ever had anything to do with". This is done with two television cameras right in our face, Larry stated back to the man, well that makes me feel good, just knowing I got to ya. As Larry made it out of the hearing room there was a great deal of people waiting to just say hi, and shake his hand, several stating good job, were not sure why your doing this but you put up one hell of a battle.

Now Larry is off to still another battle, he has drawn his battle lines, all that is left now is to act. He had hired a few small planes, and people to take photographs, was really surprised he didn't get any flack from the army, since this was 27 square mile Top Secret Chemical Manufacturing center. Larry had been checking into some bigger, and better planes, and camera types like Infra red. He found this one outfit, flew out to talk with them tied up a deal to take Infra red photos of the entire chemical weapons base. It was scary, but the Army had driven Larry to clean it up. He had his man in Denver, an expert in Infra red photography fly out and meet the person with the plane, and assist in getting exactly what we needed, then bring them back to Denver. He did and we met within a week the photographs were absolutely dazzling. It showed basin F,C,D, A,B, E, a lot of swampy sludge holes all contaminated big time. In Infra red the red hot spots stuck right out like sore thumbs. It also pinpointed forty nine (49) toxic waste pits, some as long as half a mile, and hundreds of feet wide. He was elated, and decided just to lay back for a while and let things cool off. What really worried him is did they know. When would they be coming after him.

Again, this was a secret, classified Chemical Weapons manufacturing facility. Evidently they were not aware that he had taken the photographs. Larry thought "wow" this is ironic, they had airplanes fly over the arsenal back and forth six (6) times with no notifications. Then to fly the bigger plane over several times to get the infra red photographs he now had in his possession. This really surprised Larry, but made him wonder just how secure this place really was. Since Larry had breeched the security already at least eighteen times to collect samples, and just to check out what and where stuff was stored. There was only one time they almost got him, but he had a very quiet running motorcycle, and could go where vehicles couldn't. He remembers laying out in the middle with them searching for him with spot lights can remember them going right over top his location. As he is working for the Denver Sheriff Department one of the new staff came up to him to ask him if he was Larry the Farmer that lived north of the Arsenal. This person who is now a Deputy for the city of Denver as was Larry, began telling him a few stories, and experiences he had when he was a guard at the Arsenal. Larry said oh they hired civilian personnel to guard the place. The guy told him the many times they chased some damn motorcycle out there, the person was so good we would have the spot light right on him, and he just disappeared. Larry was glad that was behind him any way and he never got caught, didn't have to go back. the Deputy even told him how they were using night vision and could not keep up with him.

Larry stayed quiet for quite some time and was clearly shocked he had not heard from the Army, or Shell Chemical Company about the fly-overs. So he finally called his first television station to let them know what he had, and was willing to pull the plug on the Army, and Shell Chemical Company, and show the World what they had done to that land. The first photograph showed hot spots in the northwest quadrant, this was some of the smaller locations. Well it hit television that evening, and hit the newspaper the following morning. By 10 a.m. Larry had Army knocking on his door, they were very cordial saying hi do you know what you have done, Larry said no what are you talking about. They said the photo you released to the news media. He said oh the photo yes, would you like a copy of it. It surely points out many of the things you have been denying existed. The only thing he would like to hear is the size of these pits, and what's in them. They stated that is impossible. So Larry ask them don't you people keep any records, there are many more photos that actually show the entire base, and every hot spot shows up a really bright red. Larry again ask them if they wanted a copy of it they said no we'll get our own. Larry said I hope you get some good ones since there will be a lot of questions that will need answers, and as we speak the first one is already in the hands of Congress. and unless something gets done about them I will start releasing more photos to the news media. Larry wiped the sweat from his brow because he figured they would have the United States Marshall with them to arrest him and hang him out to dry.

The photographs were absolutely magnificent, they pointed out every pit, dump, pond, swamp, lakes, there were even some very hot equipment areas found. Well since he had already Mailed a copy to congress, he decided to just let things chill for a bit. Larry was working two part time jobs, and one full time job, and all the work on this project.

All the contact from around the Country wanted to just hear what had happened to his cattle and to his family, but they also wanted to know why it happened. So he told them.

The facts centering around his own plight he knew by heart, but other facts he was unsure, and told the listening audience as much. He told them that he needed any information they might have. Whether the information was obtained by him, or given to him by a sympathetic employee at the Arsenal trying to relieve his or her conscience, it made little difference. If it was a documented fact, he wanted it, so he could in turn give it to the news media. Larry found the news media The armie's worst enemy.

After about six interviews, his battle with the Army and Shell Chemical Company were more widely known by employees and the general public. Information began pouring in. It was as if a value had suddenly been turned wide open and the contents within could not wait to spring forth.

Larry was rather proud to think he had issued the first Cease and Desist order, and had done all the proper filing with the State, the Governor, Attorney General, Colorado Department of Health, Department of interior, and Environmental Protection Agency. Now the State of Colorado getting more involved have also filed their first Cease and Desist order for Shell Chemical Company, and the United States Army, who had tried to reject it because they were not a resident. The Courts thought otherwise.

Larry's postage was very high at least it was a tax deductible item for a legal expense. In Washington D. C. all congressmen, and senators, and even the President of the United States, It was almost a bi weekly thing, each letter had to be addressed and sent specifically to the Congressman, or Senator. I did get two responses from President Jimmy Carter, and turned it over to our own Congressmen, and Senators. Later I became acquainted with President Ronald Reagan, and was standing behind him as he signed the bill that would prevent the manufacturing of any more poisonous gases or nerve agents in the United States.

By September several bits of information were passed along to him that he had never heard before and which left him with a sense of disgust for the people that created this monster.

It seems that part of the facility was originally leased to the CF&I Steel Corporation shortly after the war.

"It was a total disaster., a botched job", The man whom will simply be referred to as Bill said.

Bill was expressing his disgust and frustration with the CF&I Corporation, as it made shoddy attempts at the production of Benzene in a facility leased to them by the Army. The plant sat on the Rocky Mountain Arsenal property, south Center.

Bill who was a foreman over nineteen men in the little known operation, said that many of his workers "breathed in tremendous amounts of Benzene, vapors", as it was being processed. Bill maintained that the worker's shoes were sometimes "just soaked" with pure Benzene, a very highly toxic liquid that only in recent years has been cited as a suspected cause of leukemia, and/or cancer of the blood, it was a found to be a mutagen.

Bill's allegations regarding the Benzene plant were disgusting at best. "It was a very crude, hazardous and wasteful operation". Bill then stated he was ashamed to be associated with it, it was that bad. The equipment at the plant was so horribly designed that the plant never did get into production with an acceptable product.

"There was no physical examination of the plant workers, except at the time they were hired", Bill said.

He also maintained that there were spills and leaks horrific, and gigantic, that were both negligent and inexcusable. From the several one-hundred thousand gallon tanks that stored Benzene in the Arsenals tank farm, as well as leaks from the plant's processing equipment and the clay sewer pipes that led to basin A, merely a huge dirt sludge pond, where as the liquid sludge would evaporate thereby concentrating daily the toxic basin that was leaking into the under ground water stratus.

"One time Bill states he sent a man out to pump some Benzene from one of the tanks, but there wasn't any to pump". Bill said. "literally thousands of gallons of raw Benzene had leaked into the ground and the ground water just North of Denver. Larry's Farm sat about five to five and a half miles straight north of this huge tank farm. This happened in the very late forty's and early fifty's, and the underground water would travel just about a mile per year. This would have put the Huge Benzene leak under his place in 1972. Benzene was one of the toxic chemicals found in the Land wells.

Chlorinated forms of Benzene, such as were processed at the CF&I plant, tend to persist in the environment for not only months, and years, but for centuries to come. It can be taken up by plants, grass, and animals.

During a period when the Benzene plant was operating, there was little public awareness of the health hazards posed by Benzene or its chlorinated forms, used at the CF&I plant in an attempt to produce three pesticides.

Benzene and several of its chlorinated forms are now on the Environmental Protection Agency's list of toxic pollutants.

Benzene also is cited as a suspected cause of leukemia, anemia,, central nervous system depression,, skin irritations, rashes, and sores, and many changes in the human chromosomes, which contain the genes that determine hereditary traits.

OSHA has sharply reduced the limits to which workers may be exposed to Benzene.

Bill also noted that in 1947, CF&I leased other surplus facilities at the Arsenal to operate two other plants. One was to produce Caustic Chlorine, and the other plant was to produce DDT, an insecticide banned in 1972 after it was linked to cancer in animals, and humans.

At that plant, Bill said, "Raw Benzene was Chlorinated and supplied with other additives in and attempt to make pesticides, Monochlorobenzene, Orthodichlorobenzene, and Paradichlorobenzene. The last compound is used as a fumigant against clothing moths".

CF&I never was able to produce those three pesticides so as they would be commercially acceptable.

"One order of Paradichlorobenzene was shipped in cardboard drums, and was returned by the purchaser. That shipment was of such low quality that the cardboard drums almost disintegrated due to the presence of the oil that hadn't been removed in processing the material", Bill recalled.

In and attempt to remedy some of the equipment problems and health hazards at the plant, he went to the CF&I safety engineer in Pueblo, and asked for his help. The engineer, who Bill did not identify, said CF&I's pesticide activity was separate from his operation and not his concern.

Bill said he then went to the Federal Laboratory in downtown Denver, and asked a lab employee about the chemicals. The lab employee described those chemicals as "horribly toxic", Bill said.

"But when I told him who I was working for and what I was doing, the lab employee refused to have any further conversation on the subject, and simply walked out".

The CF&I Benzene plant "was just a production nightmare", Bill recalled.

I never worked harder in my life trying to get that plant to go. I was thoroughly disgusted from the production and health standpoint. No insurance company or public health agency would permit that plant today. The exposure to Benzene was so severe, that I could literally smell the stuff in my urine.

Federal Health Researchers have pointed out that some workers have shown signs of Benzene poisoning after brief, constant exposures, while others displayed resistance for long periods of time.

The document went on to say that beginning early in 1947 CF&I tried unsuccessfully to make DDT that ended up being a disastrous nightmare for them at the leased military facility on the Arsenal.

They had nothing but trouble with the plant and could never make it work quite right. By 1949 CF&I had shut down all operations of its Chlorine-Caustic Soda plant because those materials were no longer in short supply world wide.

About the same time, CF&I closed its Benzene and DDT plants at the Arsenal, the CF&I lease on those facilities was taken over by Julius Hyman & Company of St. Louis Missouri.

At that point one had to ask what the disposition of the remaining chemical stock had been. There was no record of the new lessee purchasing the bulk chemicals that had been stored in tanks or the processed chemicals that were ready for shipment at the time that the plant closed down. Further investigation offered no clue as to the final consignment of the material. One can only wonder whether the chemicals were buried or dumped into one of the many basins, or pits that are located on the Arsenal property. In any event, one can surmise that they were not disposed of in a manner that would be consistent with today's standards by which we safeguard our environment and underground water supplies. In fact, one has only to look at the evidence to realize that the fact, one has only to look at the evidence to realize that the environment never entered into their thinking at the time.

Getting the job done and making the almighty buck was the only consideration. Whether or not it was done safely did not matter, and was never the issue.

On the last Saturday of September, as he sat listening to his boys fight over which type of cereal they were going to have for breakfast, the telephone rang.

Getting up, Larry answered it with his usual good morning cheerful type voice.

“Hello,” he said.

“Is this the home of Larry Land? he asks. You’re investigating the Rocky Mountain Arsenal, the Army, and Shell Chemical Company?” A voice asked from the other end.

“Yes I am, what’s on your mind?”

“Look in your mail box, I think you’ll like what you find. Mr. Land, you’re not alone in this fight. There are many people who want you to win, because if you win we all win a little”. Click went the phone.

“Who was it?” Marie asked.

“Someone telling me to go out and look in our mailbox. Larry said, I’ll go take a look”. Marie told him to be careful.

Walking out to the box, Larry was about to open it when he suddenly, thought of what Marie had just said “Be Careful” “What if it’s a bomb?” He didn’t want to call in the bomb squad, or the police, they would just think he’s nuts, especially if it really wasn’t a bomb. Shaking his head trying to remove such thoughts, but he couldn’t. So Larry got a very long narrow stick, crouching down in the barr pit (ditch) he pointed the stick at the edge of the door pushing it open. It didn’t blow up so at least he felt safer about it being a bomb. Larry then thought what if it is a box of some kind, what would I do. Well he carefully looked into the mail box and discovered it was several documents, papers and flyers held together by some very large rubber bands and paper clips, and a newspaper story. One of the reports was concerning deep well injection, and toxic waste at the Arsenal.

Once he was back inside the house Larry sat down at the table with Marie patiently standing by his side. Nervously, almost as if h e were afraid of what they might see, they opened up and unfolded the documents, and began to read slowly at first, then finally with greater speed. Among the papers were the following documents.

STUDY SAYS LAWN
CHEMICALS MAY
POSE TERRIBLE HEALTH RISKS!

Chemicals widely used on millions of lawns across America could cause cancer, birth defects, gene mutations, or other maladies yet to great to explain, according to a study released Monday.

"To create the picture-perfect lawn, Many lawn care companies rely heavily on chemical herbicides, and pesticides", said policy analyst Laura Weiss, who wrote the study for Public Citizens, a consumer rights organization. "But these toxic chemicals may do more that just kill weeds, and insects". Many have been found to cause serious adverse health effects as well.

"Many of the chemicals are available at most hardware stores for the do-it-yourself gardeners, including such widely used brands as dursban,, sevin, spectricide, and trimec. Spectricide, although available commercially, already has been banned on golf courses and sod farms", Weiss said, "is suspected of causing birth defects, nerve and liver damage".

Weiss's study, a compilation of previous studies that is billed as the first comprehensive examination of the 40 most widely used chemicals in the lawn care industry, found that:

* Twelve are suspected carcinogens. Of them 11 have been shown to cause cancer in laboratory animals and 9 have been classified by the EPA as probable or possible causes of cancer in humans.

*21 have been shown to cause other long term health effects in laboratory animals or humans, including birth defects, gene mutations, or damage to the kidney, liver, and nervous system.

*20 have been shown to cause in short-term damage to the human central nervous system, (Brain).

*36 have been shown to cause eye, skin, throat and lung irritation, difficult breathing in humans and animals.

Of the forty chemicals, only Metalaxyl has been tested for all of its potential health effects by the EPA, the report said. Metalaxyl, and triclopyr are the only ones believed to have no long term affects.*

To explain more on DDT, known as possibly the worlds deadliest insecticide, I will refer to a book by Lewis Regenstein. His book called "America the Poisoned", tells a story of "How Deadly Chemicals are Destroying our Environment, Our Wildlife, Ourselves and—How We Can Survive"!

Larry said he would take his hat off to this man Mr. Lewis Regenstein, any time any day. What he has written is truly a gift to the world. His book, and Silent Spring by Rachel Carson are two of the greatest gifts any one could have given to us all. It was these two books that inspired Mr. Land to share with the World his experience. His hi's and low's, good times and bad, that he did all he could to make this a better, and a cleaner World for us all.

Larry is a very determined Individual, very conscientious of others, his tenacity abounds, when it comes to the safety of others. He said we don't have to live in fear because of other's greed. We don't have to live in fear for other's careless, or mischievous deeds, or in fear of other's reckless, and or negligent, ways of wrong doing for expedience, and that almighty dollar.

From the book of Lewis Reginstein, an interesting excerpt from Silent Spring, 1962., and DDT.

"Over increasingly large areas of the United States, Spring now comes unheralded by the return of the birds, and the early mornings are strangely silent where once they were filled with the beauty of bird song. This sudden silencing of the song of birds, this obliteration of the color and beauty and interest they lend to our world have come about swiftly, insidiously, and unnoticed by those whose communities are as yet unaffected."

Larry will talk of what he has learned about DDT., It's spread, it's overwhelming grip that it once held on the human race, the animals, the birds, and the fish in the sea. Virtually every American was contaminated with detectable residues of DDT.

TO ALBERT SCHWEITZER
WHO SAID
"Man has lost the capacity to foresee

and to forestall. He will end by
destroying the earth".

Even after DDT was banned in the United States, it was still manufactured right here in the U. S. right at the Rocky Mountain Arsenal just north of Denver Colorado and also else where, hundreds of millions of pounds, for use domestically, and through out other parts of the World.

100's of millions of pounds of DDT have ended up in the Oceans of North and South America. No matter what we eat, or drink it's their. It has so poisoned the seas it has affected the phytoplankton a major source of our World's oxygen.

Every person in the United States carries trace levels of DDT. in his or her body. Then you stop to think trace amounts could be life threatening, and actually lethal. The problem is the United States has not yet learned how to control chemicals. Even when they know their toxic to life, big business cannot be stopped from manufacturing it. DDT is presumed to be fully capable of causing tumors in humans. The Library of Congress estimated in mid 1980 that over 4.4 billion pounds of DDT have been used for insect control since 1940, that is one pound for every human on earth.

In THE DARKENING LAND (1972), William Longgood describes how the spread of DDT and similar pesticides to the furthest corners of the earth is accomplished mainly through wind currents: Excerpt from the book Lewis Reginstein, America the Poisoned:

"DDT circles the globe, going from one hemisphere to another, often riding on jet streams that wing along at 250 miles an hour. It is estimated that in the atmosphere at any given time, there is more than one billion pounds of DDT and it's metabolites…it is reasonable to assume that of the 126,000 tons of chlorinated hydrocarbon pesticides sold annually, more than half enters the atmosphere…Studies of British rainwater indicated that in the United Kingdom, one inch of rain would deposit a ton of pesticides….It is incredible how far and fast it travels. The man spraying his roses in Kansas City today can contaminate a polar bear in the Arctic tomorrow".

Longgood warns that "the real price of our folly "is that" once DDT is in the atmosphere, there is no escaping it. It becomes part of our environment…It becomes part of the texture of meat, vegetables, milk."

As the President's Council on Environmental Quality wrote in its 1975 annual report,

Since the beginning of the National Human Monitoring Program in 1967, DDT residues have been found in 99+ percent of all human tissue samples: the figure in 1973 was 100 percent.

DDT in the late 1950's began to bring about a public outcry, since humans' themselves were being sprayed directly with the chemical. In 1957 the U. S. Department of Agriculture began a spraying operation for gypsy moths in New York State. Airplanes spewing out kerosene laced with DDT over three million acres. Dairy Farms, Children, gardens lakes, anyone walking or commuters waiting were sprayed, sometimes more than once. It was finally brought to the attention of Congress. "Lewis Reginstein".

One quite prominent and distinguished scientist spoke out repeatedly in defense of the chemical. Dr. Norman E. Borlaug, who was awarded the 1970 Nobel Prize for his work on the "Green Revolution" in helping to develop fast-growing food plants, insisted that DDT was not only safe but necessary to prevent massive World starvation. He referred to those opposed to DDT as "hysterical environmentalists". "Lewis Reginstein".

Even though DDT has been banned for use in the United States it is still manufactured here at an alarming figure of 44,000,000 pounds, with about 900 metric tons being exported each year. DDT is slowly being replaced with other more toxic type chemicals such as Aldrin, Dieldrin, Endrin, which at least two have also now been banned for use in the United States, however still manufactured here and exported to other countries for growing the very vegetables, and fruits that could be on our tables right now. The problem is it goes with the good chain, and remains in the food until it is eaten, then it remains with that person.

Larry said 1974, 1975, and 1976, Have been the busiest years of his life. In 1975 he finally got the State of Colorado to stand behind him, and they issued to the United States Army, and to Shell Chemical Company Cease and Desist Orders, both for manufacturing and dumping toxic waste. Colorado finally came through also with Shell Chemical Company denying their requests to inject toxic waste into the ground in Morgan County. Shell will be doing a lot of battle at the bureaucratic level to push the permits they want to inject the waste, with Colorado now standing pretty strong against any more of their negligent wrong doing, it will probably be pretty remote they will ever get the permits, they will need to stay in Colorado.

In 1975 Larry experienced two more close call drive by shootings, Then once in 1974, and twice in 1975 with someone driving up behind his pick-up with a very big car, and at the right time hitting your car then pushing left or right to throw your car out of control, and they usually have something their aiming your car, at like a bridge, or some type cement structure. Larry was a licensed Peace Officer, trained driver, a Professional stock car driver so he knew what he needed to know to avoid problems usually before they would even happen. The first time in the fall of 1974 he was not really looking for anything like this to happen, but was on the alert. His pick-up had a super charged 327. He worked usually on Thursday at Brush livestock market, and would be coming home shortly after dark down interstate 76. This one night in particular early fall, he noticed a car moving up behind him rapidly so he let them get real close they were still in his lane, Larry sped up a little turned into the inside lane and slammed on his brakes. The person went on by but not before slamming on his brakes as if he thought about coming back, he slowed quite a bit then sped up again so Larry was now tailing this vehicle. That person held it right down to the speed limit and Larry stayed behind for nearly 40 miles this car neither speeding up or slowing down, then finally turned of toward Brighton, Larry went straight on home. I really think this guy wanted me but could not figure out how to get back behind me. He is probably lucky he didn't try coming back, since Larry was very heavily armed for whatever.

Larry thought about this, and what he should do if it were to happen again. He decided to put some weight in the the back of the pick-up, just in case they met again. Well nothing happened again till next spring. They tried it a little different as Larry was coming out of this small town on a State Highway, this time from an Auction, this time he had the car driving along watching the behind more than the front. When all of the sudden he noticed a car coming at him, opposite direction, a person behind the driver leaning out the window with something big and black, and when he realized what was about to happen, he swerved off the road, as he heard tow loud cracks from a gun, he was close enough to see the fire from the muzzle. At the same time a loud explosion near his head, and again everything was in slow motion as he drove the car off the road, and right across someone's yard skidding sideways, the yard had only been sodded two weeks earlier none the less he rolled up a lot of sod, and by the time he got stopped he was in another persons yard. Well he had three or four people come running up to the car where he got their attention quickly he bailed out of the car gun in hand, they ducked and he ducked, but they were so close to me they ducked right with me. They were all real nice as he explained to them he had been shot at, and had the proof right in the top edge of the door. and the roof right behind where my head was. It appeared they were using a 44 magnum, very large hole in car, reports were made and the people were given insurance information, the one guy said you can use my yard anytime.

Larry thought, this is really not fun he has more happening off duty than when on duty. He no more than got that said, it did happen again off duty, coming home from Brush down interstate 76. All of the sudden here come the headlights, he was already doing 70 mph estimating this persons speed between 90 and 100 mph. Well it was just like before the person got real close then slowed down, very close behind me Larry knew then that person was just waiting for the right moment sure enough their was a highway that crossed the freeway coming up. Larry checked the person to see if he really meant to stay with him, so he slowed down that person slowed down keeping exactly the same distance, so he sped up, and that person sped up keeping that exact distance pretty close on his bumper. He knew by now exactly what he was going to do, this was probably the same one he got last time. so he is not going to do the same thing, well not exactly. Coming up on the bridge Larry had to make his move, he sped up to about 75 mph, that person right in the same spot but seem to be riding the center of the highway. Larry could see exactly what was planned, that person was going to hit the left side bumper, and by giving it gas, he would take the traction away from my rear drive wheels, and at the right second he would turn his wheel left sending me right into the cement bridge poles, and he would go merely on down the road. Instead as they approached the bridge, the person seemed to move in closer, Larry swerved to the right slammed on his brakes, and said "See Ya". He went flying on down the road, and out of site, the last time he had his plate number but it came back on a stolen car, not his car. This time Larry tried again to out think this person, because if he caught him it could be dangerous. The way he got out of their quickly never slowing down he probably, moved to plan two, thought this time he would out maneuver Larry, by pulling off somewhere and letting him go right on down the road, then he would have another shot at getting behind him. Larry knows the Country real well out there, so he right away caught a side road leading to a different highway and back home again. He would not tell his wife of most incidents, she probably would have freaked, and wanted to move right now.

Then in late summer this time he was on duty patrolling businesses, this was his job to make sure the doors on businesses were secure in Lakewood, Arvada, and into the edge of Denver. The doors would be physically marked every night so each round after that you could see the doors were still secure. Markers were placed in the doors or windows a certain way, with each night a different way. Then with the use of the spot light. Doing Routine moves on checking the doors secure, Larry got out of the patrol car and checked the doors on this little warehouse, a loud crack of a rifle, and almost like a small explosion just above and to right of where his head was, his face was peppered with small bits of cement. None the less he was able to hit the ground fast. As six more cracks of a rifle, very explosive echoing sound each time. It sounded much like a 30-06, and as the slugs hit the cement being right there it was like small explosions. Larry worked his way over to the squad car and radio'd shots fired. By the time assistance arrived it seemed to be pretty much over. It was very difficult to figure out exactly where the shots came from. slugs were gathered, they did appear to come from a 30-06, and at some distance.

There was one more incident again coming home from Brush, this car came almost out of nowhere behind me. Remember this is a fairly lonely highway, out in the middle of nowhere just a little scary by itself. Here this car is almost as if he is challenging me. Larry sped up he sped up, Larry slowed down he slowed down keeping an almost exact distance. This time Larry had brought a friend along with him just in case, who was an excellent shot. Larry had weighted down the back of the pick-up for good traction, and so no one could get under his bumper and take traction away from the rear wheels. Tom said to Larry, It's not that I didn't really believe you, I guess I really didn't want to believe this was happening. Tom stayed somewhat down in the seat, he could feel the first jolt, this guy meant business, but he did not yet have his target. It appeared his target was coming right up as he closed in on my bumper once again. Tom said this is it opening the small window in the back of the cab. They were doing about 70 mph., Tom said to Larry here goes, Larry said make sure you hit the center of the windshield, and not the driver, or passenger side were not here to kill anyone, just scare the hell out of them. We had a 45 semi automatic, the person getting real close in line, and cement bridge ahead Tom cut loose with three rounds. This person flat slowed down, got of the bumper we kept up about the same speed watching to see if he was coming back he pulled off at the first exit, Larry just kept rolling on.

Another report stated: in 1965 the United States Army Hygiene Agency found that more than 393,000 gallons of toxic waste, most of it from Shell Chemical Company, were being dumped into Basin F each day.

Another report of information stated: In 1961, the Army began to heed its consultant's caution that Basin F should only be used temporarily. Originally when it was constructed and put into use in 1955, the Army was told that Basin F had a total life time expectancy of 15 years maximum.

Larry realized that Basin F was actually two years past its life span when his family became sick and his cattle started to die in 1972. In 1975 the basin was still in use, now five years past its design recommendation, and expectation.

In 1965 it was found that a huge part of the Basin F liner was actually missing, and in fact a great deal of very toxic waste, and chemicals were in fact leaking into the underground water strata's off the Arsenal to the tune of 25 to 40 square miles.

The United States Department of Health gave a warning in 1965 to the Army. After the inspection of Basin F, A memorandum was issued to the Army warning they must stop any further use of Basin F. Warning them if they continue further use of Basin F knowing that it's leading, then you are using it with the premise of knowing for certain it is contaminating the underground water off the Arsenal property. A copy of this was sent to the Colorado Department of Health.

The Army continued use of the Basin, then diverted the chemical waste to Basin C right next to Basin F. Basin C was noted as straw Basin because everything just disappeared. This repair took nearly a year, with all the chemical waste that was removed from Basin F, and the continued manufacturing by Shell Chemical Company using Basin C. It disappeared just like they knew it would knowingly severely contaminating the underground water not only on the Arsenal but up to 40 square miles off the Arsenal Property. The Army managed to haphazardly patch up Basin F with full intentions using it know the life expectancy was surpassed. The United States Army, and Shell Chemical Company continued it's use knowing it was leaking, knowing the life expectancy was gone, using it with the premise of knowing for certain it was contaminating the underground water off the Arsenal Property.

Early in the 1960's, with the realization that the lifespan of Basin F was limited, the Army engineers began to develop a remarkable new plan for disposing of the Arsenal's hazardous wastes. They created a 12,000 feet deep, high tech well and began pumping the toxic products into the very core of the earth. Eventually 15 million gallons of hazardous waste, produced mainly by Shell, was injected into the ground. No logs were kept as to the specific chemicals involved. All we know today, is that the Army created, what has been termed by the National Toxics Campaign, a "highly toxic soup".

At the same time, a bizarre problem began to unfold.

Soon after the deep injections began, a series of earthquakes began rumbling across the Denver area. Larry brother remembered as he was watching TV one evening, when suddenly it felt as though the house was shifting underneath them. Startled, he glance across the room to see a hanging light fixture swaying back and forth, the nick knack's and china rattling in the china cabinet, and a long growling and groaning noise.

As the earthquake phenomena became more frequent, the media and the public began to ask questions. Why was the front range of the Rockies suddenly having earthquakes? It has never happened before. Larry's father new about the injection well at the Arsenal and tried to get people to listen when he told them what was causing the earthquakes. Most of the quakes were between about 2.5 and maybe 4. In and around the Arsenal the quakes were quite intense and did cause a lot of structural damage to some homes, mostly with the brick and block homes.

As investigations began, a local geologist came up with a surprising theory. He blamed the earthquakes on the Arsenal's deep well, saying the injections of waste were causing shifts of subsurface rock. Until that time, the Arsenal's operations were relatively unknown, and not talked about to the news media, or the average Denver resident., In fact wasn't even discussed with the Colorado Department of Health until after the fact, nor was the environmental Protection Agency notified till afterwards.

The Army denied any link at first. However, with some earthquakes reaching 4.2 on the Richter scale, and with public pressure mounting, the Army found it necessary to appoint a blue-ribbon investigative panel to determine the cause.

In December 1966, the panel reported that comparisons of pumping operations at the Arsenal, well and the frequency of earthquakes provided a very suggestively, correlated statistic.

The Army mothballed the $1.4 million deep well project and went back to using Basin F knowing full well it was leaking into the underground water off the Arsenal, since 1965. Shortly afterwards, the quakes stopped along the front range.

Even after the Army was ordered by the United States Department of Health and Welfare to cease all operations of basin F. Somehow, for uncertain reasons, the Army permitted Shell Chemical Company to continue dumping millions of gallons of toxic waste each month into the still Leaking Basin F. Information made available to Larry stating that Shell Chemical has already doubled, and tripled production of pesticides. Then a statement came out to all employees for all the overtime they want, Shell was going to quadruple production for as long as they were able to operate. With this they personally Recklessly, with malice in mind, and with total negligence, tripled and quadrupled the toxic waste produced and disposed of in Basin F.

Fearful that the public would learn of the leaking Basin F the Army Commander ordered that, "All future information on Basin F and the deep well injection system would be classified confidential due to its sensitive nature". That order was later revised after Army Officials decided it would be too impractical to reclassify all Basin F and deep well records. Instead, all records of the Rocky Mountain Arsenal environmental pollution were restricted as "For Official Use Only".

Army managers displayed an even more cavalier attitude toward public health in the early 1970's, when the Army allowed more that 10,000 Boy Scouts to camp out overnight on the facility grounds.

At the time, the Army was storing thousands to Nerve Gas bombs (sarin gas). Thousands of cluster bombs containing Lewisite and Mustard agents. Plus the fact they allowed them to stay on ground known to be contaminated with deadly chemicals.

That prompted the Defense Department's Explosives Safety Board to recommend that youth groups should immediately be ordered off the site.

"since studies and calculations reveal that all of the Arsenal property can be in the Zone of potential lethal, the sanctioning of this many people to having access to the Arsenal operations, being located on the Arsenal, is questionable from an explosives safety standpoint," the board wrote in a classified memo.

Incredibly, Army officials at the Arsenal argued against the youth group ban, saying the policy change would needlessly alarm the public.

"Termination of such activities would be an undesirable impetus for public apprehension or alarm" Army inspector General, D. F. Abernethy, wrote in a classified reply. "If youth group activities were terminated, it is anticipated that significant adverse community, political and congressional reaction would be incurred by the Rocky Mountain Arsenal."

That Saturday afternoon Larry was scheduled for two interviews, one for a television station in Denver, the other a live talk show on a local radio station.

His luck so far had been good, but deep down he knew that anyone capable of this kind of environmental destruction was capable of anything, including closing the mouth of anyone who is trying to get the truth to the public as Larry was doing, They had already called him, and he defiantly was going on. The very one who was bringing public attention to these and other terrible tragedies through out the world was afraid, a little scared but not going to quit. Larry could only pray that his lawyer and the few politicians who were on his side, could get his day in Court.

By the time he was finished with the last interview it was dark. He was tired and hungry, and wanting a cup of coffee, and in no mood to be pushed around or pushed back, even though everything had gone remarkably well.

Walking out of the radio station studio to the parking lot he looked, then looked again, the stopped dead in his tracks as he stared at his truck. Even in this well lit parking lot someone had managed to slash all four of his tires, not merely cut them, but cut the side walls to ribbons. Larry said God be with me, I cannot quit now, even though this is another warning of what could be expected.

Two days later while talking with his Attorney, a plan of action was discussed and laid out on the table for their future arguments in the hallowed halls of Congress.

The arguments were based on the Attorneys opinion and judgment that everything resulted from the conduct of the Arsenal's agent and landlord, the United States Army who had set up and blue printed all the plans used manufacture, and to dispose of millions of tons of toxic chemicals. The Army was responsible for the Shell contaminants, as its own, as well as other offsite chemical manufactures they had allowed to dump into basin F, 100's of thousand's of gallons daily. That is all in the Army's Classified way of doing business without anyone knowing about it perhaps their own boss.

For these reasons the Army was considered to be the more proper defendant where a choice had to be made. Since they held sole responsibility for what went on at the Arsenal. Just like in 1965 they were warned by the United States Department of Health to stop using basin, that continued used would be under the premise they were using a basin leaking into the water wells off the Arsenal property. Remember this, and as you read the story further you will see the callous disregard for human life, the Army truly held. Please also remember the Army was using this basin, with Shell Chemical Company tripling, and quadrupling their production, and toxic chemical waste, with the "Congress of the United States Of America completely, and totally aware of this happening". I had even wrote to the president of the United States of America Richard Nixon it also fell on deaf ears. They all failed to act for the people. We'll discuss this a couple more times a little later to show how far they went.

At the beginning of their evaluation of this case, they were under the impression, that they would be able to proceed against the Army under the Federal Tort Claims Act ("FTCA"), on a "nuisance" tort theory, which would not require Larry to prove negligence on the part of the Government. They then found out that they could no longer rely on the "nuisance" theory because the United States Supreme Court holds that the FTCA does not extend to the tort of "nuisance" because it does not involve an element of negligence, which is required under Federal Law.

After you have won congressional approval, you can sue the Federal Government only under certain circumstances, and even then, there are exceptions that preclude you from having a successful case. If the wrong done to you, does not fall within statutory exceptions, the FTCA may only be invoked on a negligent or wrongful act or omission of a Government employee.

Many areas of liability are excepted from prosecution under the FTCA, and there are two express statutory exceptions to the FTCA that would appear to directly affect Larry's case, and they are:

(a) EXECUTION OF STATUTE OR REGULATIONS.

Claims based on acts or omissions of Government employees, exercising due care in the execution of statute or regulation are not cognizable under the FTCA.

It is only if the Government employee fails to exercise "due care" in the administration of such statutes and regulations, that a cause of action for the resulting harm would lie. Smith V. U. S. 101 f. Supp. 87 (Dist. Colo. 1952).

In other words, if the employees at the Arsenal are exercising due care in the execution of Federal Statutes or regulation governing the operations at the Arsenal, the Government could not be sued, or held liable under the FTCA.

(b) DISCRETIONARY FUNCTION.

Claims based on the exercise of a performance, or failure to exercise or perform a discretionary function or duty on the part of a Federal Agency or Government Employee are not cognizable under the FTCA. The purpose is to protect the Government from liability that would seriously handicap efficient Governmental operations.

whether or not the discretion is abused is immaterial. United States V. 331 f.2d 498 (10th cir. 1964).

In other words, if the Government administrators, in establishing plans, specifications or schedules of operations for the Arsenal, and are considered to have exercised discretion in establishing such, the Government cannot be successfully sued for such operations.

Many Courts have established a distinction between decisions made on a "planning" as opposed to an "operational" level with regard to this discretion," and have held that negligent or wrongful exercise of discretion at the planning level is not actionable under the FTCA. Once discretion has been exercised at this higher level, a duty of due care attaches, and those administering a decision or action under it at an "operational Level" may not act negligently.

No matter what happened, this was going to be one hell of a fight, but Larry felt real good now that Congress, and the State of Colorado were standing with him on this one, and getting stronger. Seeing other cases of haphazard methods of disposing of or storing of hazardous waste being seen, recognized will help us in cleaning up, and making a safer environment for everyone.

Larry felt that this growth that was festering on the hands of humanity, along with the current mentally dangerous officials who were responsible for making this disgrace happen, had to be stopped. The wrongs of past generations that allowed this to continue had to be cleaned up and made right with much tighter controls.

Larry was soon to begin, one of his many trips to Washington D.C. where he will eventually stand as witness before the Congress of the United States of America. Larry would not carry with him any photographs, or information regarding the Arsenal, truthfully fearing for his life, as did Karen Silkwood at the time she was killed, when she was exposing the wrong doing of the Kerr Mc Gee corporation Cimarron Facility in Oklahoma. In the summer of 1974 Karen called one morning, said hello is this Larry Land, the one who is burying the Army, and Shell Chemical Company almost daily on T V, Radio, or in the newspapers. He said well yeah you could say that, but who's this. She said this is Karen from Oklahoma. She said I work for a Company here that is really out of control, not doing the right things endangering peoples lives. I just called to talk with you a moment to find out what gives you the strength, the courage to do what you do.

You have inspired me giving me the courage and strength to do what I must, and still move forward in life. Larry said it is very simple, I love life, and I love people, what's right is right, but wrong is wrong, and I do everything in my power making those who do wrong to answer up and to "do it right". Then try to get Government officials to make sure it is done right and with safety. Karen indicated to Larry she would like to come up and see him sometime this summer. As Larry was saying good bye, he didn't realize it would be the last, he told Karen to be careful, and GOD be with you. Karen was killed in November of 1974 before she ever got up to Colorado. The Kerr Mc Gee Corporation was closed shortly thereafter when they were found responsible for wrong doing, and possibly the cause of Karen's death. It was the Congress of the United States who eventually Paid a very large sum of money to her family evidently taking part of the blame for her death. She was killed by someone coming up behind her and hitting her bumper, sending her car careening out of control heard on into a cement bridge. There was for years a lot of talk how it could have been some one from Kerr Mc Gee corporation, or possibly some Government agents who were sent out with orders to shut her mouth permanently. "She was assassinated for trying to do the right thing in life". This made Larry even madder about how things like this can happen, so he pushed harder than ever before, and began the first of what would be many trips to Washington D. C.

BALD EAGLES OF THE ROCKY MOUNTAINS DEAD IN THEIR TRACKS

"At the hands of the U.S. Army Rocky Mountain Arsenal"

CHAPTER THREE

THOSE MOST DEDICATED TO THE FUTURE ARE NOT ALWAYS THE BEST PROPHETS.

Elinor Hays, 1961

Late one morning after leaving the local newspaper office, Larry stopped at the neighborhood grocery store. He had just put a gallon of milk into the cart when a man walked up to him.

"Mr. Land, may I speak with you for a short moment?"

"Sure, what can I do for you?" He innocently replied.

"my name is Dennis Allen. I work at the Rocky Mountain Arsenal."

Caution instantly entered Larry's mind, caution that was born of desperation which in turn bred suspicion of all persons connected with the Arsenal.

"Before we go any further I must warn you that I have nothing but contempt for the United States Army, and the people associated with that place. Do you have any identification that says you work there?" Larry asked.

"of course."

At that point he reached into his jacket and pulled out his identification badge with his picture on it. It was issued by Shell Oil Company.

"So you work for Shell, Chemical, all right, what's on your mind and please make it quick?"

"I wish I could but that is impossible, there are too many ears that could overhear what I have to say. Can you meet me just past the truck stop at eleven o'clock tonight. I will be in a brown ford pick-up."

"I'm supposed to trust you just like that, with no proof. How do I know I won't get beaten or killed out there?"

"Mr. Land I have seen you on the nightly news and in the newspaper. I'm sorry that it has taken me so long to work my courage up and come to you but I am not a killer or some crackpot playing some stupid game. If they knew I was planning to meet you tonight my life would be worth less than two cents. What has happened in the past and is happening out at the Arsenal this very minute is wrong, morally, and environmentally. This will give you the fuel you need. There are two others assisting in getting all the information you'll need to help stop Shell Chemical Company in their tracks. I'm hoping it's not too late to help put an end to this."

"All right Mr. Allen, I'll take you at your word and I will meet you at eleven O'clock tonight. But for your sake nothing had better go wrong."

"I knew you would be reluctant to meet me tonight, so I came prepared. This is my Colorado Drivers License, my Visa card, and my automobile registration. With this you also have my home address, and the home of my family, with a telephone number for you to verify. Please compare the signature on all three and verify they are the same."

Examining the three items Larry saw that all three were in fact identical and clearly stated as much.

"All right, do you have a piece of scratch paper? I'll sign this paper and you compare all four signatures, verifying they are the same."

"All right, they are the same, what's your point?" Larry asked.

"Compare all the pictures, are they the same person?'

"Yes."

"These are pictures of my wife, children and myself. Take the identification, and the pictures home and give them to your wife. If anything happens tonight she can give them to the police and they have got their man. Otherwise you can return them to me, fair enough?"

"I'll meet you tonight Mr. Allen" Be There.

To say that his curiosity was aroused was definitely an understatement. To say he was scared would be even more of an understatement. The one thing he definitely was not, as the John Wayne type. Yet, Larry was excited, there could be no denying that. Was someone actually starting to listen too me, he thought.

Rushing home he told Marie what had happened at the grocery store and gave her the identification.

"Larry, could it be some kind of a trap? we've been getting a lot of threatening phone calls lately."

"I don't think so, this guy sounded too sincere, and he was scared, I could see that. Plus we have his ID and I checked the license plate number on his car against his registration when I left the store, they were the same. I honestly think I'm starting to get through to some of these people."

Marie looked at him with a mixture of concern and fear. So much had happened already.

Larry knew that she would be worried tonight and he felt guilty for causing her so much pain. Still, he knew that she understood, and would stand beside him all the way. She always had. She feared more than Shell Chemical, or Army officials would come after her and the children, but then she felt a lot like Larry did. "WHAT HAS TO BE DONE< HAS TO BE DONE."

Much later he went outside to do a few odd jobs around the farm. A sense of excitement had grasped his very being. Perhaps he was no longer alone in his fight. There were others who knew the truth and were finally willing to come forward. They had the information that he needed, but had been unable to obtain. Maybe he was turning the corner. *Like turning the tide of the battle*.

At the same time, he realized that doubt was a very real part of this. Was this man telling the truth? Was t his a setup? Larry's mind swung back and forth between hope and doubt, He went about his work, all the while arguing with himself. He thought about the questions that Marie had asked him not long ago. Did he have the right to risk their lives together? He wasn't so afraid for himself, but he did worry about his family. If anything happened to any of them, his life might as well be over. How far should he go? He asked the question of himself, but deep down inside him he already knew the answer. He had to see this through.

Again he looked over the land that he knew so well, and thought of the people who meant so much to him. Some were gone, while other's were quite ill. This shouldn't have to be. It had to stop. He could not turn his back on those that remained and walk away, not now that we had them on the move, buying up property, and destroying the homes as if they never existed.

It would be a long wait until tonight.

At ten-thirty that night Larry Kissed his wife good-bye, took one last look at the boys, as if tomorrow may not happen as he drove off toward the truck stop. As a precaution, he had brought along the hand gun that had been his fathers, a .45 caliber, model 1911-A semi-automatic, a very wicked weapon in close quarters. He prayed to God that he would not have to use it, but felt he would not hesitate if need be.

As he passed the truck stop, Larry glanced out the side window. The moon seemed unnaturally large, as if it were just a short distance from his truck. The stars were brilliant, looking like beautifully cut crystal, motionless, suspended from the sky for his benefit. Who knows maybe this would be the last night he would ever see them.

Just then the tail lights of a vehicle that was parked on the shoulder of the road, went on, then off, then on and off again. It was a ford pick-up. Larry was beginning to feel like a damn spy. Pulling up behind the truck he stopped and got out, putting the weapon in his belt as he did so. For some reason he was suddenly cold. Anxiously he waited for some signal or a move that would tell him he had walked into a trap, but it never came.

"Mr. Land, I'm glad you could make it. I brought along two other people that I work with, I hope you don't mind. Larry stated yes I do mind, you said you were going to be alone. This does nothing but put me on high alert. I would rather from now on you stick with exactly what you agreed to. Allen stated I'm sorry Mr. Land, however I believe these two people can add a lot to what is happening. Mr. Land this is Jennifer Lindsey, and Glenn Morris, this is Mr. Land." "Hi nice to meet you."

"Thank you, and nice to meet you too. All right Allen tell me what you have got."

"We have been examining records of secret operations conducted at the Arsenal and outside its boundaries. Most of those operations exposed the environment, and the public to chemical and biological agents. We feel the world has lived too long in the shadow of these chemical horrors."

"I need facts Allen, facts that are written in black and white, and signed by some damn Army General. We need documents that will hold up to scrutiny in the Courts of Law. Not hearsay or rumors from some disgruntled employee. When this crap hits the fan, it's going to go off like a bomb." Right now the lid is off the can, and they are squirming like the worms they are., Lets not let them get away."

"We have them, plenty and we can get more. Look at them Mr. Land, then get rid of them. Put them where no one will ever find them until you are ready to make your move, agreed?"

"The Army filled bombs and artillery shells with a chemical that attacks the central nervous system and kills within minutes. The chemical was 'GB' or 'GB Nerve agent, better yet sarin nerve agent, the same as they used in the subway attack in Japan. It not only kills, but can cause problems that can be absolutely devastating. Some of the employees were so afraid of leaks that they carried rabbits with them. If the rabbits died…and many did, the workers would know something was wrong, and could clear the area immediately.

"Some employees, who have since left or retired, told us that they were worried about 'miscellaneous' chemicals being handled in the area known as the 'north plants'. Those chemicals included 'VX' a nerve agent many times more toxic than 'GB'. "DIMP" DiIsoPropylMethylPhosPhoNate, and IMPA both sickening, and pain agents capable of death, the by products of the manufacturing of sarin nerve agent, and GB nerve agents. DIMP, and IMPA were one chemical away from being very deadly, it could be take from a health threatening, and injurious agent to deadly sarin nerve agent in seconds. Then we got lewisite, which is a blistering agent added to mustard Gas, once it is on you it cannot be taken off, and will burn to the bone. Of course then we also have what is called Adamsite used to make the enemy sick, without killing them, but the long term affects were later found to be deadly. Then there is phosgene gas a poisonous agent that really never worked out for the Army. This was made and stored at the Arsenal in the 1940's, and 1950's. My ability to get at top secret documents. I will tell what I read from the person who was in charge of the Army Corps of Engineers at the Rocky Mountain Arsenal, during the 40's and early 50's. He was trying very hard to get the attention of certain Congressmen, the Environmental protection agency, The Colorado State Health Department. during that time he was in charge of getting rid of toxic waste. The thought at that time is just bury it. So they buried 500,000 tons of phosgene gas sealed in 55 gallon drums, at the time all he could remember it was in a huge pit near Brighton Colorado. Like many of the other toxic waste sites on and off the Arsenal there are no records as to what is buried in what pit, or where, and there is no record of the huge pit near

Brighton, Colorado that even tells where it is located. When ask didn't they keep records the answer was no, even if there were records they were classified and then eventually destroyed. This sounds just like a catastrophe waiting to happen. I will not even venture to guess what this might mean. "Maybe this is something best answered by the ones who done it".

"Most of these chemicals have been spilled, some in large amounts, at one time or another. All of these chemicals, whether contained in weapons casings or not, are buried at different locations on and in the Arsenal. I am not talking about pounds Mr. Land, I am talking about tons. Many of the chemicals that were manufactured, destroyed, or buried, have no records, at all or maybe bits and pieces to show their existence but nothing to show the location of where they are buried. No one knows for sure what is buried out there, they just know it's there. Some of the secret projects were never mentioned in the Army's contamination assessment. The Government is even afraid to dig into the pits to find out what is there, they know the dangers. But what does one do, well the Government got befuddled so they buried there head in the sand like the ostrich, hoping it will go away, How can it?

"In one fact finding Tour we found that in one of the secret operations, the Army intentionally sprayed a chemical and biological agent across a large part of the great plains states. thus exposing millions of Americans to chemical and biological agents.

"The Army developed an agent called 'TX' it was a fungus that attacked and killed wheat, but it was only supposed to attack the strain of wheat grown in the Soviet Union. Just to be sure, the Army sprayed it over parts of the American wheat belt, again the great plains states. Kansas, Nebraska, Ohio, Minnesota, The Army reported it did not affect the wheat crop, Unfortunately the Army did not feel it was important enough to explain, or report the effects on humans. Somewhere on the north end of the arsenal, seventy tons lay buried underground". I have heard it can be life threatening, and I have heard it is carcinogenic. but I doubt the Government would want to tell the millions of Americans in the Great Plains they had sprayed them with this biological chemical fungus, you may now become ill from it, we really don't know what it will do.

"You have documentation for all of this." Larry asked?

"The answer was you bet we do, and there is much more."

signed statements by the people who actually did the work." "Regarding the TX, it can cause death, studies showing chrom-o-somal damage and mutations.

Mr. Land the pictures you just released to the press, the infra-red photographs of the arsenal. It has really got them Hot, running around trying just to figure where the records are for the pits already named. Your pictures pointed out as hot spots. I know they met with you, and understand they are going to do their own Infra-red photographs of the arsenal. Yes Allen that is probably right, they met with me and wanted the photographs from me, and I told them I have about 80 photos, literally showing up at least over 300 hot spots on the arsenal, and they were classified, and sent to the Congress of the United States. They seemed a little angry as they walked away stating they were going to get their own.

As near as I can trying to measure some of the Dump sites, there are some ranging from 1000 feet to 2000 feet in length and several hundred feet wide, 30 to 40 feet in depth, then barrels stacked in solid front to back then covered up, and no records to show what is there. Most of them who put it there are now dead, or cannot remember where they buried what. that is why we call that 18000 acre army base "Poisoned Earth" Was named by the Environmental Protection Agency "The most contaminated piece of Land on earth in the 1980's".

The United States Department of Health and Welfare, In 1965 Issued orders to the Army to stop using basin F as it was leaking into the underground water stratus, and to continue using it your doing it with the premise, Knowing it is contaminating underground water off the arsenal, after having these orders not to use the Multi-Million gallon basin for outside storage of waste. The Army continued using it even more so until 1983.

"Another one of the Army's biological experiments took place in the middle of New York City. Special Government agents carried light bulb type containers filled with bacteria into the streets of New York, then began dropping them onto the ventilation grills smashing them into the subway system. Spilling millions of germs onto the public below. The goal was to see how far and how fast the bacteria would travel, and what effects it would have on people. My God knowing it would make them ill, but exactly how ill, they didn't know, but they knew they could track it and they did.

The Peter Obero family has lived in the San Francisco are for almost a century. Peter's grandfather died from and unusual infection that puzzled Doctors. Twenty-five years later, Peter learned the bacteria that caused the infection and his grandfather's death had been sprayed over the San Francisco area by the United States Army. This also was part of a Top Secret bio-warfare experiment. This is my opinion but it is something I believe very strongly. Today the people of the United States are more afraid of their own Government, than they are of organized crime. How many more deaths have these chemicals and biological agents been responsible for that we will never know about.

"The Rocky Mountain Arsenal was the main production center for deadly weapons in the 1940's through the 1970's. The weapons produced there could and did kill within minutes, even seconds.

"In the 1960's the Kennedy Administration came along. JFK and his advisors started talking about a more "humane form of warfare. "Weapons that would incapacitate the enemy but not kill him. They called their plan "project 112". The official records of the Rocky Mountain Arsenal shows that it was involved in project 112. Records also indicate one of the weapons produced was called 'BZ'. It's an hallucinogenic drug many times more powerful that LSD.

"The Army filled bombs with more than fifty ton's of BZ, that was supposed to be enough to send every person in the world on a psychedelic trip. "BZ" isn't listed in any toxic materials or contamination records for the arsenal, or does it show whatever became of those bombs or the hallucinogenic drug that was used. I suppose like everything else it got classified and stored somewhere, and they don't know what they are. The Army learned years ago that "BZ" was a very toxic pollutant. An eyedropper full of this stuff can contaminate a large water supply, and it's very difficult to trace or remove.

"Mr. Land you better hold on to something for this next part. The way the Army learned that BZ was hard to remove from water was to dump large quantities of it and other chemicals into the creeks and ponds around military installations, but we were unable to find records showing it was done at the arsenal, just all around it. Most of these bodies of water soon flowed into the Platte river, carrying the toxic waste on out across Colorado, and into Nebraska.

"Mr. Land, take these documents and be careful, do what you can, but remember they will stop at nothing to safeguard what has happened. When we have more information we will be in touch. then one of the other's will meet you, in a different location. That way it may provide some safety for us all. Remember to conceal those papers well, good night Mr. Land."

Walking back to my truck I looked into the sky. It was as black as anything I had ever seen, with the moon at full phase, like you would imagine seeing in a horror story inside a cemetery on Halloween night.

Feeling a strange sensation in my hands, I placed the folder beneath my arm and looked at my hands. They were shaking, uncontrollably, maybe I was more frightened than I had realized. I knew before I had left the house my imagination was working overtime. Larry suddenly stopped and turned walking back toward Allen's truck. Larry's mind suddenly had everything going in slow motion, but he was handling every move very careful. As he got to the Allen's truck, Larry said you better take these, "Larry said as he handed him back his documents.

"Am I to conclude from this, that you don't think I'm going to kill you, that I am now trusted.

We can't be to careful, can we? How can I get in touch with you if something should come up?" Larry asked.

"You can't. why did you give these back so soon, you were going to give them to your wife until after our meeting tonight?"

"Perhaps I thought I'd take a chance, besides you looked pretty damn scared back at the store."

"I was, I had a difficult time controlling my composure."

"I know exactly what you mean, I felt the same way." he told him as he reached up and took the documents back from Mr. Allen.

As he was walking back to his truck,, Larry once again looked into the sky. It was black as anything he had ever seen, with the moon still brilliantly illuminated like you would imagine seeing in a cemetery on Halloween night. It was beginning to look like he would be able to look at them for a while longer, realizing his mind had put him on total alert, the adrenalin began rushing, and Larry's world had went into slow motion.

A person may think it's slow motion, so I asked a Doctor about this and he said. What is happening everything around you seems it is in slow motion but in reality when your put into a position of what could be life or death. Your mind speeds up so rapidly to allow you to handle a life or death situation, it seems as if every thing around you slows down. By your mind speeding up, and what seems like slows everything down, is why at that instant you can handle an impossible situation, and the body is given adrenalin for the strength you may need.

Rushing home like a man gone mad, Larry roared into the driveway. He wanted to tell his wife everything that was said, plus he wanted to get started reading the documents.

Dennis Allen had promised that he could get more documented facts. He was pleased with that, because he would need many more for what he had in mind.

Sitting down at the kitchen table, Larry began telling Marie about meeting Jennifer and Glenn, and about everything that was said. Instead of remaining calm, their excitement grew as they began reading and going through the papers. They had tried to prepare Themselves, but what the two of them read that night put the absolute fear of God into them. No such preparedness would have helped.

As they read each document, a strange feeling of impending doom rested in the pit of their stomachs. A feeling that shattered the earlier moments of elation. Was this feeling an omen of what was to come? Larry hoped it was not.

What they read that night was like a bad dream. The reports and memos were as follows:

-The Arsenal stores millions of gallons of toxic waste and metal laden corrosive liquids that are left over from decades of chemical weapons production.

-The manufacturing of Nerve Gas that attacks the body's central nervous system is lethal in extremely small doses. Diisopropylmethylphosphnate known as (DIMP), is a chemical compound that is the by product of GB or Sarin nerve agents. It is tasteless, colorless, odorless. It is deadly in it's own right but add one chemical compound to it, and it becomes totally lethal. Dimp. Remains entirely stable when mixed with water.

-Larry had information supplied to him, from an unnamed source that, showed in 1983 thousands of tons of DIMP, was sold to Iraq, by the United States Army, to be used against Iran in the Iraq, Iran wars, and to be used on the Kurds in the north. There were also to use with this thousands of tons of Mustard, and lewisite. Where Iraq was able to spray entire towns, killing all it's inhabitants without a prayer.

-It was also understood along with this thousands of tons of IMPA was sold to IRAQ, by the United States Army, again another very deadly agent just one step away from B-Z nerve agent. was sold to Iraq for the same purposes.

-By 1990 when the Persian Gulf war broke out Sudam Huisane had all the makings of not only sickening, and health injuring agents, to full blown "BZ nerve agents", "GB nerve agents", to "Sarin Nerve agents", To "Mustard agents", With secrets given to him by our own United States Army, with a good chemists he was set up to make just about anything. To top it all off Iraq had all the Biological warfare agents being shipped in and manufactured for him by the Germans, and the French and Russian connections.

-In an Army study, it was noted that the health effects of long term exposure to low levels of toxic chemicals in drinking water are hard to predict, or detect. Certain individuals may be more susceptible than others to chemicals due to genetics, general health, age, and lifestyle. It is believed however, that children are more susceptible than adults in most cases.

-Of the 750 different chemical/compounds used, created, or found at the Arsenal, the following 71 chemical/compounds are considered harmful to humans. The letter following each chemical denotes carcinogen classification: A Human carcinogen B-probalble human carcinogen, C-possible human carcinogen, D-Data not available to assess carcinogenity: 2, 4-D, and at least 7 types of agent Orange. Aldrin-B, Arsenic-A, Atrazine-A, Benzene-A, Benzothiazole-B, Bicyclopheptadiene-B, Cadmium-B, Carbon Tetrachloride, Chlordane-B, Chloroacetic Acid-C, Chlorobenzene-C, Chloroform-B, Chromium-A, Copper-B, DDE-B, DDT-B, DBCP-B, Cichloroethane-C, 1, 2-Dichloroethylene-C, 11, 2-Dichloroethylene-D, Dicyclopentadiene-D, Dieldrin-B Dimmethyl Disulfide-D,

Diisopropylmethalphosphonate (Dimp-D0 by product of GB never agent, DMMP-D, Dithiane-D, Endrin-D, Ethylbenzene-D, Flouride-D, Fluorocetic Acid-D, Hexachlorocyclopentadiene-D, IMPA-D, a by product of BZ nerve agent makes people to crazy,. Isodrin-D, Isopropylmethylphosphonate-D, Isopropylmetylphosphonic Acid-D, Lead-B,, Lewisite, (a blistering agent added to mustard gas)-B, Lewisite-Osice-D, Malathion-D, Mercury-D, Methylene Chloride-B, Methyl Isobutyl Ketone-D, Methylphosphonic Acid, Mustard-A, NDMA-B, NEMAGON-C, Used fro nerve agents and pesticides. 1, 4-Oxathiane-C, Parathion-D, Chloropheny (sulfone Nethy-D, P-Chlorophenylmethyl Sulfoxide-D, Sarin-D, Supona-D, 1,1,2,2-Tetrachloroethane- C, PCE-B, Thiodiglycol-D, 2,2-Thiodiglycolic Acid-,, Toluene-D, 1,1,1-Trichloroethane-D, 1,12,-Trichloroethane-C, Trichloroethylenle-B, Vapona-B, Xylene-D, zinc-C.

Note: * =Profile missing.

-The following effects have been observed on humans and animals in various government and civilian studies.

Mercury: Central Nervous system and kidney damage. Mercury enters the body through the Nerves, easily passing into the Central nervous system and doing the most damage. this can be taken in by air, water, or fish, mixes well with water, cannot see, smell, air easily mixes to intoxicate it's victims before they know what is happening. from fish large doses can be taken in that method, no matter, Mercury is so extremely dangerous. EPA banned most uses of Mercury in the United States.

Lead: Headaches, joint aches, (permanent arthralgic condition), anemia, Nerve problems, Mental retardation, Short term Memory, learning disabilities, at any age, Permanent Hypertension, Birth defects, and can cause cancer. So Very toxic and dangerous EPA banned it's use in paints in 1967, in the United States.

Cadmium: Kidney damage, anemia, pulmonary problems, high blood pressure, (Hyper tension) possible fetal damage and cancer.

Arsenic: Liver, Kidney damage, blood and Central nervous system damage, and cancer.

Chromium: very toxic at low levels, Suspected cause of some forms of cancer.

Trichlorethylene (TCE): In high concentrations, liver and kidney and central nervous system (cns) damage, skin problems, Depression of the contractility of the heart, suspected cancer and mutations.

Tetrachlorethylene (PCE): Liver and kidney damage, CNS damage, depression, and cancer.

1,1,1-Tetrachlorethylene (TCA): CNS depression, liver damage, cardiovascular changes, cancer and mutations.

Caron Tetrachloride: Liver, Kidney, and lung failure, CNS damage and cancer.

1,2-Dichloroethane: In low to moderate concentration CNS depression, liver and kidney damage, gastrointestinal problems, pulmonary effects, circulatory disturbances, and suspected cancer and mutations.

Benzene: Chromosomal damage in both humans and laboratory animals. Benzene affects blood and Immune system to cause anemia, blood disorders, and leukemia.

1,1-Dichlorethylene: At high levels, liver and kidney damage, suspected cancer and mutations.

Polychlorinated Biphenyl's (PCB'S): Liver damage, skin disorder's, gastrointestinal problems, and suspected cancer and mutations.

Chloroform: liver and kidney and heart damage, suspected cancer.

Dibromochloropropane (DBCP): Male sterility and cancer.

-GB-Nerve gas: Accidental exposure to GB Nerve gas has produced abnormalities that have persisted for a year or more in the brain waves of a large group of Rocky Mountain Arsenal civilian employees.

A substantial number of the 1,397 employees who were inadvertently exposed to minimal doses of Nerve Gas unknowingly participated in these studies.

"What they really meant was a substantial number of the 2,397 employees who were purposefully exposed to minimal doses of Nerve Gas, and were unknowingly used as guinea pigs by the Army." This was Larry's first thought of what they really meant, since the Army over many years has been so guilty of using people as guinea pigs. Larry Said he will point out if necessary as many cases where the Government has used the public as their very own guinea pigs, by releasing chemicals into the air, Schools, crowded town settings, to the subways of New York, Chicago, and San Francisco, to the air and water filtration systems.

These health abnormalities suggest that something may be wrong with the life important functions of the brain.

Classified Information: The results were so grave, These employees will never be told the truth, or the results of those studies. "Classified not for release."

-GB Nerve Gas, one of the family of organophosphonates is so potent that a tiny droplet of gas on the skin, or inhalation of an equivalent amount, can cause convulsions and death within less than a minute if the exposure isn't immediately treated with the proper antidote.

-In 1954 the Army and Shell Oil Company stopped using the unlined basins and began dumping their waste into unlined trenches while the Army constructed Basin F, a 93 acre, 240 million gallon holding basin. This basin, which was constructed to withstand the pressure of millions of gallons of chemical waste. this consisted of a rolled asphalt liner one third of an inch thick, or about as thick as the sole of a shoe, applied directly over sandy soil, no hard base at all just 1/3 of 1 inch rolled over the sand. Basin F, on a high point on the arsenal ground. It was said "OH WELLL" if the damn thing leaks it will go to the north and north west not toward Denver. Remember this is 1954 when they started putting chemicals into basin F, The gave it a maximum time span of 10 years. In 1965 the Army Environmental Hygiene Agency, along with the United States Department of Health and Welfare, recommended that steps be taken to eliminate Basin "F" as soon as possible, and thereby remove much of the present environmental hazard of exposed surface storage of toxic waste. p. App. 10 at 3-4. By 1969, there was evidence that Basin F was leaking: Indeed, a physical inspection reflected that sections of the protective membrane were absent/Id. at 11-12. Thus, by the spring of 1970, the army understood that it was operating Basin F on the premise that it was leaking. Id. at 13.

-By 1966 Army officials had notified Shell that its dumping into basin F. could not continue. In 1970 the Army notified Shell that the asphalt liner in Basin F had deteriorated, and was presumed leaking. The Army did nothing to enforce their position, and continued to let

Shell dump into Basin F for another thirteen years all the time they tripled and quadrupled their production.

In 1966 the Army started to pump the toxic was 12 to 14000 feet deep with ejection pumping into the lower rock formations. This proved fatal when they began causing earthquakes in the area, so many that people got up in arms, which caused the U. S. Geological survey to move in and put a stop to the Army's use of ejection pumping. They haphazardly repaired the part of basin F that was missing, then begin using it again. By the 1970's they tripled and quadrupled the amount of toxic waste going into basin F. knowing full well it was leaking. The State of Colorado had issued cease and desist orders to the Army and Shell Oil Co. in 1975, 1976, 1977, and 1978, it was a fact at least 30 square miles off the Army ground had been contaminated and polluted. The Army by 1974, admitted that chemicals from the arsenal had in fact been found in the Land Wells. The Colorado Dept. of Health in 1975 claimed they reported to the Lands their water was contaminated with materials emanating from the Arsenal. It was reported the Materials found in the Land wells could literally be fingerprinted to the United States Army, at Rocky Mountain Arsenal. Land's ranch stood just one mile north of the Arsenal grounds, and approximately two and three quarters of a mile from basin F. They used that basin until 1983, when the Army base for manufacturing any thing, and Shell Chemical Company were stopped for good. It was made clear that Shell Oil Company was not welcome in the State of Colorado. The Army cannot sell the Land for any type development, and can only use it as a wild life refuge, that is exactly what they used to cover up what they were in fact doing until all the animals died. They will spend hundreds of years, and many millions of taxpayer dollars cleaning up the Rocky Mountain Arsenal "known as the most contaminated piece of land on earth".

-Compounding the problem of the now leaking Basin F, there was Basin A. Basin A, had been leaking since its inception, it was an unlined 350 million gallon surface dumping site used by the Army from 1943 to 1956, Shell used if from 1954 to 1956.

-Basic C was yet another unlined 67 million gallon surface storage pond, known as "STRAW BASIN", because of its leaking bottom, and was used by both the Army and Shell until 1956. Then again in 1966, through 1970 when repairs were being made to Basin F. Basin

C was adjacent to Basin F, and was easy to pump the toxic waste from F to C., but when it got time to pump the waste back into Basin F, It had all leaked out the bottom of Basin C, living up to its name straw Basin.

-Basin E. again unlined and built in 1953 to hold excess waste from Nerve and Mustard gas production.

-By the early 1970's it was presumed that all holding areas, including trenches were leaking in excess of one-hundred thousand gallons of toxic material per day.

Looking at his wife, Larry could only stare in disbelief "HOW", he asked her, "in the name of decency could the Army, Shell Oil, EPA, and the Colorado Department of Health, knowing full well what was happening, let this happen?"

Obviously Marie could not give him an answer.

Larry said God help me, I cannot stop here. the world needs to hear the truth of what happened. He was beginning to rattle the doors in Washington D.C., and in fact around the World.

Larry!"

He never was one of those men who believed a man had to hide the truth from a woman because she was supposedly the weaker sex. This wife of his had gone through all the hell that ranching and life can throw at a person and she did it standing right by his side. When she wanted to be, she could be as tough as an old leather boot. Besides, after all the years they had been together she still drove him crazy by just looking at her. He was putty in her hands and she knew it. She also knew that he never could keep anything from her.

I've heard of this before. I think they are going to watch me for a while, and as soon as I get out of line they will set up an innocent looking accident."

"Accident! That's it we are leaving, we'll move away. I'm not going to sit here waiting for one of those bastards to kill you, or all of us. I won't do it."

"This is not the time or place to talk, lets go home," he said.

He knew what was in store for him as he walked through the door of their home. You can't live with a person for years and not know how that person will react, and exactly what she is thinking. "He feels as if he were caught between a rock and hard spot pondering the next move.

"Larry, lets pack up and leave. These people from the Army or Shell Oil Company aren't fooling around, they will do exactly what they say." By this time she had started to cry. "Why does it have to be you that fights the Army and Shell? Let someone else do it. I do not want to see you dead. I couldn't go through that. I want to leave, today, lets just go, leave everything and go."

"I can't. I won't run away. I think I know more about the Arsenal and its secrets, and what they're doing out there, than anybody. Marie don't ask me to put my tail between my legs and run away like some whipped dog."

"I am asking. Don't you care about the kids or me? Has this damn Army business become more important that having a father for your children?"

"Yes, it's for them and all the other children that I must do this. Do you remember me telling you about all the trees, literally hundreds, that my father planted when I was young? What is out there now? Nothing, all the trees are dead and gone dead, because of the poison It's not only in the water, but in the air, and soil as well. If we assume, as they have, that this world of ours is here for us to do with as we like, then we are no better that the United States Army or the Shell Oil Company.

"I have documents that have been entrusted to me showing that just a couple of years ago, in the early 1970's Shell's profits averaged between one and two million dollars per month. Shell says, "it didn't feel then and does not feel now that damage to the environment is compatible with good business operations." That's bull-shit, and what's more it's mentally sick. My God Marie, even the clay sewer pipes leaked. in 1960 alone twenty-thousand gallons a day leaked out of the pipes and into the soil. That's seven million three hundred thousand gallons a year of pure poison, and you ask me to turn by back on that and just walk away. I won't, and I can't to that Marie, You can't ask that of me," "there is a time when a man must take a stand, and that time is now. I have also learned as far back as 1965, after just nine years with basin F in use, The liner was inspected by by the Army hygiene and the United States Department of Health, they found it to be leaking, in fact part of the so called protective membrane was dissolved, completely gone. They warned the Army then they would have to cease any further use of the basin, If they continued using it. They would be using it knowing that it was leaking Millions of gallons of toxic waste into the underground water where at least 25 square miles off, the Arsenal were already severely contaminated.

"Of course, you and our children mean the world to me," he said. "That's mostly why I started this, but its more than I ever imagined. It's not just us anymore, there are more people involved. How many people have we known that have died from cancer and other strange and insidious diseases? More than we want to remember. The lunatic's that run these places, or give the order, or invent these toxic concaucions must be stopped, not only here but all over the world. We don't need these toxic poisonous wastes, and experiments that expose the public to illness and death, without answers. Just like the thirty five (35) square mile area just north of Denver, Colorado has the highest rate of cancer per capita than any other place in the world. The Rock Mountain Arsenal was so noted and named by the Environmental Protection Agency. "the most contaminated, and toxic piece of land on earth." Owned and operated by the United States Army, and the word is now out that the brass at the Arsenal was taking payoffs from Shell Chemical Company, to allow them to continue storing, and dumping toxic waste at the Arsenal nearly twenty years after the Army was warned to discontinue use of basin F. and knowing that it is contaminating the underground water, and 100's of square miles off the Arsenal property.

"It is really sad that we must worry every day about the water drink, and the air we breathe, and the soil where we live."

"What is the difference between the General's or high level people that order this poison, that kills thousands of innocent civilians secretly, like some damn coward hiding in a dark corner someplace, or even Hitler, who killed Six Million Civilians? The only difference I can see is that Hitler did it faster and was more open about it. Here the military can kill you quietly and it takes you a few years to die, but you're still just as dead, and as far as any one knows you died of normal causes."

"Marie, of all the animals on earth, man has shown himself to be the most cruel and brutal. He is the only animal that will create instruments of death for his own destruction. He is the only animal on all the earth that has ever been known to burn its young as a sacrifice to appease the wrath of some imaginary God. He is the only one that will build homes, towns, and cities at such a cost in sacrifice and suffering and then turn around and destroy it all in war." "For Glory and Power"

"We are stumbling blindly through a spiritual darkness while toying with the doubtful secrets of life and death. We have achieved brilliance without wisdom, power without conscience. We know more about war than we know about peace, more about killing than we know about living."

"If it means I lose you and the kids for fighting for what I believe in and for what is right and just, then so be it, (I am sorry, but I won't quit, not now, and not with what I know."

Several minutes went by as she collected her thoughts. he was upset and visibly shaking. In all the years they had been together he had never spoke to her in such a way.

Larry felt terrible, but he meant what he said. Now he hoped that she didn't hold him to it and decide to leave.

Looking at her, he saw something in her eyes that he had not seen a few seconds ago. The love had always been there, this was something different.

"When I was growing up," Marie said, "My father would say to us kids, 'if you're going to do something, be certain that you're right. Once you know you are, then do it the best you can. Be proud of what you do and who you are.'

This whole rotten mess has been hard on all of us, but it wasn't until now that I realized, its been hardest on you.

"I have loved you since we were young adults, nothing has changed throughout the years except that my love for you has grown even stronger. I have stood by and watched you take on jobs that would have defeated many other men and you have always succeeded. I remember watching you as you fought to save the cattle, and how you cried like a little boy when you had to destroy them. I forgot about the pain you felt then, and I'm sorry.

"I have always been very proud of you, but never more so than right now. Be proud of yourself Larry, and the work you are doing, I am."

That was his wife, smart, loving, and very sincere. Once she had made up her mind to do something it was pretty hard to change it. Maybe that was another reason why he loved her so much.

Whatever hardships were put on them by the Army, or by Shell Oil Company, he knew that his wife, for all the good reasons, was beside him.

SINCE THE ARMY, WITH ALL IT'S TOXIC WASTE HAD KILLED OFF MOST OF THE BALD EAGLES. THE U. S. IS STRUGGLING TO BRING THEM BACK FROM TOTAL EXTINCTION.

CHAPTER FOUR

With the monstrous weapons man already has, humanity is in danger of being trapped in this world by its moral adolescents. Our knowledge of science has already outstripped our capacity to control it.

Gen. Omar Bradley, November 10, 1948

On Monday morning Larry poured a cup of coffee and walked to the end of his driveway. The weekend had been a complete waste of time as far as he was concerned. He had been unable to do anything worthwhile. Some of the health problems, which had gradually become part of his life, were worse at certain times' than others. This weekend had been especially bad. His head felt like it was going to explode, and the pain that he now experienced in his bones, joints, seldom left him. The Doctors felt that his symptoms were typical of the chemicals which he had ingested through the consumption, cooking and bathing of his well water, and the very air the breathed. If it was not for the pain killers…

He was not alone. The rest of his family suffered right along with him. There were times when they were simply unable to go on with their day to day lives due to the pain. Larry's Mother at 62 years old was unable to climb up or down steps without assistance. Lifting anything became torturous as did the simple act of walking to the mail box. The headaches were blinding. There were many times that they were afraid they would die, and times that they were afraid they wouldn't die right away. They constantly lived with pain and more pain. Larry's Mother, his wife and his daughter had such rashes under the arms, in the groin area, at the elbow, the neck generally all over, they would crack open and bleed. The Doctors gave them all types of salves, and creams and told them to stay away from the water. everyone by this time had little raw sores in the mouth, nose and on the lips. The Doctors did not know what to do and of course gave us pills, salve, and creams. The breaking problems were much like that of an asthmatic, sometimes really bad. Everyone began losing little patches of hair, it would start by thinning, more and more till there was no more hair in small dime like circles, other than the Doctors checking for lice, which were none, and we were all given some type of special rinse and oil to treat the scalp with. The entire family was sick of the pain and sick of the medications, salves, and creams, it all helped a little but did not stop it's progression.

It was right during this period of time Tri County Health called all of us in for blood tests. They said it would take a few days they would let us know what they found. They called us back in with in about ten days. At that meeting there were several Doctors, The main Tri County Health Doctor got up an said you people have been poisoned, there is enough in your blood to have been lethal. All they were able to say is there seemed to be some type of Oragano Phosphate at various levels, but they were unable to put a name to it yet. However they were able to name another chemical and the toxicity, "PHENOL" It is a chemical used by the United States Army, and possibly Shell Chemical Company. The Doctor went on to say your blood samples are at or near the lethal levels for everyone. The word at that point your wells are contaminated by the Arsenal get away from the water. At that point we were informed that even using that water for bathing is dangerous. It would be similar to putting 4 tablespoons full of insecticide in your bath water each time you take a bath. A short time afterward they informed us they also found, the organo Phosphate in our water was nerve agent. "GB- nerve agent." and DIMP a Hydrolouses of GB nerve agent, and IMPA a Hydrolouses of BZ nerve agent. Two other chemicals also found in our water at the time was Nemagon, and Di-Cyclo-Penti-Dine, or DCPD, Deildrin, and Endrin.

Standing there looking across the road to three of his distant neighbors, Larry tried to imagine what they were going through. It had to be much worse on them because they were much older.

Standing there he desperately tried to remember all their faces, the faces he had grown up with, had worked for in the summers of his youth or had done business with. How many were left? He knew many had passed away, and he had just been to a funeral.

Larry knew before they died many had terrible liver problems, others had colitis, Kidney problems, yet the biggest single cause had been various forms of cancer. This whole area now had the highest cancer rate, and cancer fatalities of anywhere in the United States. All the health officials knew it, but getting them to admit, it always seemed something else.

He and his family has been told by medical specialists that they have a three to four hundred percent greater chance of contracting some form of cancer in their lifetime, than any other human in the United States.

He remembered trying to explain. He tried to tell his neighbors, and friends that Shell had tripled, and quadrupled its production just months earlier and that the Army had been conducting new experiments along with their production of gases. Everything that was produced at the Arsenal had been increased. The Army and Shell knew the Arsenal was on borrowed time and they were determined to get the most out of it. Sadly, his neighbors didn't believe, and they didn't understand. They would say things like, "our government wouldn't do something like that to it's citizens," or "How can something be in the water? I've been drinking it all my life." or "Your fighting city hall boy, and no one ever beat city hall".

He remembered talking to a widower following a funeral. He had recently lost his wife to a fast moving uterine cancer. They were at the funeral of another neighbor who had died of liver cancer detected only 2 months prior to her death. Larry told him of the things that he knew regarding our water. He replied, "I've been drinking this water since we moved here ten years now. It can't be the water." He ignored the changes which had been visible in our water. The odd odors, tastes and even occasional colors. It seemed easier for him to deny the evidence, than to deal with the facts.

As time passed, they became scared as the evidence grew to a point that could not be ignored.

Now they were dying one by one. No longer was denial an option.

When it was in their best interest, Shell Oil Company began buying up all the homes on the northern boundary of the Arsenal. Hurriedly moved the people out. some on their death bed as they are being moved out. As soon as they were out their farms and homes bulldozed down and destroyed. Larry talked to a couple he had seen earlier, and they said "it's too late Larry, we're dead already, we should have listened to you. "then". "There were a few that were holding out for more money, but died before they could sell and move. Hoping for a miracle, as though moving will save them. Some people simply walked away from their homes. Shell would find them and give them whatever for what was left, and bulldoze them under.

There is nothing there anymore. Nothing to mark the passing of fine, hard working, loyal Americans. Nothing to show where they laughed and cried. Nothing to show where their children played, where mothers planted their flower gardens, or where mothers hung clothes out to dry for her family. Nothing is left to show where fathers fixed the bicycles ridden by their children. It's all gone now, everything, like they never existed. Not one indication remains of what their Government did to them.

Larry remembered the day the chemist, who he had hired to independently test his water, came out to his house. His friends and neighbors were all gathered in the backyard to hear what he had to say.

He explained that they had found, Organophosphates, Chlorinated Hydrocarbons, Deildrin, Aldrin, Endrin, Aerozine-50, GB Nerve agent, BZ Nerve agent, Mustard gas, and Hydrozine Rocket fuel, (one of the most toxic chemicals on earth). Through the entire conversation it was like he was speaking a foreign language. They just could not comprehend, and because they couldn't, they died one by one, day after day, year by year.

The United States Army and Shell Oil Company lied. They lied to everyone, and because of their size and power, they got away with it. The army and Shell Oil have raped our countryside, devastated our environment, raped and killed our people, It is not just affecting the adults, it's killing our children too.

Chemicals and chemical waste flow from the large pipe lines that leak enormous amounts of chemicals into the under ground aquifers, then on to the waste basins such as, A, C, D, and F, some spilling from deteriorated pipes, and tanks. Some purposely injected into their 12,000 foot well, but all finding it's way into the complex system of underground water supplies known as aquifers.

Large exterior pipe in North Plants area wrapped in asbestos. Asbestos is a major problem at the arsenal since most buildings there contain it, and it must be removed before demolition takes place.

PEAK AND PRAIRIE NEWS

Many aquifers traverse the area beneath the Arsenal at different depths. Each making its own way from beneath the Arsenal across the plains of Colorado and beyond. Arsenal Chemicals have been found in the Platte River which winds out of Colorado, into Nebraska and finally into the Missouri River. The potential for eventually affecting people from across this country is great. The toxic soup, that has been created over the years, moves slowly but steadily through our underground water supply. How many people will be affected in the long run. Larry knows his family is not alone and it scares him.

He believes from the bottom of his heart, that there are some bureaucrats who know exactly what is going on, and just let it happen, while taking the kickbacks under the table. Are they paid off to keep their mouths closed? He believes that is so, even at the level of those who are in charge of the Arsenal. For what ever reason they have, he thinks it is a crime that they have let this happen for all these years. All the lies, the broken dreams, the happiness that has been lost to eternity. All this pain and unhappiness because of greed. I would like to be able to examine their finances. It would not surprise me to find enormous amounts placed in accounts for the Generals' and then there are the one who you will get hold, providing you raise enough hell you can get to a Colonel. Year by year after toxic materials was found in our wells, Like Colonel Watson, and Colonel Quintrell. Then after the army had in fact found the chemicals in the Land wells, in 1974, They were still denying any chemicals escaped the Arsenal property, five, ten, and fifteen years later.

To better illustrate his point of government cover up or pay off's lets' add one other thought.

Larry knows of a lot of men, and women who are and have served in the military. They always talk about surprise inspections by high ranking officers or political leaders flying in from Washington to look at a Federal Installation. Are we to believe that in all the years that the Rocky Mountain Arsenal was in existence, 1942-1982, that no high ranking officer, low ranking officer, a Senator, or Congressman ever came to inspect how the American taxpayers money was being spent. That not one officer or political leader ever saw what was going on? Senators and Congressmen have a security clearance, so secrecy could not have been the reason. Why didn't they ever see anything? Maybe someday, someone could answer this for him.

The Rocky Mountain Arsenal sits just north of Denver, Colorado, sort of what one would call the high plains, since Denver Colorado is called the mile high city, because it sits a mile high 5,280 feet. Denver was very lucky over the years, as the wind usually blows to the north or northwest, and the water in the aqifers run mostly to the north and northwest.

The next page you will see a map of the nearly 28 square mile Arsenal, 17,820 acres. Then looking at the map imagine the area further to the north and north west. Where over 40 square miles have been poisoned, and contaminated by the Arsenal. Place 20 more miles to the northwest and 20 more miles straight to the north and imagine how many miles that is and how many people were directly effected by the Arsenal.

Figure I

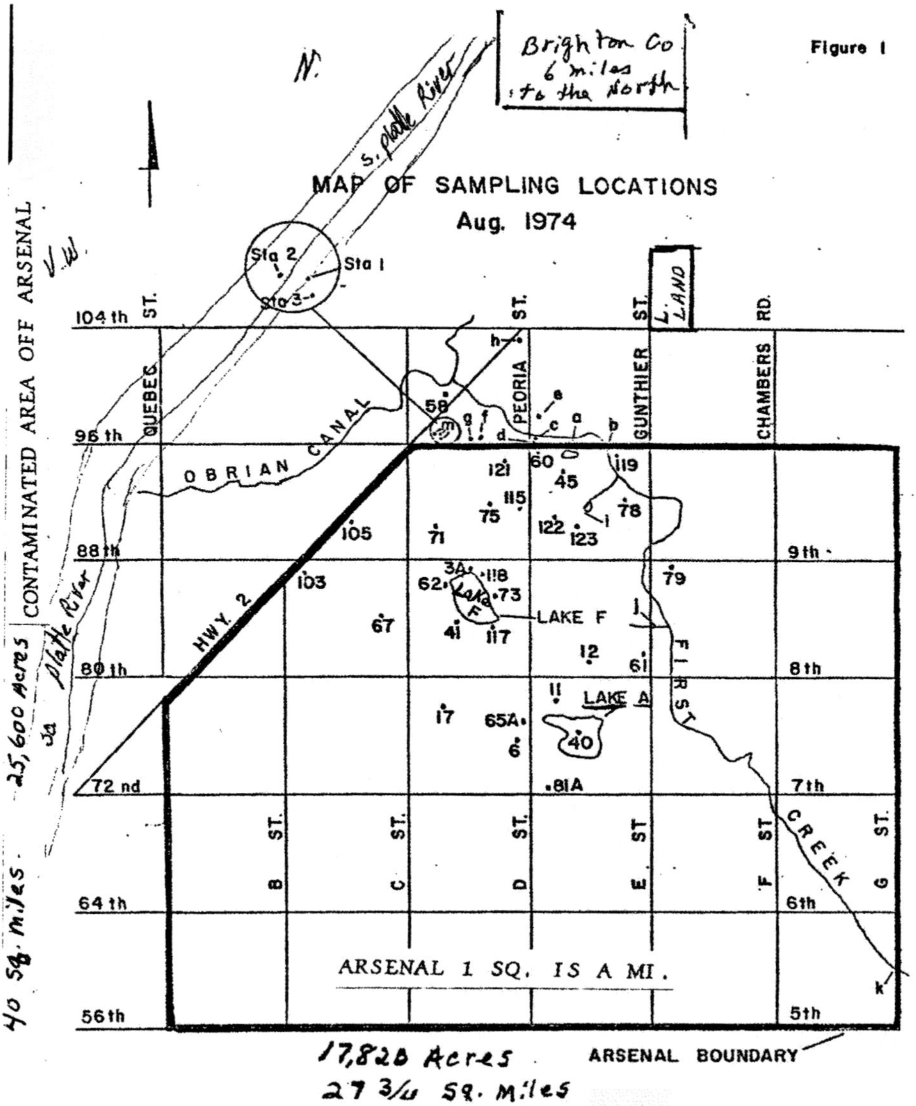

Brighton Co 6 miles to the North
N.
S. Platte River
MAP OF SAMPLING LOCATIONS
Aug. 1974
CONTAMINATED AREA OFF ARSENAL
V.W.
Sta 2
Sta 1
Sta 3
104 th
96 th
88 th
80 th
72 nd
64 th
56 th
QUEBEC ST.
PEORIA ST.
GUNTHIER ST.
CHAMBERS RD.
LAND
OBRIAN CANAL
HWY. 2
LAKE F
LAKE A
FIRST CREEK
9 th
8 th
7 th
6 th
5 th
B ST.
C ST.
D ST.
E ST
F ST
G ST.
ARSENAL 1 SQ. IS A MI.
ARSENAL BOUNDARY
25,600 Acres
40 Sq. Miles
Platte River
17,820 Acres
27 3/4 Sq. Miles

Walking back to the house Larry had just sat down when the phone began to ring. As it turned out, it was the Denver County Sheriff's Department saying he had gotten the job as a Deputy, that he had applied for a month earlier. They wanted him to come down the next afternoon to fill out some paper work. Training had been changed because he had been a Deputy in another County, therefore had prior experience. It wasn't a job he liked, but it was involvement with the criminal justice system, and he enjoyed working with people.

The thing that really made him mad, was the fact, that had in not been for the Arsenal, The Army, and Shell's cover up, he never would have lost his ability to earn a living for his family. He would have been raising, and Auctioning his livestock.

Walking into the bedroom to put a shirt on, he was just doing the last button when the phone rang again.

Walking into the kitchen, he picked up the receiver.

"Hello."

"Mr. Land, this is Dennis Allen. Can you meet Jennifer inside the restaurant at the truck stop tonight at nine o'clock?"

"Yes definitely. what have you got?"

"I'd just as soon not say over the phone, but lets just say, you won't be disappointed."

All right, tell her nine o'clock."

"Good, thank you Mr. Land, I'll talk with you later."

"Don't hang up yet. If we are going to be working together on this, my first name is Larry."

"All right, Larry. I'll talk with you later."

That night, Larry pulled into the truck stop on Interstate 76. As usual, it was full of trucks that had parked for the night. Most of the drivers were inside having supper or having a last cup of coffee, others were watching television or already asleep.

Entering the building, Larry walked past the cash register and the gift shop that was full of necessities for the drivers' and souvenirs that could be purchased by the many travelers that passed through here. Walking into the restaurant he looked around. It was 8:50 P.M. There were a few booths where drivers and some of the local residents was talking. Finding a booth next to windows, Larry sat down and ordered coffee. There was no sign of Jennifer.

9:05 nothing. He was beginning to get nervous. His mind began to play out all sorts of scenarios. What if something had happened? Fear crept into his mind.

At 9:15, Jennifer walked in, obviously very upset.

"I'm sorry Mr. Land, I ran out of gas. The gas gauge in my car doesn't work accurately. I've been wanting to get it fixed, but I just haven't had the time.

"The names Larry and it's all right, what have you got?'

"A complete listing of what was manufactured at the Arsenal, the years it was made, and what they did with it. Also a report on the trenches that were dug, along with their measurements. I also have some information on Basin A."

"Great! Are you sure no one suspects anything or you were not watched or followed here?"

"Yes positive, I took a round about way to get here, that's one reason why I ran out of gas."

Larry decided to look at the documents while in the restaurant. The documents were as follows, watch very close.

WHAT WAS MANUFACTURED

SHELL OIL COMPANY

CHEMICAL	DATE	USE	STATUS
Aldrin	1951-1974	Kills corn worm	Banned most uses 1974
Dieldrin	1951-1973	Kills weevil	Banned most uses 1974
Endrin	1953-1965	Kills the Grasshopper	Banned most uses 1979
Atrazine	1977-1978	Kills weeds	Produces tumors in Rats, still sold
Bladex	1970-1974	Kills weeds in a corn	Risk to woman

		Field	Banned 1984
Planavin	1966-1975	Kills thistle	Banned 1984
Chlordane	1947-1950	Killed roaches	Banned most uses in 1980
Nemagon	1955-1976	Kills Nematode and Worms	Banned most uses 1979
Phosdrin	1956-1973	Kills Caterpillars	Potent nervous system inhibitor
Methyl Parathion	1957-1967	Kills Weevil	Acute toxicity to Wildlife
Gardona	1967-1968	Kills house fly	Still on market
Azodrin	1965-1982	Kills beetle	Very toxic to Wildlife
Vapona	1960-1982	Kills Mosquitoes	Human Carcinogen Toxic to Humans
Nudrin	1973-1982	Kills Cutworms	Extremely Toxic to Humans

ARMY

CHEMICALS	DATES	OPERATIONS	DISPOSAL
Mustard Gas	1942-1949	Blistering agent	Waste to basin A
Mustard Gas	1949-1973	Agent Incinerated	Waste to basin A

Lewisite	1942-1945	Burn and Blister agent	Waste in A, C, D, E, and F.
M-47 Napalm Bombs	1942-1945	Fire bombs	Waste to basin A
Phosgene Gas	1943-1944	Choking agent Toxic	Waste to A, D and E & buried
Magnesium bombs	1945-1956	Explosive Device	Waste placed in Various ditches
Phosphorous Bombs	1945-1970	Explodes when exposed to air	Waste placed in Various ditches
Dichlor	1952-1954	G-B Nerve Gas Comp.	Waste to basins A, C, D, E, & F
TX anti crop agent	1962-1973	Made To destroy Soviet Wheat	Waste to basin F&C, & buried
Aerozine-50			Waste to basins
Hydrozine	1962-1982	Rocket fuel	F & C
Button Bombs	1967-1968	Troop Detection Device	Waste buried
G-B Nerve Gas	1953-1957	Attacks Nervous System	Waste to basins A, B, D, C, & F
G-B Nerve Gas	1955-1976	Weapons destroyed	Waste to basin C and F
Chloroacetophenol	1965-1966	Weapons destroyed	Waste to basin F
Microgravel Mines	1967-1968	Shrapnel mines	Waste basin F

Chemical ID	1981-1982	Kits destroyed	Waste Incinerated
Sets Adamsite	1983-1984	Nausea inducing Gas	Waste basin F
Cyanogen Chloride	1983-1984	Nausea Inducing Gas	Waste basin F
B-Z Nerve agent	1953-1975	Happy Death Gas	Waste to basins A, C, D, & F
IMPA Agent	1953-1975	Pleasant death gas	Waste to basins A, C, D, & F

The next item explained is the trenches, although a small amount of it was missing.

The magnitude of the production list almost paled the imagination when compared to the way that the Army and Shell dumped its solid toxic waste. From 1952 to 1967, Shell dug a series of trenches, 300 feet long, 25 feet wide, and 15 feet deep across from its factory.

Into these giant trenches, Shell stacked hundreds of barrels of toxic waste. When the trenches were full, Shell poured liquid fuel atop its waste and set the material on fire. The Army also burned toxic waste in a series of its own trenches. Yet Denver was unaware of where the problems was really coming from when they would have red alerts, for smog, telling everyone to stay inside, the hospitals would fill up with people, unable to breathe. In reality it was the toxic smoke coming from the Arsenal, The Army and Shell Oil Company. They didn't even care about what was happening. It wasn't until about the time they were closed permanently that the air pollution Commission finally caught Shell Oil Company operating Most of the nights through without using the scrubbing towers to help cleanse the air of toxic waste and particulate. Then bright and early just before sunrise they would quickly hook everything back up to the scrubbers. So when they were checked at sunrise they were in compliance. It was the people in and around the Arsenal that paid the price especially those in Denver. To Save Money and speed up the operations in the waning years, It is understood from employees they unhooked the scrubbers. That's how they were able to double, triple, and quadruple their production. Of course no one will admit to this.

Because of the shear volume of the material in these toxic burn pits a vast amount was never burned and was covered with soil when the trench was full.

More than once, so much smoke billowed from the hazardous waste fire that Stapleteon Airport was "fogged in" for more than an hour or two, forcing pilots to rely on cockpit instruments to take off and land. The white smoke, caused by Army's burning of phosphorous bombs and artillery shells, actually settled across downtown Denver like a blanket of snow. The smog like smoke is very irritating to the eyes, nose, throat, and lungs, and again Denver would be on the Red Alert.

The main source of this (destruction) was a waste disposal system that had largely been built as wartime expediency. This waste disposal system was the sewer, that ran to a string of unlined waste pits. They at one time thought the evaporation was just great. A large report to their upper echelon describing how effective the evaporation was in this area. It wasn't until the early 1950's they realized the toxic waste was not being evaporated. They discovered that only 15% was evaporating and the other 85% had been seeping through the porous sands down into the underground water stratus. This was seeping from some of the original basins, such as Basin, A, B, C, D, E. as early as 1953, Army and Shell officials knew that toxic waste from chemical operations was eating through joints in the Arsenal Sewers, and leaking through the sand bottom basins into the under ground water. On two separate occasions, the sewers, made of a substance similar to clay flower pots, had decayed so drastically that toxins were backing up into the Shell plant. In one case, Shell crews dug up several hundred feet of sewer line, looking for a solid pipe. They didn't find any. This is what prompted with expedience a lined basin Called basin F. Lined with an asphalt liner, leaving one to wonder how thick this liner must have been, to stand the ravages of time, earth shifting, even earth quakes, and the the strongest, and most potent acids known to mankind. That liner was 3/16 of an inch thick blown asphalt liner, and was given a life expectancy of ten (10) years. Before the ten year life expectancy came and went, 1965 they were advised any further use of Basin F would be on the premise it is leaking into the underground water stratus, and that parts of the liner were missing. They discontinued even thinking about this and pushed on Tripling, and quadrupling the waste place into basin F. Right up to 1983 when they were ordered out, and to clean the entire area, according to Colorado Standards. It had become the worst polluted piece of Land on earth. The Army will spend centuries just trying to clean it up. The half life of most the material is 500 years or more. Nearly 18,000 acres just north of Denver Colorado, can never be sold for anything, and must be maintained as a wildlife habitat. "The Army's very Own" as a reminder of what they have done.

In June of 1960, an Army Corps of Engineers study found that eleven per cent of all hazardous wastes poured into the sewers leaked out. At the time, that meant 20,000 gallons per day of poisonous, and very toxic by-products of chemical manufacturing were gushing into the soil, and when these toxins ran then into basins A, B, C, D, and E. an additional 100,000 thousand gallons or more, were disappearing daily, and it wasn't all evaporation.

From 1943 to 1956, the key to the Arsenal's disposal system was an unprotected dumping pond known as Basin A. It was a natural depression in the soil that held millions of gallons of the by-products of chemical weapons, and pesticide production.

Basin A operated under a very simple premise: Because Denver had an arid climate, liquids could be pumped into the basin and would evaporate fairly rapidly.

The problem was, the Army and Shell poured so much waste into Basin A that liquid toxins sometimes spilled over the edge of the 123-acre dumping pond. In the spring of 1953, Army and Shell officials confirmed the dangers of their dumping materials in a series of internal memos. In one letter to Shell Officials, an Army engineer wrote that overflows of toxic wastes in dumping basins was reaching "A Danger Stage."

Instead of correcting the current problem, the Army and Shell decided they needed more basins.

So there it was, more evidence as to the misconduct by the Army and Shell Oil Company in disposing of their toxic waste.

"It sure doesn't make a person feel proud of the government does it Jennifer?"

"I haven't felt proud in a long time," She answered back.

"You will, and soon, our day is coming. These people have been running and lying for to many years, they are making to many mistakes and to many enemy's. It may take us a while but we'll be there and get them in the end, and they will get their just reward."

"I wish I could believe that Mr. Land."

"Jennifer, listen to me. My mother was operated on for her gall bladder. My Attorney requested that her gall stones, tissue, and all records be kept and given back to us for testing. Her condition was very unusual so my Attorney thought this was necessary considering the high levels of contamination on her property.

"Shortly After the operation was over my attorney immediately asked for these items, which he was going to deliver to the laboratory personally. Less than an hour after the operation was complete we were told by the physician that it all had to be sent to their laboratory first, and then we could make arrangements to pick it up. The Lawyer then got the number for them, and told them he would also put it in the mail. A few days later the Attorney again called the Laboratory, no one knew what he was talking about, and their was nothing ever sent in for Viola Hollenbaugh's operation, no records whatsoever".

"The same thing happened when Larry's wife went to the dentist, severe problems with her teeth, The Attorney again had wrote a letter certified requesting that everything be saved from her operation, and teeth, They used the same excuse, by law they had to send it to their laboratory first, and he was assured they would save everything. Again after a few days no records existed as far as they knew it never happened".

Then lets back up to 1974, Tri County Health Department was assisting us with medical help. They took blood samples from at least eight of us, and promised to have results back shortly, they seem to know exactly what they were going to look for. Within days the results of the blood tests were back, Larry and Toni went in to talk of the results. They called us into a special conference room with at least three Doctors. They ask if we had ever heard of phenol, of course we had not. They said it is a material used at the Arsenal for pesticides and possibly some Government projects. They said this has been leaking into the underground water for years. They told Larry and Toni they had enough phenol in all their blood to be lethal. Again of course at this point they learned we were still bathing in the well water and told us to make sure we did not use it for anything as it can be absorbed through the skin. They called in a couple more Doctors that said there isn't a lot they could do except tell us to stay away from the water completely or we may die or do severe damage to our internal organs. They said your water is with out doubt contaminated from the Rocky Mountain Arsenal. Well when your in such a mess, you can't really think straight, we listened and left, stayed off the water completely, but didn't ask them for copies of anything., a few days later they done more blood tests, told us it looked much better. Then a short time later we were asked to get a copy of the records. Went back into the Tri County Health Clinic and asked for copies of the records and blood tests. They had nothing and nothing existed, there were no records. Well you know what anyone would have done about now, and Larry did, and it almost got him arrested.

"Now I'm not the smartest man in the world, but even I can see by now that something is going on, and I don't have to be kicked in the teeth more than once, or maybe twice before I learn.

"Any way, about two weeks later my attorney, and myself had a meeting with our Congressman. We told him that I had a lot of extensive testing done on the dead cattle at the Veterinarian college in Fort Collins.

"He wanted to see the results of the testing so he makes the necessary arrangements with the college and drives up there. When he arrives he asks for the results of the tests done on Mr. Land's cattle. The Laboratory, and clinic, both tells him that there was nothing, and that they had never heard of Larry Land or his Attorney.

"It would be an understatement to say he was just a little angry, and within a couple of days we were called into his office to explain, why we had lied."

"The Congressman started of by saying you told me you had testing done on your cattle and had talked to certain Doctors," he said. "they told me they didn't know what in the World I was talking about, and that they had never heard of either you or your attorney. Now will you please tell me just what is going on?".

"Like I had thought earlier Jennifer, I don't have to be kicked in the teeth more that once or twice to learn. This time I was one step ahead of the Army."

"Congressman, I said. "Your absolutely sure that is what they said? They had never done any testing on my animals, dead or alive, and they had no record of Larry Land or his Attorney.

"Yes of course I'm sure." That's exactly what they said Mr. Land. I don't have the time to be going out on wild goose chases, or chasing something for naught. I'd like to know what is going on! "He said"

"Larry reached down for his briefcase, brought it up to his lap and open it up. Took out a rather large stack of papers laying them out on the Congressman's desk."

"The Congressman said, what in the world is this. Larry said just oddly enough it is copies of everything that was done at the College, and look here and here, the names of the Doctor's, Lab techs, every one who had assisted in examining the animals. They told you there was nothing there and nothing had been done. Yet we have documents that are over several days as to what was seen wrong with the Cattle, and how they died no matter what was done for them. They died from a destruction of all the main organs of the body, mostly the Lungs, brains, kidneys', liver's and hearts. All which appeared black." "I find this pretty amazing since they said they had done nothing, to have all these medical records with their names on them. This thing with the United States Army is a huge Government cover up, I hope you can now see that. The sooner all you people in Washington realize the Army has lied to Congress, the people of the Untied States, and even to the President, the better off we'll all be. The Colorado State Health Department was not capable of running all the tests for them selves, so in their infinite wisdom, samples were sent to the Army and Shell Chemical Company. Of course then the Army and Shell began it's own testing program of the Land wells which was short-lived. No one was telling us anything. Shell Chemical, and the Army, along with some in the State Health Dept. concealed all the facts of what was found. No outside laboratory could check and tell what was in the water.

"All the logs for the water sampling, from the Army, Shell, and the Health Department, Certain medical Affidavits' going back to early 1970's. So you see Jennifer, in our hands. We Even have Notes found in Shell Oil Companies files regarding Larry Land stating "leave the sleeping dog lay". They have made many mistakes they cannot correct."

"You know that they are going to try and kill you don't you?." God, Larry, I feel like I'm living inside a dime novel. It never really occurred to me that things like this really went on, not in this country. Yet here we are, meeting secretly, hoping that no one will see us talking. Constantly watching our rear view mirrors to see if anyone is following us. Sometimes I hesitate to turn the key in the ignition of my car for fear that someone may have placed a bomb under the hood. Then I scold myself for letting my imagination run away with me. But that's the problem, It's not my imagination Larry. Too much has happened. The unbelievable has become a very real part of our lives. I'm afraid for all of us." Jennifer put her head in her hands in a gesture of defeat.

"They have already tried several times," I told her.

"Jennifer is shocked disbelief," I had no idea! when? where?" And that doesn't scare you? knowing that someone, probably hired by our own government wants you dead." Or more probably Shell Oil Company.

"Jennifer sat upright, Staring at him." "They've tried this before and you haven't told me."

"Yes Do you remember me telling you that when this whole thing started, my livelihood was taken from me. I got a job in Brush at the Livestock Market auctioneer and ringman, and cattle buyer. I was up there one day, It had been a busy session of buying and selling for the area ranchers. I was standing by a light post and had just finished talking to a couple of locals. They had just turned to walk away, when I heard the distinct crack of rifle fire. I dove for the ground. people milled around in confusion. There were two quick shots and then it was over. When my heart stopped pounding and I had regained my composure, I got to my feet. People were coming in my direction, concern and questions etched on their faces. I looked above me in the heavy wooden post and there a short ways above my head there were the splinters of wood where the slugs had entered. No one could figure out where the shots had come from or why. Even I don't have the answer, but only I was aware that I was pushing the Arsenal, and Shell Chemical Company situation awfully hard.

"Probably the scariest incident by far occurred, when I was coming through the Eisenhower Tunnel on I-70 from a day in the mountains, just west of Denver. I was driving home after an enjoyable day of peace and quiet. As you know, while the scenery up there is beautiful, the road demands your utmost attention. The climb up to the tunnel and the descent from it is quite steep. As I started down the mountain I noticed a pickup truck in my rear view mirror. It came closer and closer. At first, I didn't pay too much attention to it until I looked up and could only see his grill in my mirror. Then there was the first bump. He'd hit my rear bumper sharply. The car reacted, I kept it under control but decided to speed up a little to get away from him. That's exactly what he wanted. Because immediately another sharp bump came from the rear. My speed was increasing as we negotiated the steep and winding road. Part of me wanted to go even faster, but I knew that this road would not forgive too much speed. I also realized that I would only serve his purposes. I concentrated just on getting off of the mountain, when I saw that he had moved up beside me. It seemed like a monster, and not knowing what to expect. Suddenly there was a savage assault from the side into my door. Tires squealed as the car careened to the right. I was sliding on the gravel with the guardrail coming rapidly at me. There was a truck runoff area not far from me. I don't know how I managed, but the car ended up stopped in the deep gravel, and sand of the runoff, instead off the side of the mountain."

"And that doesn't scare you? knowing that someone, probably hired by our own Government, just tried to kill you. How badly they want you dead."

"Sure it does, it scares the hell out of me. Whatever they try to do to me, they've got to make it look like an accident, just like they did with Karen Silkwood. They tried all kinds of things on her and failed at everyone, until finally they ran her off the road and killed her. Her experiences and mine started at almost the exact same time in 1974, with the only difference being she died and so far, knock on wood, I've managed to stay one step ahead of them, I'm still alive. When she blew the whistle on Kerr Mcgee Corporation, she did it because she believed in what she was doing, and the safety of her fellow man. I think I have an excellent roll model to follow." All because of my fellow man, and I cannot stop when I'm doing, The only thing I live by is “it’s not over till it’s over”. “Much like the pursuit of the dragon”.

“So what will happen Mr. Land? Is it all worth it, when we are dead and gone will it have honestly made a difference?” Will any one care.”

“I plan to live long enough to write a book about what we went through, and why then and only then people will understand.”

“Looking at her with her large brown eyes looking all sad and depressed I knew what she was going through. The uncertainty, the fear of being discovered, all the emotions he had felt at least a thousand times, constantly running through her mind. I felt sorry for her, she was still very young, she knew too much for her innocent years.

“If for no other reason than the fight to save the environment, I know that I’m right, I have to believe that. That and my own hatred for what they have done is all that keeps me going. I sincerely believe that our government and corporate business have massively and abusively disrupted the environment without being aware of many of the harmful consequences of their acts, until the acts have been performed, then it was to late. The effects, which are far to difficult to understand, and sometimes irreversible, are now upon us. Like the sorcerer’s apprentice, we are acting upon dangerously incomplete knowledge. They are in effect, conducting a huge experiment on the people of this country.

“Because we depend on so many detailed and subtle aspects of the environment any change imposed on it for the sake of some economic benefit has it’s price. Sooner or later, wittingly or unwittingly, we must pay for every intrusion on the natural environment. Do you have any children Jennifer?”

“No, I wish I did but right now with my husband in school we don’t think it’s the right time,” she said.

“When the time comes and you have children, what would you like them to see, to feel, to experience in this land?”

“I guess everything that I have and more, up until this all started, and more. I would be like other parents and want my children to have all the things I didn’t.”

“That’s exactly the way you should feel. If we do not make a common stand, if the people of the United States do not get their heads out of the sand and wake up in time to tell this Government, and big business, enough is enough, I doubt very seriously whether or not there will be anything left for our children or grandchildren. What were doing her today is for them and their tomorrow's, and even our own tomorrow's. We all live in our own private world that we have created for ourselves, and in our hurried existence to make a better living for ourselves, we watch the world pass us by. If it does not involve us, we do not concern ourselves. We must open our eyes to the future of this country and stop those people or companies, that insist on poisoning it just for the almighty dollar. I know that you are depressed, but hang in there with me, "WE SHALL WIN".

"You sound so sure of yourself, so positive," she said. Jennifer looked as though she was almost to the point of tears. Larry hoped she wouldn't do that. he could not stand to see a woman cry. It turned him to instant mush."

"I am sure of myself because they have tried to kill me on several occasions now and failed. They are in the wrong, and they know it. that tells me they are afraid. It also tells me there is someone, or some force that wants the people to know the truth, in time so something can still be done to lessen the effects. Call it God or whatever you like, but I honestly believe he wants us to tell the people of this country what has happened and continues too happen, before it's too late, It is called getting a grip on it.

"You sure have a way of working a person up."

"My wife says its my charming personality," Larry said as he started to laugh.

They stayed for another hour or so talking and then he had to leave. he told Jennifer about his new job at the Denver Sheriff Department, and that as soon as he had a number where they could reach him he would let them know.

The next day Larry drove to the Denver County Sheriff's Office and filled out the necessary paper work. He had a steady job again and that made him feel good, and he was looking forward to it now that he was feeling better. It was going to be working the midnight shift, he liked that cause it gave him the days to himself which he wanted to continue his struggle.

There were still many things bothering him, not the least of which were his children. For some reason the toxins that the Army, and

Shell unleashed into the ground water, affects children faster, and he believed harder, at least in the early stages than it does adults. His wife and he had spent many sleepless nights at the side of their children's beds while this industrial vomit flowed through their veins, and the only two things the children were capable of doing was holding their hands and puking their guts out because there was nothing that would help except time and the good Lord. You look at your children, with large splotches of hair missing from their heads, seeing them so sick and in so much pain, not knowing what to do, was like a living horror, Larry felt strong about pulling through this.

Remembering that, he reaffirmed his intentions that he would never quit as long as that poisoned hell hole was in operation. Even if it took his last breath. Just to make sure something like this would be stopped and never repeated. That's when Larry called it "DEATH FACTORY U. S. A."

On the way home he made a slight detour and drove past the main gate of the Arsenal in Commerce City. He was surprised to say the least, at what he saw. There were men carrying picket signs outside the gate. The employees of Shell Oil Company were on strike.

With some thought, he decided to stop. Maybe someone would give him some information that he could pass on to his attorney.

One man in particular seemed willing to talk. He told Larry that when they did the main part of their production of pesticides, herbicides, and any one of the seven types of agent orange being manufactured for the Army under contract. The stack scrubbers (filters) would be shut down to a bare minimum. So that no one could see the cloud of contaminants being blown to the wind, they could double or triple the nights production, then just before sunrise production was slowed down, and the scrubbers in the stacks were again put into operation about 4:30 a.m. so the air would clear from the nights operation. There would be days it would not dissipate as rapidly as they would like to have seen, but no one at the Air pollution unit began work before 8:00 a.m. by then it would be mostly gone, or blown with the every days pollution being blown up against the mountains to the west. The Air pollution people would be testing during the day and give all good reports, never did they test at night. I was told their supervisor's would not allow that at all. This would slowly stack up over the City of Denver until there would be that daily cloud hanging over Denver. Shell would watch to make sure the wind was blowing to the East North East before its nightly operations began. There were a few times the wind would shift back to the West and would build rapidly to the critical point as the pollution would build up against the mountains like a dome over the city. where everyone in Denver would be warned they are on red alert for smog, anyone with any problems advised to close all windows and stay inside and wear your respirators, and have oxygen available. That particular person also told Larry You won't have any problem knowing what happened if you look up and see a reddish haze in the sky, Shell got caught with the wind shifting. The reddish material nitrous acids, was used in the manufacturing of Many of the herbicides, including agent orange, and that is where it got it's name. "AGENT ORANGE". Mostly all 2-4D products, 2,4,5-T, Silvex, TCDD, and DIOXIN-is an unavoidable by product in the manufacture of certain herbicides, as was discussed. DIOXIN and TCDD, is believed to the most dangerous of all. Larry could remember on some clear moon lit nights as he was watching the stacks at Shell Oil, around eight to nine p.m. could see clearly this blackish red, almost puke orange pollution start bellowing out of the stacks. Larry knew then whether Denver was going to have a bad pollution, day or just a normal day, while everyone in the East were severely poisoned probably as they slept.

As they were standing there, an elderly gentleman drove up, walked up and joined into the conversation. He listened quite interestingly. "He began talking about other parts of the arsenal he pointed out he had worked for the Army as a civilian employee for thirty (30) years. He quickly asserted if you want to see some real information on this place, I could give you an eyeful. I've got documents and records at home you wouldn't believe he said.

This old man definitely had all of Larry's attention. He knew all the people there at the gate, and they knew he liked his beer, they quickly directed us to go to a bar about a mile down the street and have a beer. There you would better be able to talk, cause right now you know were being watched. Licking his lips like a thirsty dog, the man eagerly agreed.

I followed the old fellow he said he knows exactly where to go. We arrived at this small bar had just sat down this guy started to open up as to what he was going to be able to get for me, when two M.P.'s Military Police walked in and stood, one on each side of the door as if awaiting for something.

Turning towards the door, the old man saw the two giants standing there, and literally turned an off-white, and definitely forgot how to talk. Suddenly he got up and said he had to use the restroom. One of the M.P.'s made the same choice and followed him into the restroom. The other one just stood by the door staring directly at Larry, as if he was waiting for him to try something.

Several minutes later, both the M.P and the old man emerged from the restroom. The old man looked even paler than before. Walking over to his stool, with one M.P. right behind him and without another word, he picked up his jacket, and left with the M.P.'s.

No good-by, no acknowledgment, no nothing as if Larry did not even exist.

Larry had the old fellow's name and address, he had scribbled on a piece of paper. About a week later he thought things had cooled of by now so he went over to Westminister. A north westerly suburb to Denver. When Larry arrived the house was locked, no one around. He talked with one of the neighbor's who had come over to see what He was doing, He said they had moved, real rapidly and no one knew why, everything seemed to be ok.

Larry had not realized that the Army had a way of knowing what they were talking about, until one day when he was working for an auctioneer herding cattle into a ring to be sold. He happened to see one of the guys he was talking with at the front gate of the Arsenal. Larry said hello, and walked up to the man hoping that they could talk. The man looked at him and said, "Stay the hell away from me." I don't want anything-to do-with you. They heard every damn word we said at the front gate. I am to have no further contact with you. The front gate was bugged, they heard every word we said."

"How could that be?" Larry asked.

"I don't know, but you stay the hell away from me or I swear I'll let you have it."

"Look, Larry said I'm not trying to start anything, that's already been done, when I saw you standing over here I just thought we could pick up where we left off, and maybe talk a bit more.

"Ya sure, then one day I learn everything I've said to you has been past on to them. I've got a job and a family to think about, I can't afford to get mixed up in anything with you, so get lost!"

"Are you telling me that what is going on out there doesn't bother you? What about your family for Christ sake."

"I…I can't be the savior of the world or be responsible for it, and one man is not going to make any difference. You come in here huffing and puffing about saving humanity and the environment and all that other bull shit, well let me tell you something. You have no idea of what they are capable of or the power they have. You're not going to win, and do you know why? Because they're not going to let you. All it takes is one snap of the finger by some big shot and things start to happen. Nasty things like people disappear, people die. You don't think for one minute that I don't know what I'm talking about, well I sure do. That Arsenal will be here long after you and I have turned to dust. It will be here and in operation because there is too much power and money involved. Now please just stay the hell away from me!"

"Larry says, well look here I don't give a damn who you are, right now I consider you one of there puppet's, and he ask him to take off, that this is his sale, not yours.

"Larry said You know it's strange, but I feel sorry for you." "What goes around comes around, while right now your runnin round happy as a boar hog in mud. I feel sorry for you Mr. because you're already dead, you just haven't realized it yet."

Turning, Larry walked away, and started to make his way back to his pickup. As he walked up to his vehicle, he happened to look down at the car parked next to his. A bumper sticker in bright red letters jumped out at him. It read: Every man is a fool for at least five minutes each day. Wisdom consists in not exceeding the limit.

"Before Larry could leave the sale area, the same man walked up to him, and said I didn't mean to sound so irrational, but we should stay shy of each other at least for a while, keep up the good work I know what your doing is right.

CHAPTER FIVE

The objector, and the rebel who raises his voice against what he believes to be an injustice of the present, and the wrongs of the past is the one who hunches the world along

Clarence Darrow, 1920

By the time the summer of 1977. rolled around, the health and financial well-being in the Land family had improved enough that they were again able to get away for an occasional weekend with the kids.

After returning from one such weekends, Marie and Larry entered the house and Larry was immediately stuck by the feeling that things were not as they had been. It wasn't anything that was obviously out of place, it was just a feeling, a feeling of intrusion. Walking through the house he quietly observed its contents, careful not to alarm the children. He did not see anything out of place but the feeling of intrusion persisted and was as strong as ever.

After unloading the car and getting the children to bed, Marie went to the kitchen to make a pot of coffee. As she walked into the kitchen, with Larry behind her, she turned and looked at him. He could see the worry on her face.

"Larry, they've been here again, haven't they? Somehow I can always feel it. Things have been moved, dust lines have changed" she said.

She had noticed the small things that only a woman in her own home would instinctively see. Marie saw the things, what were to a man, taken for granted, unimportant things that a man was insensitive to seeing. Drawers that had not quite been closed, keepsakes that had been picked up, examined, then returned to their normal place, only now they were slightly ajar.

Immediately they went to the room that held boxes and file cabinet full of the material and documentation that had been carefully collected over the years. As they looked closer, it became evident that everything had been inspected by someone unknown to them only a few days before. Larry had suspected that something was going on many times in the past, but had always thought it was from tension or lack of sleep. The thing that always made him think that it was his imagination was that nothing was ever taken. When you know this, you then know it was professional. However Larry knew by now exactly how the Army, and Shell Oil Company collected things, so he left nothing at chance for them to collect. Usually if they wanted something they just took it. Larry kept nothing that was new and different, or any thing as to how he was collecting his information, the only thing kept at the house was what was already public information, for which he also had copies made of all of it and placed in a vault. Any information that Larry had received, or had put together, or the Infra Red Photo's he had taken, that effectively spelled out every hot spot or toxic waste dump used, or created by the Army and Shell Oil Company, from Denver to Brighton. The Army was already told to do their own damn Infra Red Photos. This when Larry had already took the photo's. Which he did release two Infra Red Photo's showing the exact location of two toxic waste sites on the arsenal property, using Infra red the hot spots show up red in color and showing the exact locations, and how hot it is. The photo's Larry has do show at least a potential of up to fifty (50) locations. It was in the news the following day, with the stories to match, putting the Army on the hot seat. The army was at his door almost immediately giving him some warnings. He thought here I go, he truly thought they were going to take him away, and they were also prepared with U.S. Marshals. Larry advised them, anything I may have, had already been prepared, and released to Congress, The Army then did go to a great extent of doing their own infra red photography, and then prepared all the excuses as to why it was done in this manner. They stated the arsenal was set up during World War Two, and their were not always times to do it all correctly. and so they thought they were prepared for Congress. They only added a little more to their own demise when Congress asked them then why the hell did the United States Army, and Shell Oil Company continue using basin F for an additional twenty five (25) years as a toxic sewer after they were advised by the United States Department of Health to stop using it, any further use will be under the premise it is contaminating the underground water both on and off the arsenal property. This is the demise that finally ended their existence. Now he wasn't so sure. Perhaps he had missed something. In any event, there was nothing that they could do but to speculate as to the possible presence of an unknown visitor in their lives. Larry looked at Marie, put his arms around her, turned out the light, and left the room with its secrets.

During the period of time, when the Democratic Party Headquarters were broken into, Larry was working as a Deputy Sheriff for the City and County of Denver, He worked mainly with U.S. Marshal prisoner's being held. During the time Larry was checking the prisoners, a particular man who had been at the Jail only a few days, was standing at the door looking out the window. "He stated L. Land, would your first name be Larry".

"He said yes how do you know? Larry asked.

"A friendly warning, watch yourself, they are really after your ass now for doing that last interview." "Do you realize you trespassed on a secret Military Base when you photographed the Rocky Mountain Arsenal. This is why the F.B.I., is after you."

"The F.B.I. wants my ass, what are you talking about?"

"The F.B.I, the Government, you made some very important people unhappy. Ever since you started doing this whole arsenal thing, I've been in and out of your house on numerous occasions."

"You've been to my home?" Larry said, sounding suspicious.

"Sure many times, starting around 1973 or so. I am burglar, I work for the F.B.I. specializing in hi pro file cases. Got accused of being involved in the Democratic party headquarters burglary, they will have me out of here fairly quick. We have been looking for documents of facts, names, and or pictures. They think you know too much, and your talking to every one that will listen, In their eye's your talking to all the wrong people, and that is a bad combination.

"I'm suppose to believe all this right? well alright how many times have you been to my house?"

"Right off hand I could not give you an accurate number. Guessing I would say probably twenty times. You live out on 104th and Potomac, a four bedroom ranch style home, a natural red wood fence with a large rock and flower garden, with a flagpole, flag at half staff in the front yard."

"You could have gotten that information anywhere, or simply driven past and made the connection."

"Still don't believe me, is that it? The last time I was there about two weeks ago on a Friday night. If you walk in through the kitchen entrance the table is off to the left, cabinets to the right, appliances are almond colored. Your wife has different size woven baskets hanging on and around the soffits in the kitchen. Your bowling ball is in the kitchen closet. In the living room you have a television, stereo, the usual furniture, and a small grand piano sitting in front of the picture window, shall I continue?

"Sure you could have learned that somewhere too."

"They were right, you are a stubborn cuss. Very well, we'll do it the hard way. In your bedroom you have a gun rack hanging on the east wall. You have four rifles and two shotguns with trigger guards on all six., and I want you to know I took a real hankering to your Winchester 30-30 lever action, with an octagon barrel, The Buffalo Bill Commemorative. On the west wall is your bed with the foot of the bed facing east. Your wife's jewelry box sits to the right on the dresser if your facing it. In the jewelry box are several pictures of your parents, an old wallet and pocket watch. There are two class rings in there that I assume belongs to you and your wife. Seeing that your wife came up, she has the cutest pair of soft pink flannel pajama's with tiny red hearts all over. In case you're wondering, I have total recall, a photographic memory. Would you like me to go on?"

"No, you made your point. Why are you telling me all this?"

"Under normal circumstances I wouldn't, but you must admit these are not normal circumstances. Me here and you there, a chance in a million meet. The People behind your cause, lot of media, and even now the State Department of Health, and sounds like you've now moved into Congress. I read a lot of your papers, letters, I believe in what your doing, although I can't help you in any way, nor am I fully convinced. It's true, what they are doing at the Arsenal is wrong, I just wanted to say to you "Bravo" you have woke the sleeping giant World wide." I probably won't be here for more than a couple more days, Keep it together he says, just call me Tim, I wanted to give you a friendly tip, watch your step. You will know next time if I ever get out your way, I'll leave something you'll recognize.

This man had just admitted breaking into his home and admitted he was a Government burglar. The only reason that Larry could think of for the Army to do this, is that they must be getting pretty damn worried about this little rancher stirring the pot. "so to speak"

Roughly twenty minutes later a U.S. Marshall walked in to the correctional facility, asking for the release of this man. On the way out Larry heard him say, "Jim lets go get an early morning breakfast, I'm starving."

"All right Tim," the Marshall said. "I could go for that myself."

So his name is Tim I'll have to remember that until we meet again.

Throughout the whole ordeal with the Arsenal, aside from his family being poisoned, what happened next was probably the most bizarre and painful. Larry really couldn't explain it in words because it was more a feeling, almost like friendship, that he had for this man. He had never been good at describing his feelings. For what ever reason it affected him in such a way that for a long time afterwards he could not easily talk about it.

In April, Larry received a phone call from a fellow officer who worked at the jail with him. A new prisoner had been brought in, and he was in appalling shape.

Early the next morning as Larry sat reading the booking sheet he saw a man listed under John Doe, and no charges were showing "Just Hold For U.S. Marshall's. No identification of any kind had been found on the person or in his property. It was as though he was dirty bum picked up for vagrancy, yet his suit was quite expensive, but absolutely filthy no jewelry, no D/L credit cards or money. He had been dropped off with only a Hold for the U.S. Marshall. They sometimes won't give any information on the subject, Just a hold with no property. If we have any problems such as death, or medical questions Call the Marshall.

Before leaving, Larry's Supervisor advised him to keep a real close eye on this guy, highly suicidal, and the day before had wiped feces on the wall, the floor and himself. Mentally this guy was a few bricks short of a full load.

Several days passed before Larry could even begin talking to this man, and Officer Land was very good at getting people with mental problems to open up and talk. His mental balance and physical appearance was simply awful.

As the days past this man's demeanor, and manner was slowly improving, as he began to talk, clean himself up, and settle down. It also appeared he was trying very hard to remember who he was, and where he came from.

The next day was Larry's day off, but when he saw him next he told him he lived just outside of Dallas Texas, and what his name was. He asserted he wanted to call someone but could not remember the names or telephone numbers. He wanted a Phone book but the only one Officer Land could get for him was Colorado. He stated that won't work. As days passed by the more he would talk to Larry the more he was remembering. He wouldn't talk with any one else because they wouldn't listen or laughed at him, and would not let him call anyone or even try. Larry tried for him several times but got no answer or was told wrong number. It seemed a basic trust was being developed between Larry and him.

"Officer, Land could you check my belongings and see if I came in with any money? I am desperate for a cigarette."

"The only belongings you had when you came in was the clothes on your back, nothing else."

Looking at him Larry saw a deep frown come across his face. Even this had been taken from him. He was confused, disoriented, and at the same time scared to death. He was locked up and he could not understand why. Larry Couldn't tell him anything because he didn't know anymore than the man did. They usually leave of some money for you guy's when your going to be here for any length of time.

The man was again asking for a phone call, Larry felt kind of sorry for the man since he already tried and was unable to get a hold of anyone. He just did not appear to be the type of man to be behind bars.

"Larry, said sure I think it would be a good idea for you to try again, and to hear a familiar voice, it might make you feel better." Larry got the clip board for all phone calls had to be listed and to whom, then approved by the shift sergeant.

"He listed John Edwards, CIA Home Office, Langley, Virginia."

"Larry stated "C'mon you gotta be kidding me, I'm not going to play any silly games. If this isn't so there will be no more phone calls." "DO YOU UNDERSTAND."

Looking at his name tag he said, "Mr. Land, My name is Phillip Brown. I am employed by the Central Intelligence Agency and have been for the last seventeen years. Before that I was an officer in the Marine Corps. I have no reason to lie to you. Would you please make this call for me, no one else will."

To this day Larry hasn't been able to forget that experience a person hears of horror stories like this in other countries. We live in the United States of America could something like this really be happening. He didn't want to believe what was happening but he couldn't help knowing somehow Brown was telling the truth, and in that sense he was thankful for what little he could do.

It was time to get people out of their telephone calls, the one's that were pre approved. Brown's phone call was pre approved to a John Edwards at 703-482-1100. These phone calls were made sure they happened and to the specific person by the officer. After about four rings the phone was answered by a nice sounding woman announcing politely that Larry had reached the Central Intelligence Agency. Brown stated at this number they'll put me right through to him, Larry asked that a Phillip Brown would like to speak with Mr. John Edwards, she said one moment please.

"It seemed as if several minutes had passed, and just as Larry was about to hang up the phone a man's voice came on the phone this is John Edwards, May I help you. The call was confirmed so Larry then told Mr. Edwards, this is Officer Land with the Denver County Sheriff's Office. I have a man here by the name of Phillip Brown, he says he works for the CIA, and he is asking to talk with you, personally, Larry said."

"He sad put him on."

Larry worked in the Maximum Security part of the Jail, The inmates were brought into the Officer's office, handcuffed to make those calls in the officer's presence after the call was confirmed as to his request, and the person on the other end wanted to talk with him.

Larry handed the receiver to the prisoner he heard, "John what is the meaning of this, why the hell are you doing this to me? I have worked for the agency for seventeen years with never a blemish on my record. What is going on and why am I being treated like this?" Then I heard him say OH no he's out doing something else. "I believe that meant me.", and he just covered my A__.

What was said on the other end he had no idea, except the prisoner said, "what was it that I was suppose to have seen out there?"

One thing, I felt certain this man was a CIA agent, who done something, or seen something he wasn't suppose to. Larry couldn't even begin to believe what he was hearing and seeing. "This was. one of their own without doubt. Now I feel I'm right in the middle of something I don't want.

"Get me out of here John…ok very well, two hours, I'll be waiting. "OH" John after this little incident my trust isn't what it used to be, so if I do not hear from someone, my next call will be to the newspapers. Thanks John."

After hanging up Larry walked him back to his cell. He could see that he was deeply troubled.

"As officer Land was taking of his cuffs, he said thanks a lot, and I really mean that. I don't have any idea why I'm in here, and in this condition, He wasn't even remembering the first few days here and what he had done, nor what kind of shape he was in. Maybe it's all just a big mistake and Edward's can straighten it out."

"You said maybe, why maybe?" Larry asked.

"Because I've seen things like this before. One minute an agent is there, the next he's not. Some times they see, hear or do things they shouldn't." I was sent out here to investigate certain allegations about a place called Rocky Flats, which you probably heard about it's located out west of Denver. They make plutonium for the atomic bombs. It sits in Jefferson County, and there has been a lot of people protesting it, and blocking the tracks, a hundred people were taken into custody just the other day for blocking the tracks."

"Mr. Land, I hope it doesn't backfire on them and me, but I have a pretty good idea what is going to happen to me. I shouldn't be saying anything to you about this. I don't want to get you involved in anything I may have said, but I want you to know that Rocky Flats is leaking highly carcinogenic plutonium, and it has escaped several miles now from the plant. Just remember I didn't say anything to you." "Huge cover up."

About an hour later a well dressed man was here to pick up Phillip Brown. He showed his credentials showing he was a Deputy U.S. Marshall, and Brown was released into his custody.

Three days later Phillip Brown was back in lock up, and as mentally deranged as he had been the first time that Larry had seen him. This time however all (personal) identification even some money had been left on his person.

"Larry Knew the Rocky Flats was getting a lot of Television, and newspaper coverage, especially with all the people blocking the rail road tracks, by sitting and lying right on the tracks. It was making big news about it's leaks, and it of course being an ultra hazardous activity.

"Larry had by now seen enough, and learned enough about those using Heroin, Meth Amphetamines, Black Tar Heroin, even various acids, was coming onto the drug scene. They had a special place in the jail just for people coming down. Fume sniffing was now a very big thing with the young people Some really fried their brains."

"Larry could see readily the problems Brown was having was probably unwilling or possible BZ Nerve agents heroin use. He had tracks up and down his arms just like before, some were terribly bruised. Orders was left he was to have no visitors, and no phone calls to anyone."

With in three days the U.S. Marshall's were here to pick up Brown. He was just beginning again to make sense and start taking care of himself. Larry Checked him back out of the jail, and hoped he would find an end to his problems.

Larry was home that night trying to forget about Mr. Brown, and what happened it was constantly on his mind, just wondering what happened to him. Some time much later he dozed off. He was restless and sleep was the last thing he wanted.

Larry awoke from a tormented doze off, It's about 2 a.m. he's thinking about Brown, and what they done to him, and why. Maybe it was about some Government secrets since it was Rocky Flats. The most Classified and confidential and secret Army installation in the Country That makes the heart of the atomic bomb. There is no proof as to what Brown was doing, but the way things happened in one sense you probably do. Yet, coincidences like that, do not happen in every day life. It's like something played out in some horror movie, or a horror story. They use drugs in brain washing techniques, and I believe this was one that I saw first hand. To string someone out on heroin, and Meth amphetamines, for many days, maybe even a week or two. then to put him in some place like the jail with cement floors, with a round brass hole in the floor for a toilet. To let him come down the hard way, where he has lost control of his bladder, and his bowels at the same time cannot eat, each time he tries he will vomit it back up, Gut wrenching cramps, while the top of your head is blowing up, and every bone, and muscle in your body hurts, like in a living hell, where he is aware of the pain and suffering. Then when he comes down after a week or two and starts talking, they pick them up to see if it works, and if not they'll go through the same routine again, and maybe again. UNTIL!

Larry asked himself then, and now, what type of people can do something like this to another human being without even blinking an eye. Maybe until next time. He felt used, almost the clean up in brain washing with drugs. caught in the middle, "Do what your told, No questions asked, keep your nose clean."

Larry had a few cups of coffee, and heavy on the brandy to help him sleep, and get a few things off his mind. He curled up in the blankets and soon dozed off into a deep sleep.

Around dawn his eyes sprang open to the sound of the telephone ringing. Rushing into the kitchen he picked up the receiver and was surprised to hear Dennis on the other end.

"Dennis asked Larry if he would meet three of them at the café in Brighton in thirty minutes? I've got an idea, I think you'll like." "This'll blow the Army's socks off." Larry then said let's make that thirty, with a lot of ish. (that's the Colorado excuse for not being on time.) Larry said "All right, I'll be there in thirty-ish minutes." Larry answered.

Larry thought for a moment about Mr. Brown, if he blows the army's socks off what is it they may want to do to him. The Rocky Mountain arsenal probably second most secure, and secret Army bases in the country, manufacturing Saren, GB, BZ Nerve, Arozine Hydrozine, Mustard & Lewisite agents, and who knows what else. A whole lot like Rocky Flats who manufactures triggers for the Atom Bombs, Atomic Bombs, Hydrogen Bombs, near Denver, They were coming under a lot of criticism and fire from all over the Country, England, France, Canada, & Washington D.C. Regarding the manufacturing process and the use of plutonium. With the Government pushing their weight around, stating nothings wrong. Larry said he would blow the whistle on them any time. he just don't want to be another Mr. Brown. That really worried him. he will take every precaution out there.

Forty minutes later he was seated in a booth with Dennis, Jennifer, and Glenn, along with a good strong hot cup of coffee.

"OK Dennis, what's your plan?"

"I was thinking last night that what we need, along with everything else, is emotional evidence. Something that the public can not read or listen to, but something they can see, evidence that is visual. You've still got all those Infra Red Photo's you had done of the entire arsenal, right."

"Larry says yeah all but the one that was already turned over to the newspaper. He said there's lots' photos of that place, and sixty (60) hot spots." It had a big splash alright, the Army was at my door the next morning saying they wanted the rest of the photo's I had. He told them all the photo's had been given to the United States Congress. They stated they were going to do an extensive set of their own. There had been bits and pieces from the Army to the News about certain trenches. Nothing on the open basin storage like lakes A, B, D, C, and F. Some of the trenches had no records as to what may be there, however, they are showing up as highly toxic, and very hot in the photo's."

"Dennis said well what do you think of dropping a couple more on the press and television."

"Larry said let me think about this for a while, we just got the State officials working with us, the Health Department is coming around where there acting like they know what they are doing at least now, since they haven't done anything for over thirty (30) years except hold the Army's hand. The doors in Washington D.C. have just begun to rattle, and we've now got some interesting things happening in Washington D. C., and he said we sure don't want to screw that up. We just got Senator Hart sounding somewhat interested in what's going on. Bill McNichols the Mayor of Denver even is sounding in a little assistance. Representative Patricia Schroeder's were out to see what was taking place. Several of the State Representatives are now asking a lot of questions of the Army and what their doing out there. To top it all off the Attorney's ask if I would Lay low for a while. Larry said he spent a little over eight thousand Dollars, ($8,000.00) on the photo's so far, we have to lay back some since things are really starting to move. (It's called stirring the pot, stir but don't over stir.) I've heard wind the State of Colorado is about to issue Cease and Desist orders on the Army and Shell Chemical Company, both with a lot more teeth than the first set.

"Suddenly Jennifer put her hand in her purse and came up with a long white envelope, handing it to Larry."

"Inside the envelope was fifteen fresh big ones, one hundred dollar bills."

"Larry somewhat startled said what is this."

"Glen said we want you to know your not alone, and we knew you've spent a lot of money so far. Were in it with you all the way, and wanted to help with the photography you've already done.

"Thanks to all of you Larry said."

"Glenn said he had one more bit of information, an internal memo in the Army hierarchy." "It was intended as a Press release, but some higher ranking officer decided against it, and it was never aired. This is still another good example of the recklessness that has gone on at the Arsenal. The bad part is that there is no date on it so we have no idea how old it is." But don't forget there are ways we can fix that too, give me a little more time.

"Thanks Glenn, let me take a look at it and you maybe will have a date for it by then," Larry said.

The article read. "A deadly Nerve Gas bomb and eighteen Mustard Gas Shells, still carrying explosive charges, were unearthed just beneath the surface of the ground, an Arsenal official disclosed late Wednesday."

"Were not sure at all what is buried out there," the Arsenal Official said. "I know for a fact that many such devices were buried, and tested in the past, how many unexploded devices will be found is uncertain." "There were no records kept on this."

"The Nerve gas bomb known as an M-125, was severely corroded and contained 2.6 pounds of Sarin nerve agent. If spilled, this amount would be enough to kill anyone within 300 to 900 feet. However the bomb's contents actually are liquid until exploded, at such a time the substance vaporizes into a poisonous cloud. Depending on the wind speed, it's updraft or downdraft could be carried for miles." One drop will kill within minutes, when you think of acute, "Saren redefines the word with a short course to certain death."

"The Mustard gas shells, range in size from 75 to 155 millimeters, contain a persistent, powerful blistering agent that was common in World Wars I and II. Mustard is a solid or liquid, depending on the temperature, until it's exploded. End of Release. Comes down over it's victims, severely burning their skin, and suffocates a helpless victim within Seconds, or up to a few minutes."

"Glenn, do you think that there are a lot of bombs and shells buried out there?" Larry Asked.

"I know there are! There is not one bit of doubt in my mind," Glenn replied."

"That's about what I thought. All right I'll be at home, I want do some more thinking about what to do with the rest of the infra Red photos. I think we first need to track the Army on what they do and don't do when they release certain information about something, and when they don't we'll be there, a continuous thorn in their side forcing them to tell the truth.

Most things in life, Larry learned the hard way but Marie was something special. as she had more old fashion common sense that anyone he had ever known, and usually when she said something she was right.

He often wondered how she felt about all this. She had always been there for him, and yet he knew that she carried her own fears and pain with her. Watching their children suffer had torn at her very being. He remembered visiting her in the hospital following her miscarriage. He would walk quietly into the room only to see her lying there, staring out the window with tears in her eyes He thought there was a part of her that wanted to leave this place so that she could protect her family, The other part of her wanted to stay and fight this as much as he did. After all she watched as Larry and Herself also fell victims to this while losing the livestock, and everything they owned. It left her with an internal struggle within herself, a battle that was impossible to resolve There were times that he shared those inner battles. Times when he too, yearned for peace and a safe place in this world. But then he wondered, does that safe place really exist? Maybe that's what gave him the inner strength to do what he does best, helping to make this a safe place for everyone.

Larry began thinking where this all started. The Rocky Mountain Arsenal, He remembered flying over the arsenal on several occasions.

knowing the lay out he was able to pick out exactly what he was looking at. All the basins and trenches stood out like a poisoned sore thumb. The soil in many areas had a yellow, reddish, or blue-green tinge to it that resembled a rainbow laid out on the ground. The so-called fresh water lakes looked more like a dirty translucent yellow instead of sparking blue water.

At certain angles the ground below looked as if it wasn't even part of this planet, instead it resembled some barren landscape from another world. There were very few signs of wild life anywhere.

Larry then tried to remember back to the Infra-Red Photos, not only did the photo's show in bright red and yellow all the hot spots containing contaminated heavy metals, but they showed where the Army had buried weapons containing hazardous material. Bombs, Barrels, Artillery Shells, mortar rounds, some exploded and some unexploded. Steel thousand pound nerve gas containers, and mustard agent containers. Anything the Army or Shell Chemical Company did not know how to destroy safely, they just buried it, "Much like the ostrich," and much of that was still capable of exploding.

There were two photo's that distinctly showed reddish colored lines running from Basin A, B, C, D, and Basin F, and some highly contaminated leaching toward the Land ranch, that showed a lot of high red ridges.

AS Larry was thinking of all this he then began to think of all the devastation for his family, thinking of his parents who died from drinking their water, and his and Marie's child born dead by spontaneous abortion caused drinking the well water, his neighbor's that were having all kinds of health maladies, the Doctor's could not explain, the ones they could explain were dying of cancer. Then there was Phillip Brown, he gave Larry a lot of information about Rocky Flats, once he started talking, I would bring my tape recorders down. Then I began to wonder maybe I'm hearing some stuff here I don't really want to know. Larry pretty much kept most of the information under wrap.

Larry didn't know the proper term for the human condition he was experiencing, but he was remembering once when he flew over the arsenal, and he was looking down from the plane. He thought of all of those who had paid the ultimate price for the existence of the place, and it made him happy, and almost light hearted. All of us, the living as well as the dead were laughing at the Army's stupidity. Because of laziness or ignorance they had lied to the public and soon they would be paying for their arrogance through the Court system of our land.

"As long as the world shall last there will be wrongs, and if no man objected and no man rebelled, those wrongs would last forever."

Clarence Darrow
Chicago 1920.

He was right. It was true, Larry wanted to see the Arsenal closed, but he also wanted to correct the wrongs that had been done by the Army and Shell Chemical Co., to the people who lived in the area.

Justice was what he wanted. Justice for the survivors and for those who had not been so fortunate. They had to be held accountable.

Was it just human nature that made people the way they are? he didn't know the answer, but he wished someone would tell him. Why must everything that centers around the so called progress of man be dragged down to its lowest element? Why must we constantly pay for this so-called progress in the number of human lives lost?

How many people have died of disease due to the experiments conducted by the Army in New York, San Francisco, Minneapolis, Louisiana, Atlanta, Chicago, and the entire wheat belt of the Midwest?

"The Army admits having done these experiments on the people so they could see if what they had developed actually worked."

The Army very quietly after being blamed for these experiments, and proof supplied to the Government have many excuses as to why something like this was done to our own people. Young Children always seem to be a big part of these experiments, and our Government done very little to assist even in medical surveillance.

"Is it so wrong for us to love one another instead of always trying to destroy each other?"

I think that when man reaches a certain point in his life, he begins to forget the beauty and warmth of his Country. In much the same way that he forgets what it was like to be caressed by a beautiful woman or the memories of his youth. Monetary concerns always seem to take precedence.

When he reaches a position of authority that has taken him a life time to achieve, he no longer fully realizes the long term effect of his decisions. He knows that his time on earth is limited. He cares, but it is much in the same way as he cares for his lover. As his age advances, the closeness that was so dear to his youth, grows more distant. Closeness is no longer as important as it once was. He has forgotten the meaning of companionship and understanding, because he thinks it is no longer relevant to his life, and his job, and others, who have consumed him.

There is no defense for the behavior of our leaders. How else could this have happened at the Arsenal, and hundreds of other places like it, have gone unnoticed and untouched over the years, causing distress, injury, and Death to many, while still remaining untouched.

It was at this time that his wife walked in catching him by surprise. He had been concentrating so hard on the pictures that he had failed to hear her come back from the Doctor's.

She and the kids had been to the Doctor for their weekly tests and check ups, and he could tell by the expression on her face nothing had changed…it seemed as if it never would.

"The Doctor wants to see you on Friday. He's mad because you haven't been in for your tests lately," Marie said.

"All right," said Larry.

Since they had started using bottled water and not bathing in well water, the effects of the poison in their systems had diminished dramatically. The pain and swollen joints were still very much there but the constant thirst, hair loss, the rashes, and sores in the mouth and nose, darkening of the gums, and teeth, the vomiting, and diarrhea, the horrible dizziness, and headaches, difficulty breathing, the flinching and dropping things, shaking, drooling, all seem very slowly improving, and may take years of improvement. All though Doctor's can't do much since they have not developed treatment for Nerve agent Poisoning and the United States Army have kept all tests and other data secret from them. The pain is somewhat tolerable with the strong pain medicines.

Looking down at the photo's he suddenly decided not to say anything just yet to Marie. As casually as he could, he began putting everything back into the package.

He had no idea why he made this decision except, That after seeing those photo's, he realized that the Arsenal was much more contaminated than he had even thought. His wife had more that her

Share of things to worry about. The last thing she needed now was for him to add even more.

Suddenly he wanted to get away. He didn't know why, he just did. Looking up at her he said, "Honey lets get the Kid's go out for an early supper, I know we can't really afford it but lets treat ourselves anyway. Besides I have to drop this package off at the Attorney's office."

"Is anything wrong Larry?" She asked.

"No, not really, I just wanted to get out for a while. Everything feels like it's closing in on me."

"OK, I'll get the kids."

While Larry waited for Marie to gather the kids and to clean them up, he placed the other package, the one he had opened, in a metal box he had installed under the floor of the kitchen sink in between the floor joists. He made sure that all of the household cleaning supplies were replaced not only to cover the hidden door, but also to look as though they had recently been used.

After delivering the unopened package to his attorney's office, he decided on a quiet restaurant north of Denver.

Unfortunately, with all his well thought out plans, that were subject to change. His plan was over ruled by his War Department, "the wife and kids" who were in favor of Mexican food. Without saying a word, and after a careful reflection, he had to admit it was a better choice. He had been soundly defeated. Besides, discretion on his part, at a time like this was better than an all out mutiny.

After an early supper they decided they would go over to the shopping mall and walk around for an hour or so.

Larry always enjoyed walking. It offered a time for reflection which was sometimes hard to come by during the days hurried pace. A little exercise wasn't a bad idea either.

Several hours later they returned home, relaxed and very much content, just itchen to stretch out.

They had been home no more than half an hour when Marie began to notice things. A file drawer in the bedroom that was not closed completely. A gap in the clothes hanging in the closet that was not there earlier, Things in the drawers in disarray, and the shoes lying on the closet floor were arranged differently than what they were left. It's all these little things that leaves everything out of balance.

"Larry, they've been here again. It looks like they've gone through everything. We looked for some kind of bugs, camera's listening devices etc, the kind of thing used to intrude upon another. We had talked about it was odd that they haven't already tried using something like this.

Walking into the kitchen he saw a fork that had been carelessly dropped on the floor. This he thought was very strange considering the fact that they were suppose to be professionals. Walking over to the cupboard he bent over to pick up the fork, and instantly thought of the metal box.

Opening the cupboard doors beneath the sink, he sank to his knees and began moving paper towels and cleaning supplies out of the way. The first thing he saw was the latch to the hidden door had been undone. His heart felt like it was skipping beats as he lifted the door and placed his hand into the metal box. It was empty. The package of infrared photo's was gone!

Larry was mad, mad enough to rip someone apart. Not only was he mad at the people who had done this, but also at himself for not having a more secure place here at home. This is the first time they really breached one of my secure storage area's. How ever mad as he was, he wasn't really mad at himself that much, because Larry always quadrupled any thing he had and only kept the working paper's secured at home. He already had three more sets of the Infrared photo's, one went to his attorney, the other two sets were taken out of state.

Larry thinks this will only make them roll over a lot faster because they know what I had, and don't know I still have others. What Larry is using them for now, is to make sure the Army tells the truth of what has been polluted, and what the sources from the Arsenal are.

Just then Marie called him into the bedroom again. By the sound of her voice he could tell she was upset and had been crying for the last few minutes.

"What's the matter, what is it?" Larry asked.

"C'mon lets go into the kitchen," he said to her. "I can think better if I have a cup of coffee in my hand."

Sitting at the table in semi darkness, his mind was a total blank. There were no thoughts of the Rocky Mountain Arsenal, or of the people who had broken into his home, or thoughts of his family,

Nothing entered his mind. It was like God had intentionally shut his mind off. It was as if he no longer existed. He just sat in the chair no longer connected to reality.

"Larry, are you all right, Marie asked?"

"I'm all right, I'm just depressed. Something has been bothering me ever since we realized someone had been in the house. It was as if who ever it was that was in the house wanted us to know who he was, I just cannot put a finger on it."

Standing up, Larry walked over to the five (5) gallon bottle of water that stood in the corner, and rinsed out his cup. Refilling his cup with hot coffee, he stood there, a moment the same obsessive, irritating thought jumping from side to side in his mind.

As he turned to go back to the table it hit him like a cement block. The FBI, but more than that, a name…Tim.

The man who had claimed to be a professional burglar for the Government, and told me that he had broken into his home fifteen (15) to twenty 20) times, looking for anything to do with Government operations, mainly Rocky Mountain Arsenal.

Well Larry sighed, not really feeling too bad that he was again burglarized, because they at least know what he knew, without realizing Larry was one step ahead of them, and still had what they stole from him. It will just be more fun as he irritates the hell out of them forever, while keeping them on their toes.

The bald eagle was once a common sight throughout the United States and most of Canada, but now it is in grave danger of total annihilation by two of our nations largest polluters, the Rocky Mountain Arsenal, (Army) and Shell Chemical Company, by their mishandling, and disposal of toxic pollutants, just outside the city limits of Denver, Colorado.

BY "LARRY LAND"

CHAPTER SIX

WHERE HAVE THE BALD EAGLES GONE THE AMERICAN SYMBOL OF FREEDOM

GRAVE DANGER OF EXTINCTION BY POLLUTION

Although I'm sometimes pessimistic about man's future, I don't believe him to be innately evil. I'm more worried about his insatiable curiosity than I am about his poor character. His preoccupation with the moon is disturbing to me, particularly since his own rivers run dirty and his air is getting fouler every year.

E. B. WHITE, February 25, 1966.

Why would this man from the FBI, break into Larry's house, and steal his pictures, and then give him a clue as to who did it? Larry could not be mistaken. It had to be the same man who was in jail months earlier and then mysteriously walked out with the U. S. Marshall, saying he was hungry, and wanted to get something to eat.

Was this his way of telling Larry that he still believed in his cause, but that he had a job to do and orders to follow? Possibly. He did say that if he got out that way he would let me know, a clue can be found.

"If you're right," Marie was saying, "and he was letting you know that it was him, you can't think that he would help you in any way. He's not going to jeopardize his position, or his future, or that of his family. He knows what these people are capable, and he doesn't want to be buried in some jail for many years because he crossed someone. He walks a tight wire every day of his life.

Larry looked into Marie's eyes and saw a frightening combination of anger, fear, and concern. Most of the time she was able to mask the feeling that were inconsistent with her desire to support him, and to stop some of the madness in this world. However, he knew that those feelings were part of her. How could it be any other way? She was, after all, a wife and mother and her first concern was bound to be for their safety, and well being. The break-ins seemed to be the vehicle that always created the cracks in her otherwise calm exterior. Perhaps she had reached her limit.

"I haven't forgotten," Larry said, "but maybe if I could talk to him, just the two of us, he might change his mind and decide to help."

"I think..."

Just then the phone rang, it was Jennifer calling. He wants you to speak at a meeting scheduled for this coming Friday, concerning toxic chemicals, and the disposal of the same. The opposition will be there defending the Arsenal."

"Well Larry said I do know how to speak to large groups of people, and get my points across, of course that's speaking as an Auctioneer.

"Jennifer said, "just stand up and tell them what you have learned about the chemicals manufactured at the Arsenal, and the horrible methods used for disposal."

"Jennifer, I would like that, I may not always win but I usually sell what I have. I'll just tell it like it is, on what I have found out, and the information I have pictures of, and what happened to me, and the Colorado State Health Department stating they notified the Land's by Mail that their drinking, and hygiene water was tested in June of 1975, and did contain chemicals from the Rocky Mountain Arsenal. A couple of other things I could talk about is the Honest John Missile program at the Arsenal, on dispersing Sarin Nerve agents. also the part on basin F. when the Army Hygiene Department, and the U. S. Department of Health, ask them to stop using basin F. and in fact warned them in 1967, that basin F was found to be leaking, an inspection done showed large parts of the protective liner were missing, any further use of basin F would be on the premise it was leaking and polluting the under ground water off the Arsenal property. In fact over twenty five thousand (25000) acres to the Northwest and to the North of the Arsenal were already contaminated. After hearing this warning they continued using Basin F increasingly more than before.

"Write down some notes on what you want to talk about, I know once you get going you cover every point necessary to explain something. Larry the word of what is going on out there has to be delivered to these people before it's too late. Dennis feels you're the one to do it. The people will listen to you, they know you've been hurt by this just like they are being hurt. Dennis says everybody from the Health Department right on up the ladder could have their hands in some one else's pocket, under the table and getting rich. Larry will you do it for all of us.

"Can I have a little time to think about it, Larry says."

"Jennifer said OH NO! I know you better than that. I want your answer right now, this very minute, while I have you on the phone. You're not scared are you," she said laughing as she goaded him."

Larry said, "you're sure a high pressure sales person today and you know what I do with high pressure people., but you talked me into it I'm glad you're on my side."

"You can do it Larry, you have nothing to lose, and maybe everything to gain." 'I know you'll make a lot of new friends, and there will be some able to help.

"How many people do you think will be here?"

"That I don't know. I should think probably two or three dozen. Depends on how many have heard about it, and how many have been damaged already from it. Meetings of this nature usually aren't very big."

"All right Larry said, I feel I'm backed into the corner with no where to go but take a stance. I feel the time has come for me to put up or shut up. I'm a leader and cannot shut up, so I will take that stance, I have so far, sometimes feeling like I have other's behind me, but when I need them there hard to find.

"You have two days to think of something. Don't worry you'll be great. I have a lot of confidence in you. The meeting is at the Ramada Inn on Quebec avenue, near the airport at seven-thirty Friday night. Conference room will be posted right inside the front door. I'll meet you at the front door at seven P.M."

"You're a big help Jennifer. This whole thing was probably your idea." he said.

"Well I did help a little, I volunteered the both of us," she replied laughing.

"I knew it," but as Ben Franklin once said, "Well done is better than well said." "we'll have to make sure it's well done," Larry said see you there.

For the next two days Larry poured his heart out trying to come up with something intelligent for these people. Finally on Friday afternoon, the day of the meeting, he had something that he thought would be appropriate. Looking at it, he could only shake his head, Larry had been over the data, so many times he knew most of it right of the top of his head. all he needed was little reminders of each subject and a direction of how they would be used. His notes read.

1. United States Health Department warning the Army of using basin F. 1965
2. The Colorado Department of Health stating they had warned Larry Land by Mail of Arsenal contaminants being found in his well water. "There was no letter ever sent as claimed."
3. The Tri-County Health Department also claiming they sent a letter to Larry Land warning him of the consequences of what was found in his wells emanated from the Rocky Mountain Arsenal. "There was no letter ever sent as claimed."
4. The "Honest John Missile" and the payload.
5. The M-125 ward head filled with Sarin nerve agent.
6. 75 to 155 millimeter shells containing persistent Mustard agent, this was common in World War I, and II,.

Larry was feeling as good as he did in 1974 when he appeared at the Hearings for Shell Chemical Company, when the were trying in a last ditch effort to stay in business. They had purchased a large amount of property in Morgan County near Brush Colorado, where they had already drilled several wells to the Lacota sands, and the Dakota sands, several thousand feet deep. where they were going to haul the waste from the Arsenal, about 65 miles to the north, and force pump it into these layers, or stratus under hundreds of pounds (PSI) of pressure forcing this into these sealed stratus.

"Larry was able to prove to the Court that doing this would breach every upper water strata in the area to become polluted with deadly chemicals being forced down their wells, under hundreds of pounds pressure it would force contamination back up for miles around, where back in the early 1900's, 100's of dry holes were drilled to the same levels and below, looking for oil. These huge and very deep holes when found to be dry (no oil), they would be mudded shut, meaning filled and sealed to the surface with mud. With Shell Chemical pumping this under pressure would allow chemicals to breach these holes that had been mudded up down to the same levels, and below. This under pressure would breach the seals then forcing Contaminants, and pollution back up these mudded holes, leaching into the water wells for much of Brush, "Ft. Morgan, and into surrounding areas." This would then flow in the direction of Nebraska, Kansas, Iowa, and Missouri, contaminating domestic wells hundreds of years from now. Life time expectancy of at least some of these toxic chemicals is 500 to a 1000 years life expectancy.

Mr. Land had met all requirements of the Courts to cross examine all witness, and he proved to the Court the expert witnesses for Shell had only dollar signs in their eyes, and lacked knowledge and common sense of what they were about to do. They were to show their feasibility study to the State of Colorado, and they had very little, to none. Larry was well prepared for most everything, he had spent day and night pouring over information both legal, and about the search for oil in the early 1900's, and for safety sake how was these huge deep holes covered and sealed. He had several attorneys' assisting him in this endeavor to prove the danger of allowing Shell Chemical permits to do any more devastation in Colorado, or elsewhere.

The last day of the Trial and when the Court had adjourned, news cameras were all around since this was big news. When one of Shells Attorney's walked up to Larry, and said "you are the most radical little Son of B_____, I have ever had to deal with," As Larry looked down upon him, His mostly bald head was a glowing red. With news cameras running right in their faces, Larry said to him, "Thanks for your support, I know I was wrong once, but I was only mistaken." "not for a moment putting himself on this man's same level." This story did hit the evening news, beeping out what Larry was called.

As Larry stood outside the Court room to see if anything might happen. The people from Brush, and Morgan County filed out of the Court room. Stopping to shake his hand for the interest he had taken for them up north. He said each hand shake was a good vote of confidence, and loaded with thanks. He became a "Hero" to many at the Health Department, and the folks up north.

Shell had also been involved with the U. S. Army in pumping toxic waste, under pressure, to the 12 to 14,000 foot levels. They had got away with this for nearly four years, until the pressure was put on the Army and the blame was their's for causing serious earthquakes, in the Denver area and to the North. Most of these quakes happened in 1964, 1965, 1966, they would register on the Rich-ter scale of 3.5 to 4.0. Lot of people were showing damage to their homes, and the

Army denying and taking no responsibility. In 1966 they were caught red handed, and under the table. They were immediately stopped in this endeavor that was done secretly and with no actual feasibility study for what they had done. The Army is use to doing things this way and if it causes a problem, they just deny it and hope it goes away.

To cut this short, Larry was a hero to many, and a radical to a few, however he managed to get his point across to everyone. With in weeks news came from the Court, Larry had only one thing to say. “Whoa we have won” “Shell Chemical has lost.”

“Happiness lies in the joy of achievement and the thrill of creative effort.” His favorite motivational quote by.”

President Franklin Roosevelt

Larry remembers back now how difficult this seemed, but really when he put his mind to it, the information was plentiful. He schooled himself, and filled up with the knowledge that was there all the time. He felt as if he had at least two college degrees. He was told by several attorneys, chemical and geology experts. “You’re a man of wisdom beyond most, You have become a Geologist, Hydrologist, Chemist, and Attorney all in one, and a hero to many.” To help build his self confidence, Larry remembers another motivational Quote he had read…

“All things are difficult before they are easy.”

“John Norley”

Larry as he got closer to the Inn was becoming more confident than ever for what he had to do.

“He began thinking of that little train that could, as it’s steaming up the big hill saying ‘I think I can’ ‘I think I can’ As he nears the top of the hill he said “I know I can” “I know I can” As he rounds the top of the hill, and on the way down he began saying with confidence “I knew I could” “I knew I could.”

Larry remembering what it takes, “Real leaders are ordinary people with extraordinary determination.”

As he drove up to the Inn, he was shocked to see the parking lot completely full of cars. The familiar building that he had passed on his way to Denver many times. He finally found a parking place clear to the back of another Inn, their lots were also full.

Walking up to the main entrance of the Inn, there were a lot of people just milling around talking. looking around he soon seen Jennifer walking toward him, and Marie, they were sure glad to see her as she was good support too Larry. She then led him and Marie to the huge room where the meeting was going to take place. Taking one good look he was surprised as to the size of the room. Chairs lined in rows from left to right, and front to the back. The podium was rather large, and I hoped they had a real good P A system. Larry is thinking he has a good strong voice, and is used to speaking through a P A system. Its always nice not to have to yell what your trying to say, so they can hear you in the back of the room. It looks like there is going to be several hundred people here, half the seats are full, and many out front with more coming in.

Larry had just turned his head when Dennis saw them standing there and escorted Marie and the kids to seats in back of and to the right of the podium. Larry was given a seat to the immediate left where the other speakers were seated.

One of the guest speakers was the head Chemist for the Colorado Department of Health. He was a person who Larry had many disagreements with many times in the past. It was this man, Bill Punt, who had stated, and informed the Health Department, the People North and Northwest of the Arsenal, and to the Congress of thee United States. That it was impossible for any chemical to leak from the Arsenal and find its way into the underground water supply with out the Health Department being aware of it.

Larry could tell that he was not happy to see him there, much less, ready and willing to speak to the people he had been lying to for many years, he knew his days were numbered, and the days of taking it under the table were about to end.

With a straight face and void of all emotion, Punt got up and addressed the audience. He stating his position for the Colorado Health Department, that of the Army, and Shell Chemical Company. He stated the people's or the city well water is not contaminated, that people were not being poisoned by toxic chemicals from the Arsenal. We at the State Health Department have tests we run that will tell us if the chloride is over 200 ppm that it is toxic with chemicals from the Arsenal. Although these tests haven't been done on any wells now since since 1965 The Health Department can assure you there is no way in Gods Green Earth could pollutants be leaking from the Arsenal. However these test are now going to begin again, we'll see that no pollution is leaking from the Arsenal.

One by one Dennis introduced the speakers and called them to the podium. Larry was Dennis's biggest gun, He was saved until last, so he could hear what others were saying. Dennis knowing some were here just to white wash what is happening, and this would give Larry a good start.

Finally it was Larry's turn and as Dennis introduced him to the audience, he began walking up to the microphone with a lot of confidence, and some anger, and a lot of courage. He looked better than other's before him.

As Larry leaned over a little adjusting the microphone, almost as if he knew exactly what he was doing. "Ladies and Gentlemen, Twenty five hundred years ago, a Greek lawyer by the name of Demosthenes said, 'Every investigation that can be made to safeguard the well being of all citizens should be made.' That an Orator should inform the citizens and be held responsible for the facts he gives, to honor his duty of public confidence. What are those duties for which an Orator should be held responsible? To discern events in their beginning, to foresee what is coming, and to forewarn others.

I'm not an Orator, nor am I a physician, or an Attorney, I am a simple rancher, I am one of you, with certain intrinsic values, but what Demosthenes said so long ago, I shall do here tonight., I shall tell you the truth.

"You are gathered here with your friends and neighbors tonight, in this place of learning, to hear the facts. unlike the other guest speakers here tonight I am not a speech maker, I do not draw a large salary by doing nothing, and unlike them, I do know the facts.

"During World War II, the Nazis created a holocaust by systematically hunting down and murdering a selected group of people, not to matter of what Country.

"The people who lived here and around the Rock Mountain Arsenal have been, since 1942 in their own holocaust created by the Army, 1948, Julius Hyman Company, and 1954, Shell Chemical Company. They are manufacturing chemicals that kill people with toxic chemicals that seep into the water, toxic emissions that fill the air with tovic gases and pesticides. Some are so deadly they have been banned in the United States of America, because they have entered the food chain.

"They are banned in the United States however they are still being manufactured by Shell Chemical Company right here at the Rocky Mountain Arsenal. Then sold and shipped to other Countries, who use these chemicals with out direction, and ship the vegetables and fruits right back into the United States Where probably have set right on your tables. Aldrin, and Dieldrin. Aldrin, a pesticide used to control termites in tree nurseries and buildings, insects in grain storage, soil pests on corn fields and possible control of tsetse flies. Aldrin is very toxic and carcinogenic. "Probably more toxic than the nerve agents." Dieldrin, Similar uses as those of Aldrin, Deildrin is very toxic, A carcinogen, and harms the reproductive system.

The "No Pest Strip" Shell right here at the Arsenal was manufacturing them in the early 60's, and until they were banned in the late 60's. These were strips people would buy, and could be hung up in the middle of a room, and it would be absolutely void of insects. Until they discovered it was also getting rid of peoples pets, dogs, cats, fish, birds. This finally made them say Whoa, what is this doing to humans. Last I had heard it is a known carcinogen, damages the liver, the lungs, brain, kidneys, reproductive system, immune system. consequently this left Shell with millions already produced and all that was shipped back. AS I understand they dumped thousands of tons into large pits and some were burned where the clouds of smoke hung in the air over the farmland and small towns to the North. Henderson, Commerce city, Brighton, clear up to Ft. Lupton. People complained of headaches, nausea, vomiting, burning eyes, difficulty breathing, and last I had heard not published that it was a very big carcinogen, affects the Central Nervous System, and reproductive system.

"The Nazi way was sometimes fast and painless, and sometimes not. The United States Army's, and Shell Chemical Companies way is guaranteed to be painful, slow, and causes a variety of diseases, and as I have said before it causes death.

I compare this to the Holocaust for several reasons, when you tell someone you got serious medical problems, they think your crazy, don't treat ya, and the United States Army just denies any wrong doing exactly as the Nazi's done, while they were killing millions. People you had better think positive they are killing us right here.

"My parents didn't smoke, do drugs, drink alcohol, they died from drinking the water, and breathing the air contaminated by the Army and Shell Chemical Company. Their internal organs had been completely destroyed and they were on borrowed time. My Mother died screaming in sheer pain., My Father died also in severe unbearable pain.

I sit by as the neighbors have begun dying, trying to do something but the Army denies being at fault, even when toxic nerve agents were found in the well water. For the population north and northwest of the Arsenal, we have the highest rate per capita of Cancer any where in the World. We all have been poisoned by chemicals known to cause cancer, by both the United States Army and Shell Chemical Company.

"It's time we wake up and smell the roses."

"Sine 1973 I have been trying to sue the Army that is protected by the Federal Tort Claims Act. FTCA., which protects the Government from being sued for negligence or wrong doing. If I could get a Congressional Reference, I could get my day in Court in what is called the Court of Claims, It has been repeatedly shelved in Congress. It must go before the House Judiciary Committee, then the full House.

"Since 1973 I have been gathering facts. Facts concerning the operation and the handling of Ultra Hazardous Chemicals at the Arsenal. What I have learned would make most grown men cry.

"For all practical purposes, the people living west, northwest, north, and northeast of the boundaries of the Arsenal, are in their own concentration camp waiting to die. If you don't die, you will soon recognize it from all the pain and suffering you'll have to endure".

"The voice of the people must be heard all the way to the House of Congress in Washington D.C. The Rocky Mountain Arsenal, and Shell Chemical Company must be closed permanently, forever and the land purged of the toxins poured into it. The half life of most of the man made chemicals is five hundred (500), to one thousand (1,000) years."

"I would like to lay out for you just a few what has been touched upon in the last few month months and 3½ years since my cattle were destroyed, our lives forever changed." Larry said.

A. "IN 1965, The Army environmental Hygiene Agency recommended that steps be taken 'To Eliminate Lake F' as soon as possible, and thereby remove much of the present environmental hazard of exposed surface storage of toxic wastes.

a. "By 1969 there was evidence that Basin F was leaking; indeed for quite some time, A physical inspection reflected that sections of the protective membrane were absent. There was no time frame or records kept for how long this protective membrane had been absent. Thus by the spring of 1969, The Army understood that it was operating Basin F on the premise that it was leaking.

b. "These toxic chemicals have already Polluted more than twenty five thousand acres, off the Arsenal to the north and north west thereof as far north to Brighton Colorado, and through out the Commerce City area and many thousands of acres in between.

c. "Larry Stated his ranch was only 1 3/4 miles from the basin F storage of Millions and Millions of gallons of toxic soup."

B. "The Colorado Department of Health, and he said I would like to direct this to Bill Punt, the Department's Chief Chemist, who has been denying for the Government that this was even happening. On July 11, 1975 they had by correspondence notified Larry Land, of the presence of chemicals in the house well that did in fact emanate from the Arsenal. Bill You have been denying this was even happening, are you the reason I never received any correspondence from the Colorado Department of Health about my water being contaminated by the Arsenal."

a. "Larry also learned that the Tri County District Health Department on July 7, 1975, claim they also notified him by correspondence, they informed Larry Land of Chemicals in his well water. It also indicated the Rocky Mountain Arsenal, as the source of contamination."

b. "Larry stated he had not been notified of any toxic contamination in his water, by the State, or Tri County

Health Departments. Mr. Bill Punt, since you're the one in charge of making sure those chemicals did not leak off the Arsenal, and you openly claimed "there is no way in God's Green Earth Chemicals could be leaking from the Arsenal, and into your wells. My question is will you answer the questions, where are those letters, and why have you been saying this could not happen, and over the years it has been happening, and you said nothing. I don't want to hear any of your excuses, In fact he said he wanted to see some answers in your report."

C. "There was once stored at the Rocky Mountain Arsenal enough GB-Nerve agent to wipe out the entire city population of city of Denver, Colorado, many times over. It was stored in 1,000 pound and ton barrel containers. They also had missiles with cluster bombs, and the M125 containing 2.6 pounds of GB nerve agent, remains liquid until exploded vaporizing into a poisonous cloud.

Ton barrel containers which were filled with nerve agents at one time are stacked in North Plants area.

D. "Now we'll talk about "Honest John" Missiles, made else where, the bomblets it would carry were made, loaded with lethal gas, and explosives, right here at the Rocky Mountain Arsenal, and tested here also 1954 right into the 60's. You must realize that Shell Chemical Company, the Army have continually refused to provide any documentation on anything any thing, so what happened here, was pieced together little by little. So this kind of information can be brought to the public as quickly as possible."

a. The "Honest John Missiles", "were designed to carry 368 bomblets to an enemy target. Disperse the canisters overhead in all directions. Each bomblet containing 1.3 pounds of deadly Sarin, or GB nerve agents, would then fall to a lower altitude then explode. On detonation of each canister, the explosive charge would spread the sarin. They were designed with the force of the explosion to change the liquid into a gas, a more deadly form of sarin, GB gas, covering a far larger area, it actually doubled to tripled the saturated area as it showers the enemy with a liquid agent capable of causing respiratory paralysis, convulsions, and death within minutes of contact. Just a single drop will kill humans on contact. To kill everyone in an entire city, without any damage to the structures, would be a city with no life." The electrical impulses of the neurotransmitter molecules of the nervous system, are halted by Sarin, or GB gas, leaving only convulsions and death."

b. "The small canisters are about 6 inches in diameter, filled with 1.3 pounds GB nerve agent. Much like the Honest John Missile, the deadly Sarin, and GB nerve agents were developed in Germany in 1938. Four canisters per square mile would allow total saturation, death without warning. The missile carrying 368 would give a total saturation of 92 square miles. Cities such as New York, Boston, Washington D.C., Minneapolis/St. Paul, Chicago, Atlanta, Dallas, Houston, Los Angeles, San Francisco, Portland, Seattle,. Can you imagine cities with no life. Very Close to the Neutron Bomb that was being talked of at that same time. Could there by a connection? I truly believe North

Korea, and China probably would have been no more, had these been used. Imagine with in seconds the people with just minor exposure would be, sweating, muscular twitching, pin point pupils, running nose, tightness of chest, with shortness of breath, blurry vision. Then those with severe exposure, headaches, cramps, nausea, vomiting, losing control of the bowels, and involuntary urination, jerking motions, staggering, involuntary shaking, and trembling, lethargic, convulsions, drowsiness, coma, respiratory failure and death." Injury even with small does, leave a trail of permanent injury, and death.

E. "Then for the smaller areas, or towns they would used the M-125, or for the war heads, which contained 2.6 Pounds of Sarin, or GB Nerve agent. Each one upon detonation for spreading the deadly nerve agent, The force of the explosion would change the liquid into a gas, a far more deadly form of Sarin or GB. This could kill everything within a four square mile area, leaving all structures in place.

a. How Sarin or GB affects the body. Developed in Germany in 1938. It is colorless, odorless to most. It disrupts the normal workings of the nerve cells, and it's pathway into the body seems to be the nerves, causing immediate injury to the Central Nervous System, (CNS). GB, Sarin, and BZ nerve agents are all very stable in water, can last for hundreds of years with it's same deadly effect. DIMP. was found in the Land wells, which is a by-product of both the manufacturing and the destruction of GB and Sarin Nerve agents. Dimp is stable with at least 3% GB at all times. IMPA, still another by product of another nerve agent that was manufactured by the Army at the Rocky Mountain Arsenal, the by product of BZ nerve agent. IMPA stays stays stable also at 3% BZ nerve agent. Both these items were just recently found in the Land water wells by the Army, and Shell Chemical Company, and the Colorado Department of Health. If they tell you DIMP, or IMPA are safe, it's a lie, they contain low levels of the deadly nerve agents, and if it don't kill you right away, you will suffer, from the symptoms listed above for a life time.

F. Lets talk about the 75 to 155 millimeter shells that were also filled at the Arsenal with Mustard, and Lewisite agents. Was more common during World War I, and II, It was a persistent, and powerful blistering agent. Mustard is a solid or liquid, depending on the temperature, until it is exploded, if it gets on you you can't get if off, burns and blister's to the bone. The Army during the 1960's pretty much stopped filling shell's with Mustard, or Lewisite, agents although they had a huge stockpile of the agents.

G. The Army at the Arsenal made the Nerve agents, and was involved in loading many different devices with the agents. They were still manufacturing and loading various devices, while also doing a large amount of testing and experiments on same, from the early 1950's through the 1960's. They ended up with huge stockpiles of GB, Sarin, and BZ Nerve agents. During the 1970's the Army made many botched attempts at disposing of the GB, agent and during these attempts GB agents had leaked out into the air from their stacks, to the North of the Arsenal. Their monitors in the stacks would show leaks, but they continued their disposal of the GB, and Sarin, and BZ nerve agents. It was just blowing out into the open farmland, causing many illnesses, muscular twitching, sweating, pin point pupils, runny nose, tightness of chest, shortness of breath, blurred vision, No one knew what to do.

We are probably pretty damn lucky, they kept the Honest John Missiles away from the Army, and at a different place than here at the Rocky Mountain Arsenal, and not giving the Army any chance to experiment, or test. If they had the missile available, they may have experimented and tested, and "POOF" we all disappeared, and of course without explanation. Denver and surrounding communities population is gone. "HEADLINE NEWSPAPER's" United States Army has developed NU-WAY to lower the population anywhere in the world. The question would be without explanation. Where'd all the people go?

Larry said "I had told you earlier, the deadly effects of Sarin or GB nerve agent. How it halted the neurotransmitter molecules. Then I said is there a possible connection? In the 1950's and 1960's we all heard about the bomb that would kill everything with in a city, with no structural damage. "The neu-tran bomb. I'll spell out again what is halted, or paralyzed, and brings on death within minutes. Neu /tran/.

"Neuro Transmitter" in the brain. The Honest John Missile, loaded with 368 explosive Sarin/GB canister's, this would have obliterated all life form, New York, or Los Angeles.

"A quote of one of the Worlds greatest men. (THEODORE ROOSEVELT)

"The credit belongs to the man who is actually in the arena; Whose face is marred by dust and sweat and blood; who strives valiantly; who errs and comes short again and again; Who knows the great enthusiasms, the great devotions, and spends himself in a worthy cause; Who at the best knows in the end the triumph of high achievements; and who at the worst, if he fails, at least fails while daring greatly...."

THEODORE ROOSEVELT

"Ladies and Gentlemen, I believe this nation is facing an unparalleled crisis due to the reckless irresponsibility of the Chemical Industry, the foot dragging of Government Agencies, and the lack of realization by the general public. Our air, our water, our food and the work place have become progressively more contaminated with a wide range of toxic chemicals that produce reproductive, neurological cancer, and behavioral diseases".

"Today we are dealing with an epidemic of cancer. In 1960 one in four of us would get cancer and one in four of us will die from it. Many Doctors give a conservative estimate that by the end of this century one in two will be affected by cancer and one three will die from it. The plain and simple truth is that industry, with an excessive preoccupation for the short term interest, is indifferent, and has been indifferent, and continues to be indifferent, to the health and welfare of the nation. It remains preoccupied with only the short term economic interest".

"I am very concerned about the direction that this Country is heading. Large Corporations would not be able to get away with the things that they do if it were not for financial pay-offs made to our political leaders! We allow our elected officials to make irrational decisions that affect us all. We must, at all costs, eliminate the kick-back schemes between big Corporations and our leaders."

"Environmental Policy's are set by which ever way the major campaign contributions go. The only way those people or the Corporations are going to be stopped is by revolution, a grass roots revolution. A revolution not by Political leaders, but by the average hardworking American, that demand these executive leaders and their Corporations be held responsible. More reliable reporting methods of what they are doing, or what they may be contemplating doing.

"For the last few years, Larry, and his wife have written, and filed with the courts cease and desist orders against the Arsenal, and Shell Chemical Company. They mailed them to the Governor, the Attorney General, the State Supreme Court, and the Colorado Department of Health. We have repeatedly sent them to various offices in Washington D. C. Including the President of the United States Richard Nixon."

"unfortunately most of our pleas for help go unanswered. How ever we did hear each time from a few who cared. Someday more will hear our cries help, and close the death factory that now is beginning to haunt us all. We have got Colorado going real well, and I understand they have or will issue cease and desist orders on behalf of the State of Colorado."

"Because of what they have done to my parents, my family, my friends, neighbors, and their parents, I will not quit until justice is served."

"I have in my possession, documents stating what was manufactured out at the Arsenal, what was manufactured, the by product, what was by product of the demilitarizing, what they had made. What was done with it, for example how many heard of basin F?"

A long pause followed. Larry said "I didn't think very many of you had. Basin F is a Ninety-three acre, two-hundred forty million gallon holding pond for literally everything the Army ever made, or tried to make. Nerve gas, (BZ, GB, Sarin,) Mustard agents, Lewisite, Phosgene, Phosphorus, Napalm, aerozine, Hydrazine, seven types of Agent Orange. All the pesticide and herbicide produced by Julius Hyman Co., and Shell Chemical Company., along with anything else were not sure of. It has been leaking into your drinking water for the last thirty eight years.

"Larry just uncovered another one that had been kept very quiet for years. The people in Brighton were told it was nitrate in the drinking water, and advised "Do not use the water for infants and toddlers' It could cause brain damage. Well my friends it is overwhelming, to know it has been covered up very well by not only the Colorado Department of Health, but by the Army, and Shell Chemical Company. Diisopropylmethylphosphonate (DIMP), and Isopropylmethylphosphonate (IMPA). This was found in Brighton city water in 1974, these are the byproducts of GB, BZ, and Sarin nerve agents this will always remain about 3% of the nerve agents. Small amounts like this will and did cause sweating, muscular twitching, headaches, Nausea, running nose as if one has a cold, tightness of the chest, shortness of breath, cramps, vomiting, and hurting at every joint in your body."

"We, the people of the world, are not here to predict the future but to change it for the good. We are not here as helpless creatures but as son's and daughter's of Adam and Eva, capable of affecting our own fate."

"We are not here to avoid decisions but to make hard, decisive choices between good and evil, right and wrong, by using an ethical system not invented by man, but by our own creator. A framework of truth and moral guidelines through which we can find deliverance from despair."

"We are not here to glorify in ourselves, but to glorify in that which he gave us, and in he who made us all, and who will eventually judge each of us on how well we did on the journey we all take but once."

"There can be no excuse for what happened at the Rocky Mountain Arsenal, and to the people of the United States Army and Shell Chemical Company, and the pentagon in Washington D.C., who made the decisions to let this happen, there is no excuse, God will judge on ones merit."

"The time has come, that we, the people who live in this land, must band together for the common good of all people living in these communities. If the Arsenal can be closed here, then other factories of death elsewhere around the World can be closed. This madness must be stopped and it is up to us to stop it. We must wake up our Government, and Industry leaders."

"Our own United States Army has done chemical warfare testing over the cities of New York and San Francisco, Denver, Los Angeles, Minneapolis/St. Paul, the entire midwestern wheat belt. Colorado, Nebraska, Kansas, Iowa, Minnesota and the Dakota's. Chemical warfare agents used on men, women, and children alike, so the Army could record the effects of what they were using.

"We do not have a Dachau, Auschwitz, Buchenwald, or Bergen Belsen concentration camps here. Instead, we have Denver, Brighton, Commerce City, Los Angeles, San Francisco, New York, Minneapolis/St. Paul, Chicago, and the midwestern wheat belt. They have the technology available, an entire city could be made void of life in one Neu-Tran Bomb."

"Since this nightmare of death that struck me in 1972, the Investigations, and legal actions that have ensued, My home has been burglarized no less than twenty five times, for the purpose of stealing incriminating documents that I have been turning over to the News as needed. The people he believes responsible for this crime have repeatedly failed in their attempts. The people responsible for this crime work for the United States Government (Army), more to the point the FBI, is believed also involved."

"I have been shot at on more than one occasion. They have, on numerous occasions, attempted to run me off the road much in the same manner that Karen Silkwood was murdered. In my case they attempted to run me into a bridge support column, a retaining wall, and a piece of heavy earth moving equipment, and twice they attempted to run me off a high mountain road. (Loveland Pass)

"Because of the Ultra Hazardous toxic chemicals they have dumped by the millions of gallons, in Basin A, B, C, D, E, and F, these open basins for storing toxic waste have not only polluted the ground, the ground water, but the air itself for miles in all directions of the Rocky Mountain Arsenal. Basins A, B, C, D, and E, are all unlined basins never meant to be used. It was a matter of using large earth moving equipment in the lower area's between two huge hills, and damning it up creating one toxic Lake after another. Absolutely no Hydrology, or Geology study of the area, no percolation test were done then or now. 1954 they created Basin F, much in the same way, then they called it a lined basin after spraying an asphalt liner. It was given a life time expectancy of ten years. Well he thought this might be ok, he many times wondered how thick this liner must be, and since the toxic waste that was to be placed into Basin "F" for evaporation, is so acidic and totally caustic, until he learned the asphalt liner was less than 3/8 inch thick, less than that of a thinly soled pair of shoes. It was made to hold 240 Million gallons of toxic soup, and as they began the use of the basin they discovered part of the liner was already missing., but they said they repaired it and kept right on using it. In 1965 they knew full well the basin was leaking, by 1969 an inspection of the liner showed clearly the toxic soup had breached the liner, and was leaking into the underground water. The Army was warned then they were using Basin F on the Premise it was leaking deadly ultrahazardous chemicals into the Country side, if fact by 1970 it had already polluted over 25000 acres off the Arsenal property, and they continued using the basin more than ever by over filling it at times. Then using a "spray raft" trying to get better evaporation by squirting millions of gallons of Ultra Hazardous materials contaminating the air for miles around the Arsenal." At times we could see a yellowish, bluish, red haze in the air, almost like a rainbow, we would have difficulty breathing, red eyes blurred, runny nose as if you just caught a terrible cold, nauseated, muscles twitching, disoriented (as if being drunk), sweating in cold weather or hot, not realizing it was deadly and toxic chemicals, and we were experiencing everything close to death.

"The United States Army has now been charged with, trespassing, failure to repair, A failure to follow recommendations, willful negligence, fraud and deceit. Also the flushing actions used by the Army in a memo by damning up a mound of natural water, and pumping into basin C Irrigation water in an attempt to flush the Contaminants from beneath the arsenal."

Larry said. "Because of the fraud, deceit, negligence, wrongdoing, failure of the Government to exercise the required skills in its' operation of the Rocky Mountain Arsenal, and willful trespass, onto my property, both from the air and the water. All my livestock destroyed."

Larry said "I was a cattleman, a cattle buyer, had my license for both the State of Colorado and Bonds to purchase on a Federal basis. The first problem I had was the State licensing, coming right into the auction markets inspecting my license and permit, and reminding him that he could ship none of the livestock he purchased out of State under 31 days. reminding me I would need a Federal license. Larry brought out his Federal license, and stated this is just to let you know I am well covered. Then he began to be hounded by the U. S.

Marshall's right at the Sale Barns. At that time I had all his required $30,000 dollar bonds for his Federal Licensing as a cattle buyer. meaning I could buy sell livestock any where in the United States. I had customers mostly in the west, California, Arizona, Nevada, Idaho, Utah, New Mexico. That was my status, I was a buyer, seller, Auctioneer breeder, I had and could get the best dairy cattle in the country, and my customers knew this from past service. Some I had raised, and some I knew where to get them, Yet some I purchased at the sale barns in Brush, Ft. Collins, Greeley, north Denver, Be able to buy them one day ship them the next.

Larry stated. "In 1972 nearly 650 head of my dairy cattle were killed and destroyed within a very short period of time. They showed signs and symptoms of cramps, (difficulty moving their necks, front and hind quarters), coughing and difficulty breathing, vomiting, slobbering, defecation, and urination, twitching, jerking, staggering, drowsiness, drooping ears, running eyes, (red irritated), runny nose, sores in, and on nose and mouth, eyes, unable to stand, unable to get up, comatose, convulsions, and death, with a 3 days of starting on the water the bred heifer's aborted their calves.

"The Army, the Colorado State Health Department, the veterinarian, both State & Federal veterinarians, the college Ft. Collins sent a person, both State and Federal health inspectors' Any testing they could run came back negative, no one had any idea other than it is something unknown, we don't know. Even after showing them the water would run red. Was later found to be Resin beds of the 300 to 400 mesh size that would move any where water run. The only thing they were not natural, man made and used to trace leaks of contamination. The Army would not admit to this at all. "No one would even indicate, it may be the Arsenal."

"Larry's Veterinarian stated there is something in the water, it is very toxic and we are getting some signs now of secondary infection, and that we can treat. However it could be very contagious to any other livestock should it be pneumonia. He said I don't think for a minute it could be pneumonia, and he said I would like to put your remaining herd in with a completely healthy dairy herd, because if they don't get sick we'll know then that it is the water. About 285 head that were left were moved in with this other heard, watched very close, & was able to see improvement within 3 to 4 days, within 2 to 3 weeks there was a definite improvement to what the problems were, & there was no disease or sickness brought into this other herd, he had not reported a single animal sickened. He was elated, saying I knew it was the water all along, & what we have done here proves that point. He said Larry, there is no doubt in my mind your Cattle were destroyed by the water they drank, & the air they breathed.

Larry Said. "The part with the U. S. Marshall's office didn't stop, Larry was notified because of his financial loss in livestock, they would now require a $300,000.00 Bond. First of all they get quite expensive in that range. I had just got covered with the bonds and a U. S. Marshall met me at the entrance of the Brush Livestock Market to see if I was covered. I think they would catch me buying without bonds, however I had Bonds and he seemed quite frustrated with me. Within a week I got a letter in the mail from the Department of Justice, Office of the U. S. Marshall telling me that it has been determined by the Department of Justice, because of the financial loss in live stock in 1972, they would nullify any bonds to buy and sell livestock shipping them across state lines. That any livestock I purchased I would have to prove I had kept them in my possession for a minimum of 31 days before it could be resold. within the State of Colorado, absolutely none could be moved across any state lines. For each head you purchase and sell you will be fined $5,000.00 dollars, and 6 months in jail. The Justice Department literally put the straight jacket on me. Taking any and all means of making a living from me. I asked why how can they do this, to my senator's and to my Congressmen, "No one but No one would touch it." even though they would agree they can't do this. They were putting on the final brakes for me to even raise hell, make a living, support or take care of my family.

Larry said, "Well, what they done is really piss me off. Your gonna see how one man can literally wage war against our own United States Army, and Shell Chemical Company. I made up my mind I will close them down.

"friends and neighbors you must open your eyes, do not merely look, but see what is happening around you. What flows from the Arsenal, into our water, and floats through the air at night, is at this very moment killing you."

"How many of you have smelled that strange putrid odor of death at night? That odor comes from the smoke stacks at the Shell Chemical Company when scrubbers, and their filters are shut down while being cleaned. The reason for this is two fold, one, so the stacks are in compliance each morning in case of any inspections, Federal or State, Number two, so Shell Chemical can remain in full operation during the night time hours with out the use of filters. On a moon lit night you could see the reddish brown clouds moving across the sky, early in the A.M. 4:30 to 5:30, to the northeast, and the north, Then sometimes to the west right over down town Denver Colorado "Smog city U. S. A." with the mountains locking in all this pollution over the city of Denver, and as the day goes by the heavier the air the more it settles down. I could almost tell before hand when the City of Denver was going to call the Red alert because of the smog, warning everyone to close their windows, stay inside, and if outside wear a respirator. you can see as the smoke stacks begin getting lighter and lighter, until almost clear, leaving Shell Chemical in complete compliance, for business for another day."

"Likened to the slimy, sneaky night creatures they have become, they use the cloak of darkness to conceal their heinous crime. The we have the United States Army operating under the cloak of darkness, deceit, fraud, and willful negligence to conceal their heinous crimes."

That is not all, the U. S. Army is doing the very same thing here with what they call Project-Eagle, Phase II, Nerve gas formed in the exhaust stacks, and released out into the atmosphere after it had reacted with sodium hydroxide. The reformation would be much more pronounced where conditions were less controlled when either DIMP or Nerve gas was washed down the chemical drain to Basin A. DIMP is formed in the same chemical process to form nerve gas. The Arsenal used distillation as a method to separate the DIMP from the Nerve gas.

Distillation is an inefficient method of separation. It really means concentration. You will always have low levels of Nerve gas in DIMP. These stacks were leaking low levels of nerve agent out into the farmland night after night.

"You and I, are today, engaged in a great battle. A battle that is not fought with guns and bombs, or against a foreign enemy. Our enemy is our own brother, our fellow countrymen, the policy makers of the United States Army, The Pentagon, and Shell Chemical Company based out of London England, and Holland. This is our true enemy!

"We the American public are the victims! We are at the mercy of crooked politicians and large corporations, graft in the Army, and at the pentagon. The American public is a power to be reckoned with only if they get their heads out of the sand and say enough is enough."

"Ladies and Gentlemen, the price for liberty is eternal vigilance, and until the American public decides to take control of their own fate, nothing will be done. The Government and large corporations have repeatedly shown they cannot be trusted, and therefore not relied on. If we are going to protect this planet, the planet we all call home, we must join together in harmony, and jointly defeat this monster within. We must leave a legacy for our children and their children, we must learn to rely on ourselves, and be able to rely on one another.

"Ladies and Gentlemen, Abraham Lincoln once said and I say to you now, that 'with my own ability' 'I cannot succeed' 'Without the sustenance of divine providence, and you the people' 'I cannot hope to succeed' "with them" 'I cannot fail.'

"Before I leave here tonight there is one final piece of information that I would like to share."

"In the late 1940's the Army Corps of Engineers buried five hundred thousand tons of pure phosgene liquid in steel containers in a huge pit, all he could remember was near Brighton Colorado."

"This man who wrote the letter to a Congressman, was the man in command of the Army Corps of Engineers. He became very concerned after nearly 25 years of, possible leakage. Somewhere near Brighton Colorado is all he could remember. phosgene was a nerve agent that was made at the Arsenal during the 40's that never worked out. the Army had learned the secrets of what Germany had, it was far superior to phosgene. Phosgene is a colorless liquid (gas) very unpleasant odor, and a severe respiratory irritant. The new secrets for the manufacturing of GB nerve agent it would killing within minutes, far outweighed Phosgene. The Army was now stuck with figuring out how to dispose of several hundred thousand tons of phosgene. They done the only thing they knew dug huge pits, and buried the phosgene steel barrels and all. At the Arsenal, and around the area they have approximately sixty (60) huge pits that were filled and buried with unknown materials. Infra Red shows some are so hot they are actually afraid to even try digging it up to find out what may be there. The Army kept no records as to what they buried."

"During the 1950's a major disaster struck the city of Minamata, Japan. The first sign of impending doom was the death of the City's cat population, as the animals underwent convulsions and often leaped into the sea. Over time, some 700 people were killed directly, and 9,000 others were permanently injured, and disabled. The City's economic base was in ruins, as fish died and the fishing industry was forced to shut down. Minamata's population gradually declined, and now less than 5,000 remain of the 50,000 who once lived there. This disaster was traced to a single cause...the dumping of a mercury-based compound into the bay by a chemical company."

"The Minamata incident has become a landmark as the first widely publicized confrontation between the citizenry, and the manufactured hazardous substance poured into the environment. Many other incidents have followed since then. Frequent reports of abnormally high numbers of Deformities, Cancer, Prolonged illnesses, miscarriages, and retardation. To include the disappearance of local animals, bird, and insect life. Cries for help have come from around the World. Often such health problems can be traced to manufactured hazardous chemicals released into the air, ground, or water we drink, the air we breathe, and the food we eat."

"Can you remember the terrible cries for help, during the late 30's and 1940's when Hitler was rounding up people putting them in concentration camps, began shooting them, but he complained that was a waste of ammunition, they began using pesticides followed by the nerve agents. killing people with chemicals and then bury them in open pits, or cremated them in the huge crematoriums. Of course after he plundered and stole every thing they owned, and then stored in it Switzerland Banks and elsewhere."

"you can tour the Smithsonian institute, and museum in Washington D. C. See the pictures, and the cries for help, but no one heard or listened, no one helped, until far too late. It will bring tears to your eyes. It is as if guilt takes hold of you gently saying open your eyes. We all must jump in and together get control of the Government and the huge Corporations who are running our world today with no one to answer to. The Constitution states "for the people and by the people, not for the people against the people like they are now doing.

"Before we become another Minamata, or we have them everywhere, Simply put, 'Hazardous or Ultra Hazardous substances' include any substance that poses a threat to human health or to the environment. The term usually encompasses a broader range of materials than does the term "Toxic", such as. Phytotoxins, Nerve agents, Insecticides, Abnormally dangerous activity, Ultra hazardous activity, which generally indicates a substance that will be directly toxic, or lethal to humans."

"Tonight I ask that you join me in defeating our common enemy. Write to your Senator's, your Congressmen, demanding that the Rocky Mountain Arsenal be closed permanently, and a full investigation be started. Our own state has now started and that took nearly 3 years before anyone would listen. We have our first cease and desist order now served on the United States Army, and to Shell Chemical Company. They have been effectively cut off in the north for pumping anything under the ground. This has appeared in the news, from Los Angeles, to New York, to England, France, Australia, South Korea, and all the way back to Washington D. C.. We have rattled the doors in the right places. It has opened up to investigation several other terribly polluted sights across the world, several right here in the United States. It's time to clean House, Together we cannot fail." "Thank You."

Larry said, "Well there it was, for better or worse I done what I could to inform the people around this area who cared to come here tonight. Larry said to his wife, kids, and friends, saying lets go out front where we can talk, and catch what I hope will be fresh air."

Later as everyone stood outside the main entrance of the hotel, people came up and shook Larry's hand and expressed their personal feelings about the Arsenal, mostly belief, yet there were some disbelief, which he was glad to hear the feedback, "good or bad." Again he was hearing what he had heard during his early visits to the television, newspaper's, and radio talk show, and it sounded good. At least it was getting out there.

Unfortunately, there were still people out there who had their doubts concerning the Arsenal. They had noticed the odd happening around their homes, such as the strange tastes in their water, the eggs they gathered with very thin shells or no shell at all, or the horrible odors in their air at night. Unexplained deaths amongst their neighbors, pets and livestock were common but yet, they refused to believe. Instead they believed the lies put out by the Government and Shell Chemical Company.

Some talked of illnesses amongst their close knit family members, and close friends. Cancer was always a main concern. Still they were afraid. Their fear never went away and how could Larry blame them after all of his experiences. When you fear for your family's safety, it is a very sobering experience. In Larry's case however, he not only had to worry about their health, and safety, he also had to worry about the possibility of one of their so-called, "accidents" that he had been warned about. Why should anyone want to overtly join him in this fight. They were all behind him in spirit, but at certain times, that just didn't seem to be enough somehow. Every one agreed they were behind him 100%, whenever Larry ask for some help, he was told your fighting city hall, The Government, you'll never win, I don't want to be involved, and you know, that can be the truth if you quit, you'll never win.

The simple truth was, that while it would have eased his burden to have had just one person stand with him, some support, he understood why they could not. He would continue for all of them, even though it would take him years longer to get to the place that he longed to be in, the Congress and the Courts. Perhaps it's just the letters to Congress, that will open the doors.

"If I were asked to give what I consider the single most useful bit of advice for all humanity, it would be this: Expect trouble as an inevitable part of life and when it comes, hold your head high, look it squarely in the eye and say," "I will be bigger than you. You cannot defeat me."

ANN LANDERS

Dennis, Jennifer, and Glenn all said that he had given an impressive speech, and gave out a great deal of knowledge. He only hoped that it would serve its purpose.

On the way home from the meeting, Marie and the boys were as excited as he ever seen them, and it made him feel good about himself. There was no mistaking their pride, So to celebrate the evenings big event, they stopped and ordered ice cream at of course what else "A Dolly Madison ice cream parlor."

As he sat looking at his children, he again felt the familiar stirring in his heart that what he was doing was right, not only for them but for all children everywhere. If they were to have any kind of a future, if they were to ever enjoy the land as he and many others had, then the raping of the American heritage, and their land must be stopped. The misguided belief of a few, that the Government is under some mystical belief, they own the land, and the American Citizenry are merely temporary occupants is wrong. The American Government owns nothing more than what the American public say they own. This land is owned by all American Citizens, past and present. "The Government are merely managers.

The next morning Marie said she was going out to get the morning paper. When she came in she was bursting with excitement as she opened the newspaper for Larry to see. It stated in bold print; LOCAL MAN WAGES WAR WITH THE UNITED STATES ARMY! Said he will shut their operations down in Colorado.

The article went on to tell how Larry had spoken, the night before at the large hotel on Quebec avenue, to a very large group of people, releasing a lot of good information about the Army, and Shell Chemical Company at the Arsenal and what they have so negligently done has affected this area. Larry said they have filled me with tenacity and I will never let go. Larry said before I'm done I will know exactly how they have made the most contaminated piece of Land on earth.

This was done by fraud and deceit, to the people and Congress. by misrepresentation, to Colorado, the people and Congress. The United States Army, and Shell Chemical Company violated every standard of reasonable care when working with ultra hazardous materials. Continuing to use Basin F. after being warned in 1969 any further use will be with the knowledge Basin F. is Leaking, and had already contaminated 25,000 acres off the Arsenal property to the north of the Arsenal. This is far beyond negligence, wrong doing, careless, or reckless. this was a deliberate act willful endangerment to humans. This is a Federal Human Rights Violation. The United States Army had a duty to contain and confine the Ultrahazardous Chemicals to its property, and failed to warn those in harms way.

The radio, television, and newspaper coverage had picked dramatically after giving his speech. The way Larry looks at it, If the Army and Shell Chemical Company can't take the heat, then they had better damn well stay out of the kitchen.

It had been on television, and newspapers in New York, Los Angeles, Salt Lake City Utah, to Dallas Texas, and at one time Larry was on 69 different radio stations. Everyone wanted to point toward the death of all the sheep on a Utah ranch. They had planes going over the pasture land, seemed to be spraying something, anyway thousands of sheep ended up mysteriously dead. This was probably another on of those tests the Army was involved in without telling anyone, and damn well not taking the blame.

Larry had his first bill before the Congress of the United States merely asking for permission to sue the United States Army for their mishandling of Ultrahazardous materials. Legally it is called fraud and deceit, he calls it plane lies. It had gotten shelved and never made it to the House Judiciary Committee. When a Bill is not past at the end of a Congress, it must once again be done all over for the next Congress, and until it gets heard before the House Judiciary Committee. Congress only lasts for two years at a time., then everything must be redone, and those same attempts to have it heard are then repeated. He did not plan to stop here there would be more, many more if needed.

Larry knew exactly what he would have to do now, and not giving up, he got out the Bill that was placed before the House subcommittee, and began reading it over, planning his next moves. adrenaline soon filled his system, he felt himself shaking within, it was overtaking him, he felt confined, and trapped, and he knew he had to go for a walk, it seemed to give him a chance to think, and better direct his energy. He felt deserted, as though no one cared. Even the people he was trying to help weren't there. He was beginning to resent them. They were all sitting back safely waiting for him to come up with the answers, to make everything right again. What was wrong with them? He needed help. Maybe he couldn't do this all by himself.

Larry thought of Eddie Rickenbacker, and what he once said. "Courage is doing what you're afraid to do. There can be no courage unless you're scared."

EDDIE RICKENBACKER

BEING HONEST; Before us lie two paths, honesty and dishonesty. The ignorant embark on the dishonest path; the wise on the honest. For in helping others, you help yourself; in hurting others, you hurt yourself. Those who remain honest know the truth; trust rises above fame. And honesty is still the best policy, and with courage you will win.

Larry was thinking of the type of things you might see at a movie. He wasn't dishonest, or some waco nut, and to him his credibility and honesty were everything.

As Larry was nearing the house, Marie came out to meet him knowing something was wrong. They sat down together looking up at his Flag still flying half staff, and thinking of what that meant. It was for his parents, and their lost son who will never feel the love of life.

"The bill was tabled and never made the subcommittee," he told her. "They want me to wait and try again in the next Congress." You see a bill is defeated that has not been heard when a two year Congress ends."

"I'm so sorry Larry, I know how hard you had worked on that, and the promises that you were told."

"He said I'm so sick and tired, of being sick, and tired, he said I wish the Doctor's could do something more, maybe it's something that is gonna take time, a lotta time, he knows there has already been a lot of improvement with all of us."

"She said I'm Sorry I had hoped things would be cleared up by now. At least the boy's hair is growing back in, the nausea and vomiting has mostly diminished, everyone's eyes, and noses are pretty well cleared up now. We still have a way to go on some of the rest."

"It's not your fault honey, you don't have anything to be sorry about. You've done more than your share in helping me along. I'm just tired of waiting, and waiting, everything takes so damn long. C'mon princess, lets take a short walk, and we'll all feel better, besides it is good for us."

Taking her hand they started down the road, he said let's walk the half mile he used to as a kid to the old school house Cactus Ridge, he said we called it two hoots and a holler, and we'll be there, unless his father would let him ride the horse that was fun and quick. The old school still has the horse stable, and the teachers-house (that was where the teacher lived). That old school house was built sometime around the turn of the century.

The sun was just getting ready to set behind the mountains, Bands of yellow and red filled the sky, and the last remaining light filtered through the clouds. The green on the distant slopes seemed to be magnified, as if looking at it through powerful binoculars. Larry thought of the warm summer days when he had stopped to look at the wheat fields. There was always that short but magical time, right before the wheat turned golden, when it was tall and green. The gentle summer wind would move across the field creating emerald waves against the gray thunderheads that would soon deliver a shower to cool the afternoon. It was moments like this that he allowed himself to absorb the beauty around him and calm the rage that burned within. This was what his mother had always referred to as "food for the angry soul." Later the moon would rise and cover the land like a blanket in soft light. God, how he loved this land, this is where he was born and raised and went to school, almost his whole world as a young boy.

As they were walking, Larry expressed to Marie again how he felt about this country and even though he had been married to this woman for many years, there were still private thoughts and feelings that remained deep in the recesses of his mind.

Turning to start home, the evening silence was suddenly shattered by the sound of a vehicle as it slowed to make the turn onto their road. Turning back around to look,, the headlights momentarily blinded them. A feeling of impending doom was suddenly on him. instinctively he drew Marie behind him.

"If anything happens, get into the ditch as fast as you can and hide yourself in the brush. If anything happens to me, take the kids and go to Minnesota. Jimmy will take care of everything that is necessary."

"Larry I'm…"

The car was about three hundred feet away, Larry pulled out his 44 Magnum, and said I think this will stop him, he told Marie to get back and out of the way. He was slowing down, Larry thought it's now or never, he had yet to bring his weapon up, the driver is the only one that they could see, as he saw them, and the man waved, as the car continued to slow it went passed them several lengths, stopped, then began backing up. Larry watched very close as the guy stretched out of the car, Larry ordered the man to place his hands on the top of his car, with his 44 drawn he had no problems with this person.

The driver did exactly as ordered stayed put, with his hands on top of the car right where he had gotten out. "Mr. Land? I thought that it was you." He then said I know you have the upper hand, but you've got nothing to worry about with me. I'm not armed, and that is not the reason I am here. I have a package for you, and I realize, I should have done this a little differently not on the roadway.

That voice, it was familiar, Larry had heard it before, but where. Then he ask what's your name. Larry had this man do the whole routine of being frisk searched.

"Mr. Land, my name is Tim, and I have a package that belongs to you. In good conscience I can no longer follow my orders." I truly felt guilty after I had went through your house again. You really do a pretty good job of keeping things gone, and secreting information.

"Larry said you got that right, I don't often keep anything here any way.

"These are the package of pictures that was in the little steel container under the kitchen sink, But I also have brought out some more information that I had borrowed from the Army, thought you would find it quite interesting."

"Why the change of heart, why are you helping me?"

"When I took the job with the FBI, it was either take the job or go to jail, so I took the job. It was to serve and protect the Country, but I soon learned mostly it's protecting a few high ranking Officials who, out of greed, or their own sense of loyalty, made a lot of mistakes in how they would like to run the Country."

"Ever since the day when I met you in jail, and you made the statement about the children of this nation having a decent place to grow up, and not to worry about pollution and poison gases, nor any similarity of what Hitler had done in Europe. I have been thinking why in the hell did I do this, Mr. and Mrs. Land, I'm sorry." What your doing needs help, and I got excited and wanted to tell you I'm ready to help, it may not be much, but I don't have to find things either."

"Thank you. Would you like to come up to the house for a cup of coffee, we do use bottled water, it isn't the contaminated stuff."

"No thank you, I'd better not be seen here, one never knows when their watching you, every move you make, right now were both good targets."

"By the way Marie, this is the man who so accurately described your pink flannel pajama's with the little red hearts, as he was describing the interior of the house, he jokingly said maybe this will jog your memory."

"I gathered as much. It's nice to finally meet you, face to face that is," Marie replied.

"I've been sitting on those pictures ever since I took them, trying to decide what to do with them. Tonight after talking to my wife, I knew. She believes in you and what you're doing. You know you're a pretty high I-CON for the Government. Any way she reads everything written about the Arsenal and about you. Diane was at the Hotel the night you made our big speech. She thought it was horrendous and actually drove the point across. When she came home she was ready for war, and came right out and told me The Army is wrong, the Government is wrong, and big business is wrong noting Shell Chemical Company, some of this needs to be stopped."

"There is one thing I don't understand is why all the cloak and dagger business, what did they hope to prove?" Larry asked, hoping to get some useful information out of someone on the inside."

"They were not out to prove anything. It was your credibility and the many lawsuits, that's what they wanted to destroy and stop. When this didn't work they took other steps. They will continue taking whatever steps are necessary to stop you, including and not limited to assassination. Be damn careful, with Shell Chemical Company, and the Army your up against some of the most powerful and ruthless men in the country."

"We've got to cut this off or you'll be in bigger trouble than I." said Larry.

Tim said." Your house will be inspected again, if it's not me, make sure you keep up the way you have been doing with all the information you've been getting, your damn well a big step ahead of me, Thanks we'll talk again. Here is your package I didn't let them have at all, so you can make sure you put your information against there's should they try to white wash what their doing. Let me know what you think of the information I borrowed for you, and you know how to use it I'm sure. There are some codes to let you know what we need to follow, and never talk about anything here at your house, or on the phone. There are bugs about.

They all said their mutual good bye's, Larry just says see ya, he doesn't believe in good bye's. Ms. Land said "Thank you for coming out, it was more than I'd ever expected, good bye again.

"Larry was rather excited and couldn't wait to get home and look at some of the new information, Tim had given him."

"Since the loss of Larry's livestock to the pollution from the Arsenal, he continued working with his Auction business specializing in dairy cattle. One thing about him he was an expert with livestock, he could take a look, and walk through a pen of cattle maybe 40 to 75 head he would walk to the front of the pen take out his pencil, and paper and figure out the exact weight of each animal, the gross weight of them all then almost exactly what they would bring that particular day. He was quite unique in his own way, their were some that didn't believe that could be done. Larry made a bet with several cattlemen in that area, it cost each of them $1,000.00. Larry walked through a pen of mixed breeds, there were 69, he then wrote on a piece of paper the Judge would keep. 1st paper 69 head. 2nd paper Gross weight of, 46,575 lbs., 3rd piece the average weight of each animal is 675 lbs. He said he would not miss the average weight per animal more than 10 pounds + or -, the gross weight the same way + or – 690 pounds., that particular day according to the market on the radio was right around 41.50 to 42.00 cents per pound, he then offered the man 41.00 cents per pound based on leaving him a very slight cushion. The man accepted the offer, and within two hours the semi was there to pick up the animals and transport them to the Brush livestock market where Larry would have them sold. He went to the barn with the semi, at the Market this again is where Larry shines, he sorted the animals into 8 different groups to go through the ring, the trick here is to make the animals in each group appear to be very much alike, and same breed. trying to sell two or three different breeds together will virtually cost the seller 10 to 15 cents per pound. Larry had 15 Ranchers who had placed a bet Larry would lose. He was pretty happy with what he had done, this would be one for the Guinness book of records. although he was still a little scared. Just think if he misses the weight by 690 lbs. he would owe a $1,000.00 to each of the other ranchers. This is something held very serious by those who participate, and ya don't renege but Larry was good.

The next day at the sale, as each group came into the ring for presentation, they were looking good, the total weight of the group would flash on the screen, then on another screen right beside that on would flash an average weight. In most modern day auction markets the ring itself is a balanced computerized scale. You as a buyer or seller immediately know what the animals weigh, and the average weight, then if you studied your market just a little for the last few days, you would be able to quickly calculate how many cents per pound you would pay, and when you should quit bidding. This particular day just kinda left everyone in shock at this point, the first group just sold for 56 cents per pound. The next group came into the ring and the Auctioneer would say this is some out of the same lot. One group after the other was selling hitting 56 to 57 cents. We had 3 ranchers bidding, all looking for exactly what we offered for sale. The average after all were sold was 69 animals, average per animal was 682 lbs. for a total gross weight of 47,058 lbs.

46,575 –47,058= a difference in gross is 483 lbs.

675 - 682= a difference per animal 7 lbs.

Larry had just won the $15,000.00 in a matter of minutes

cattle sold for 56.75 cents per pound.

sold $26,05.42 paid $19,095.5 = $7,609.67

All Larry could say is, I guess someone knew I needed a little help about now, because it all came in one day and I don't even know yet exactly what made me do it, but to look at $27,609.67 right here in my hand makes me feel, damn I'm good.

Marie's Parents had called and wanted us to come back to Minnesota for the holidays, and we got nothing holding us back.

Marie's parents still celebrated a traditional old fashioned Italian Christmas complete with all the trimmings.

This would be a joyous occasion for their family and just what the doctor ordered to relieve some of the tension around the house.

That night they broke the news to the kids and began describing what their grandparents did on Christmas. They asked if the would be able to work on stringing the popcorn, and cranberries, together and they could make a few different paper ornaments too, and bring a picture of each of us to hand on the tree. Grandpa will have a lot of special lights.

At first they thought it was funny, but soon became eager and excited about their upcoming adventure. Nine hundred miles away from home in the middle of winter, and Minnesota can be extremely cold.

Larry began looking through the paper's Tim had given him a picture describing the Arsenal south plant located in sections 1 and 2, and what was manufactured there during World War II.

1. 3,566 tons of Levinstein Mustard.
2. 158 tons of distilled Mustard.
3. 4,553 tons of Lewisite.
4. 2,689 tons of Arsenic Trichloride.
5. 789 tons of Sulphur Dichloride.
6. 480 tons of Thionyl Chloride.
7. 31,680 tons of chlorine.
8. 155,000 tons of incendiary munitions.

The South Plant also contained it's own sodium hydroxide concentrator unit. It chemical sewer was basin A location is section 36, and to the Sanitary sewer also located in section 24.

Basin A was constructed by placing an earthen dike on the lower portion of a natural surface drainage feature. Basin A was unlined, and constructed of natural loamy and sandy soils formed in wind laid deposits. The records does not reflect any studies made prior to construction of the Basin A, in order to determine either the hydrology under the Arsenal or the rate of seepage or percolation of water from the basin. Water placed in basin A raised the water table and increased the movement of water and pollutants north, and north west toward the home place.

PESTICIDE, AND HERBICIDE PRODUCTION;

The south plant was leased to Colorado Fuel and Iron Chemical Company, in 1946 to operate the chlorine unit, and the sodium hydroxide concentration unit, and to produce chlorinated benzene, and DDT.

The South Plant lease was transferred in 1950 to Julius Hyman Chemical Company which commenced the manufacture of dieldrin and aldrin. Two of the deadliest of all pesticides, 'even more deadly than DDT. which had been banned', and were finally banned for use in the United States of America. How ever Shell Chemical Company was allowed to continue it's manufacture, It was then shipped to other unsuspecting developing Countries outside the United States. This pesticide was known as the Chlorinated Hydracarbon.

The lease on the South Plant was then transferred to Shell Chemical Company in 1951, Shell continued the manufacture, of Dieldrin, and Aldrin, and commenced the manufacture of Nemagon, as well as many other pesticides and herbicides. Shell dumped untreated pesticides, and herbicides into Basin A.

EXPANSION FOR NERVE GAS;

The North Plant located in section 25 was completed in 1953, and manufactured nerve gas. Liquid wastes from this operation were discharged into Basin A which was enlarged for the purpose. Basin A proved to be inadequate and its over flow went into the natural drainage area. So the Army created more basins merely by placing dikes on the lower end of natural surface drainage features. They created basins C, D, and E, located in section 26. These basins were

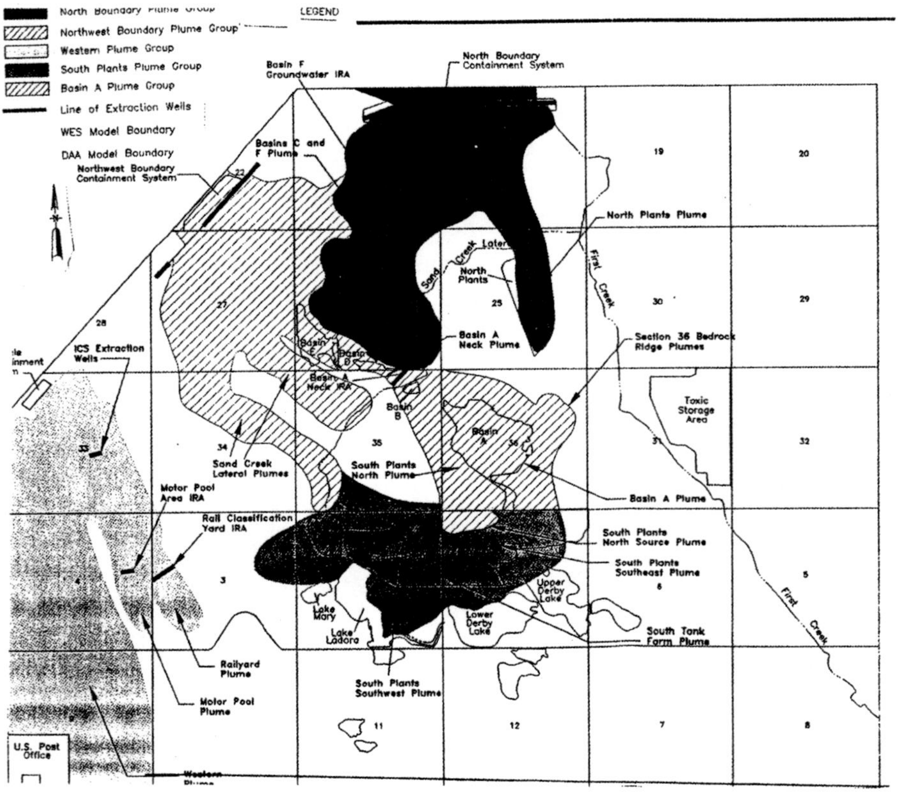
LEGEND
North Boundary Plume Group
Northwest Boundary Plume Group
Western Plume Group
South Plants Plume Group
Basin A Plume Group
Line of Extraction Wells
WES Model Boundary
DAA Model Boundary
Northwest Boundary Containment System
Basin F Groundwater IRA
North Boundary Containment System
Basins C and F Plume
North Plants Plume
Sand Creek Lateral
North Plants
First Creek
Basin A Neck Plume
Section 36 Bedrock Ridge Plumes
ICS Extraction Wells
Basin A Neck IRA
Basin B
Toxic Storage Area
Basin A
Sand Creek Lateral Plumes
South Plants North Plume
Basin A Plume
Motor Pool Area IRA
Rail Classification Yard IRA
South Plants North Source Plume
South Plants Southeast Plume
Upper Derby Lake
Lower Derby Lake
Lake Mary
Lake Ladora
South Tank Farm Plume
Railyard Plume
Motor Pool Plume
South Plants Southwest Plume
U.S. Post Office
19
20
30
29
32
5
7
8
11
12

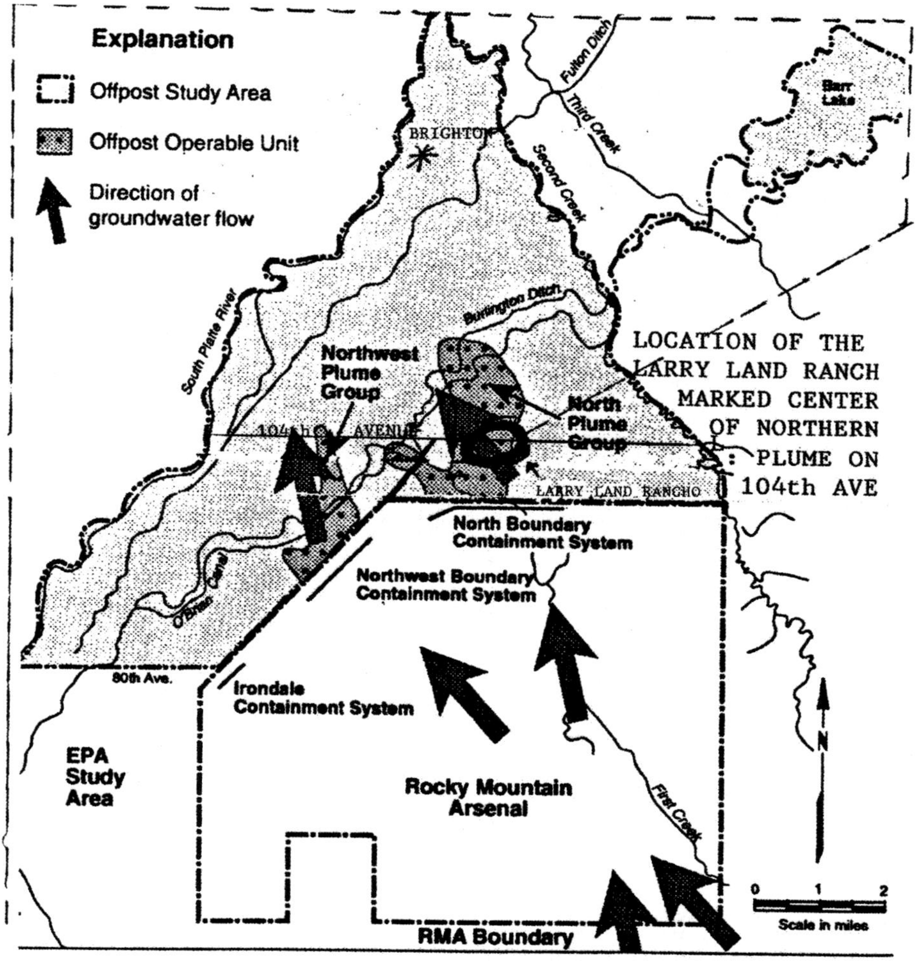

Figure 1: Rocky Mountain Arsenal and Offpost Study Areas
BROKEN LINE REPRESENTS THE 35000 ACRES OFF THE ARSENAL PROPERTY THAT IS SO BADLY CONTAMINATED

unlined, no Engineering studies were Conducted prior to the construction of the new basins. Basin C is the most permeable of the basins. It's natural drainage was straight into a layer of gravel, allowing the seepage of contaminants into the underground water that left the Arsenal. They poured lots of contaminants into basic C, because everything disappeared much quicker than the others, (THE ARMY KNEW EXACTLY WHAT THEY WERE DOING). In fact they couldn't keep it full, 100's of thousand's of Arsenal chemicals being flushed from the Arsenal daily. Now we're talking of at least 4 different nerve agents, mustard agents, lewisite, pesticides, and herbicides, rocket fuel, the worst the worst Ultra Hazardous chemicals known. This continual seepage into the drinking water, made the saturation level higher and higher every day. They called Basin C, straw basin, and they knew exactly what was happening, they are the ones who done the bulldozing, and seen first hand the layer's of sand and gravel they were in.

RESERVOIR FLUSHING;

In 1957 the Army began putting Irrigation water in basin C, the purpose of the water was to "flush" and "dilute" the polluted aquifer. Such waters also had the effect of driving pollutants beyond the boundary of the Arsenal and increasing the height of the water table. The record does not reflect when or whether this practice ceased but does reflect that water was in evidence in Basin C as late as 1974. A blocking off the water coming in by the South plant creating a mound of fresh water, and was used for the same purpose, to purposefully dilute and flush the aquifer, under the Arsenal.

1950's CONTAMINATION;

The first complaints of contaminated water were made in 1951 by farmer's irrigating land northwest of the Arsenal. Increased pumping of ground water for irrigation in 1954 resulted in an increase in complaints. By 1960 thirty-three claims for ground water contamination, totaling $1,769,466.00 were filed by landowners northwest of the Arsenal. Records of these claims are not available. By 1965, $165,000.00 had been paid on 27 claims, and 6 claims were still pending. The Army admits that it acknowledged liability for the contamination, and made settlement on those claims.

Several days later, Marie came into the spare bedroom that they had converted into a den, While Larry was working on some tax papers.

"Larry, I've been thinking, instead of flying back to my parents, let's drive. I think it would do you a world of good to occupy your mind with something else instead of always worrying about the Arsenal and Trucks."

"He agreed on the idea. and said We could take three or four days, and make some sort of vacation of it. he said he would call Dennis, and see if he would watch the house."

"I think it would be good for all of us," Marie said.

Again his wife saw through the misty veil of despair that was over shadowing his otherwise good disposition.

It was decided that they would leave here on the eighteenth of the month and leave Minnesota on the third of fourth of January.

With everything that had to be done the time flew by. The house had to be checked over, the trucking business needed a plan, and the drivers had to be briefed as to what would happen on those days. He had already put in for his vacation at the Sheriff Department. As for the Auctions they are always planned well in advance.

His darling wife must have misunderstood something he had said during one of their conversations, because she packed enough clothes to dress the entire Third Marine Division. She not only used every suitcase they had in the house, but went out and bought two additional ones, and borrowed two more.

This was when his wife was at her finest, he thought, the caring of her family. A lot of men scoff at a woman's work in the home. Fortunately Larry had never been one of those, and learned very quickly a long time ago to appreciate all the things she did, and to be able to understand when to keep his mouth shut. He knew how hard Marie worked to keep her family and the home running smoothly, and quite honestly he didn't think he could last out the day if he had her job.

Looking at her he realized that this trip would be one of the rare times in their lives that would contain many wonderful memories for them to look back on. He knew how hard life had been on her and the kid's and this delicacy of time would not be forgotten, remembered by all.

After many days of waiting, and running around, the time had come for them to finally leave. Larry finished loading the last of the Christmas gifts into the car, and Marie rechecked the house to make sure all doors and windows were locked securely.

After looking around to make sure they hadn't overlooked anything, everyone piled into the car.

At the end of the driveway, just as the car was backing around, onto the road, the historic last words were spoken by his daughter, "Daddy, I have to go to the bathroom."

Laughing to himself he was firmly convinced, that where children are concerned, this is the most communicable disease known to modern man. It inevitably happens that when one child has to go to the bathroom it spreads to the remaining children like a prairie fire out of control.

After several minute's deliberation, and knowing that both bathrooms would be occupied, Larry shut the car off and made his toward the house, as he stood outside the door with his legs crossed waiting, Marie came out and said why is your legs crossed, she began laughing. "Larry said I'm in line."

Nearly three days and a thousand miles later they arrived a Marie's Mother, and Father's home.

The days spent with his wife's parents would be long remembered by their children. Wonderful memories of Christmas from another time, enriching their lives with traditional values and customs, that were meant to be passed on from generation to generation.

The time spent getting back home, after a week at the In-laws will be long remembered also. As they packed every thing back up and ready to leave it began to rain, he said "OH MY GOSH" the temperature is only 30 degrees, and the rain is falling nearly like slush, as we left we thought it'll be ok we'll just drive slow, and we did. all the way from St. Paul, Minnesota, to Des Moines Iowa, then West toward Nebraska, all the way through Council Bluffs, and around Omaha, was just solid ice like a skating rink on the highway, There were many times he had to stop the car along the highway, chip the ice of the windows. We did not make good time but we were averaging about twenty miles an hour, the highways were really lonely, very few cars and no trucks at all on the interstate. It got a little better toward Lincoln Nebraska, and then after Lincoln it was clear sailing, no ice, and was dry. Then where Interstate 80 goes into Interstate 76 into Colorado, he got to thinking, that road sure looks awful shiny, and when he let off the accelerator, they began going in slow circles right down the middle of the freeway, very scary until it was over, and we were sitting in the middle of the freeway, They had completed twelve circles right down the middle of the highway, and very thankful that experience was as safe as it was. They got home later that day, unpacked and just sat around and relaxed.

Larry was glad, and at the same time thankful that they had made that trip back to America's heartland. It gave him the opportunity to think of his position in life and the place he held in the lives of his children. It seemed to him that both life and children required him to make a difference. He knew that what he was doing was hard on all of them, and hard for them to understand the magnitude of it all, but he knew, without a shadow of doubt, that he was right. He often wondered why it was him that had to fight this battle. It seemed that somewhere there must be a person more articulate or more educated, and more able, someone with more contacts or at least more knowledge of the political world and make the difference. Still, he looked around and there was only himself. Then he got to thinking well he already had taken on a very large corporation, Shell Chemical Company, and stopped them in their tracks from any more polluting in Colorado. He'd already caused the Cease and Desist orders to be served on the United States Army, and Shell Chemical Company, and now it's only a matter of time. Washington D.C. is now getting involved in this matter, turning up the heat on them.

Mother nature had a grip on the Land that winter and was slow to release her hold. As Larry stood in front of his living room window, he saw acres of barren land spotted with patches of snow that were left from the rampage of the blizzards that often visit this country. The trees were bare and seemingly lifeless, yet he knew that there was always the hope of a new awakening to be realized with the coming of spring. March was here in all its power and glory, he was again feeling the bonds of confinement. Winter never seemed to bother him until this time of year. The novelty of the change of seasons had left him, and the deadness of winter always closed in on him. He needed to get back to the land back to his livestock. He needed to feel productive. His struggle with the Government dragged on and on much as the coldness of the barren scene before him did. Sometimes he felt that the coldness that was in him went all the way to his soul. Then, as if someone had felt his feelings, he received a phone call from his lawyer.

"Larry, today at 11:30 A.M. a cease and desist order was again issued to the senior officer in charge at the Arsenal. As of May 31, 1982, The United States Army, at the Rocky Mountain Arsenal will cease all operations and will be permanently closed.

Larry just sat there for a moment. For the first time in his life he was nearly speechless. He sat there motionless with the receiver pressed to his ear. Land had stopped them in their tracks. It took ten (10) years. Shell Chemical Company was also issued the last cease and desist order. They had both received orders in 1975, then again in 1978, and now 1982, each time has been with heavier demands an consequences for not following the order's. As I understand the time frame for Shell Chemical Company was set in 1982.

The Army will continue It's operation of the existing ground water containment and treatment systems, and internal groundwater treatment systems. The Army would also construct and operate an additional ground water containment, and treatment system in the south plants and Basin A area. All contaminated structures will be demolished, the rubble will be placed in an on site hazardous waste landfill. There has been talk about removal of several million cubic yards of soil, placing it in a hazardous waste land fill, The Army would then place an impermeable cover (cap) over these areas to eliminate any exposure to humans or wildlife.

Larry said, "Hope sees the invisible, feels the intangible and achieves the impossible"

AMERICAN BALD EAGLE SYMBOL OF PIECE AND FREEDOM

[*This caption was placed above the picture in the original.*]

CHAPTER SEVEN

THE PICTORAL BACK GROUND ON THIS PAGE IS PART OF THE WELLS, THAT MAKE UP THE WEST BOUNDARY GROUND WATER CONTAINMENT & TREATMENT SYSTEM.

The most alarming of all man's assaults upon the environment is the contamination of air, earth, rivers and sea with dangerous and even lethal materials. This pollution is for the most part irrecoverable; In this now universal contamination of the environment, chemicals are the sinister and little-recognized partners of radiation in changing the very nature of the world, the very nature of life.

"Rachel Carson, Silent spring"

By the early 1980's a new awareness of the contamination problem had emerged and was being talked about by both, the Government agency's who had held silence before, and by the American public.

By 1970's and 80's a ground water containment/ treatment system, and internal groundwater treatment system was installed to stop the flow of pollution to the North of the Arsenal. The next ground water treatment center is on the drawing table for the South Plant area where Shell Chemical Company manufactured the deadliest of all pesticides. The water containment system, and internal groundwater treatment system for the West, and Northwest were begun in the late 40's 50's 60's and 70's.

Shell Chemical Company manufactured the deadliest and most lethal of all pesticides. They were eventually banned for use in the United States. Again the only problem with that Shell Chemical still sells this to foreign countries who in turn sell the produce back into the United States. These chemicals are systemic, and came back to the tables of "America the Poisoned."

"Lewis Regenstein:

Systemic; Is a pesticide used in crop planting to eliminate pests that would tend to eat part of the crop. The only problem is when it is sprayed on the seeds, ground or the plants, and then the plant passes it on to the vegetable or fruit. Deildrin, and Aldrin were two such pesticides. These were manufactured using Chlorinated hydrocarbons. Exactly like a Nerve agent, yet even smaller amounts can kill humans.

Another pesticide manufactured and distributed across the across the United States, Canada, and many other parts of the World, was the Vapona No Pest Strip. Hang it up in the center of a room from a light, light draws the insect. They were fantastic, a wave of the future but, deadly. Some people hung one in each room of their house, each one was prepared for a 12' x 12' room. unbeknowenced to any one each of you had hung deadly nerve agent in every room of your house. People started to notice it was great on fly's, moth's, spider's, but they soon started finding their pet fish dead, their cats dead, their dogs dead, the birds dead, and to think for a moment we were sleeping with that in our house. People began having bizarre, health maladies, and was soon recalled taken off the market, and things began getting back to normal. If Shell were going to market something like this for every one's home. One would think they had to do a lot of testing a lot of research on the effects to humans, since this is going to be hung in their house. Larry has yet to this day found any thing telling him of any research that was ever done by Shell other than they knew it killed insects. Nothing was done to say it would be safe for animals, fish, birds, or humans. Shell Chemical Company began getting truck loads, all boxed of Vapona strips, from all over the country back, then they didn't know what to do with it, so as usual they dug large pits, and began dumping it into the pits. Then a couple of times they were caught by the fire department trying to burn the whole pit, People in Denver were getting pretty ill and many bizarre health maladies., No one but no one was notified, even generally it was hush hush. Yet Shell got by with that also. I know this sounds like a horror story from hell but it is truth, and it was a failed product of Shell Chemical Company, at the Rocky Mountain Arsenal, just North of Denver, Colorado. It's now laying in huge pits, where it has contaminated the soil and probably the water below.

Shell Chemical Company, and Shell Oil Company is as I understand a Dutch Monarchy owned Corporation. Why are we in American allowing this to happen, and the only reason they are here instead of there is as Larry understood the Dutch will not let them plunder and pollute their Country like they have done here, and with no restraints, and when it takes a little money under the table, I can imagine they are number one.

Supposedly the Army had totally stopped dumping at the Arsenal in 1981 according to reports.

In late 1982 Shell finally turned the keys to the pesticide manufacturing facility back to the Army. All most trying to sneak out with nothing owing for what they had caused. The United States was not that naïve, and they were settled in now for some long Court battles not only with the United States, but all the many private insurance Companies, that had insured them, one of them being the Lloyds of London. Trials as I understand will be held in California.

In 1983 a massive legal struggle, To be the biggest and most complex in the history of American Jurisprudence, began on October 5, 1983 when Shell filed a lawsuit demanding that two hundred fifty insurance companies pay it's share of the Arsenal clean up.

Two months later, on December 9, 1983 the Army sued Shell Oil Company for $1.8 billion as compensation for the environmental devastation the Oil giant left behind. It was as though the snakes that had so long inhabited the same space, had suddenly turned on one another.

** Three years later on November 14, 1986, the state of Colorado sued the Army for natural-resources damages under the Federal Superfund hazardous-waste law.

** At the center of the dispute was the future of twenty seven square miles, owned by Federal taxpayers just outside of a large metropolitan area Denver Colorado. The Army and Shell Officials said repeatedly that it was unrealistic to think that the Arsenal land would ever be fit for homes, buildings, warehouse area, or even gardening or farming. Nor would the water ever be safe to drink or its lakes be suitable for harvesting edible fish. For years it was talked that Denver Stapleton International Airport would one day take over the 27000 acres to expand the Airport. Since Stapleton bordered the Arsenal on the South, already had huge run-ways running over top interstate 70 into the Arsenal where it had already expanded. There was talk at one time to seal the surface with some type cap, but that they soon learned was not feasible. The new Denver Airport is now out east of Denver about twenty five miles near Bennet Colorado. It has bragged about being the largest Airport in the world.

** Larry has always maintained that the Army and Shell Chemical Company have been conducting a massive public relations campaign in the hopes of soothing ruffled feathers (covering up a terrible mistake), or waiting for enough time to pass so people who were witnesses no longer exist, rather than dealing with the problem itself. The fact is the scope of the problem is so large, and the tunnel is so huge with no light at the end, it may take hundreds of years, and billions of taxpayer dollars, for the Army and Shell Chemical Company to pay for the haphazard bungling, and loss of control of the Worlds worst Ultra hazardous materials, polluted and ravaged the country side from Colorado, to Nebraska, and Kansas.

** Our Country had been producing chemicals that do not belong on the market. There are chemicals in the Arsenal that have been banned, removed from use due to the Ultra hazardous Conditions they produce. They have ended up in our drinking water, The toxins are so dangerous you cannot bathe or even wet your skin, as it will kill by absorption.

** The problem farmer's were having west of the Arsenal in the 50's, it was they discovered 2,4-D, had never been manufactured at the Arsenal. This prompted the University of Colorado science Department to surmise that 2,4-D had been created spontaneously, when wastes from Army and Shell operations were dumped into basin A, forming a witches' brew of new poisons'. At that time they figured that 7 square miles had already been contaminated by the Arsenal.

** Arsenal manufacturing operations became so great, that the 240 million gallon Basin "F" was filled nearly to the point of overflow at least three different times in the 1960's, and 70's. Please note years Basin F was overfilled, to the point of over flow is grossly over filled. This is during the time the Army was totally aware Basin "F" was leaking, and were using under the premise it was in fact leaking. The Army facilities director George Donnelly, and General Peter Olinchuk, of the pentagon. They found the asphalt liner, built to protect groundwater from Basin "Fs" toxic chemicals had liquefied, and it was filled with 130 million gallons of highly toxic substances, and was in fact leaking. This is of record by the Army Department of Hygiene, and the United States Department of Health and Welfare. This is total irresponsibility on the part of the United States Army, and the pentagon of the United States, and it's Officers in charge. This was filled with the most Ultra Hazardous toxic poisons in the World. It would make the most contaminated piece of Land on earth.

** Please notice paragraph above states Basin F was filled with 130 million gallons of toxins, Then please note the top of the paragraph of how they overfilled it, knowing it was leaking with still another 110 million gallons of Ultra Hazardous Chemicals.

** Then the Army banned release any further information on Basin F would be classified confidential due to the sensitive nature. "So they didn't get caught up in their cover up."

** The Army tried to make everyone believe they were not adding any waste to Basin F. Yet it is of record they were using Basin F as late as 1982, Shell Chemical Company, were also said to have quit putting waste into basin F, 1978 but again records will show they were still dumping into basin F as late as 1982, and the United States Army was totally aware of this, and allowing it to take place. It's entirely possible that money was changing hands under the tables.

** Army managers displayed an even more cavalier attitude toward public health in the early 1970's when they allowed more than 10,000 boy scouts to camp overnight in the facility. The following day the camp out was ordered stopped immediately.

** One other new development in the area Shell Oil Company is buying all houses, where there have been problems, moving the people elsewhere and bulldozing the homes. This I believe was a quick methodology to eliminate people joining in any law suits.

** The Army officials met with Shell Officials early 70's claiming that 85 percent of the wastes being generated, and dumped into Basin F. is from Shell Chemical Company, and they want Shell to pay 85 percent of the cost for a new basin.

** In 1973 the Army, Shell Chemical officials met with the U.S. Geological survey, Chief Hydrologist and Geologist. They reached a conclusion that the Land wells were in fact contaminated by the Arsenal with chemicals emanating from the Arsenal.

** In the early1980's the Army still concerned of saving the Bald Eagles, finally learned the Eagles were so contaminated and dying was because of what they ate. Prairie Dogs at the Arsenal was abounding with thousands, even though they were so Ultra toxic they killed the Eagles. The Army Biologists ultimately decided to exterminate all the prairie dogs, to keep the Bald Eagle's from praying on the rodents, since they were ultra hazardous from the toxic wastes. The Army says they exterminated the prairie dogs because they thought they might have the plague fleas. I still think it was because they were so contaminated and was killing all the Bald Eagles, and that didn't look good in the news.

** We have here at the Rocky Mountain Arsenal, and the surrounding area the worlds worst nightmare for pollution. these are Ultra Hazardous chemicals were dealing with, the deadliest on earth. If we were to compare the United States Army, Rocky Mountain Arsenal, with the Nuclear reactor accident at Cher Noble USSR. The Arsenal would far outweigh the death and destruction of Cher Noble. North and Northwest of the Arsenal we are seeing the highest rate of cancer per capita of anywhere in the World.

** This Country was able to produce food long before pesticides came along, and people have fought, and put pressure on Government and Industry for years in order to reduce their usage. The key is that we are using a very wide range of toxic and cancer causing chemicals and are placing them in our food supply.

** For some time Alar was used on our apples in order to give them greater eye appeal. It was the public who, when warned of the dangers of Alar, in turn forced the food industry to discontinue its use. Apples still look good, but we are not ingesting the chemical. So much of our food has been laced with coloring agents or chemicals meant to give the food greater shelf life that we often do not even think about what we are eating.

** Until we realize that we must reduce or eliminate these artificial ingredients from our environment, we will be talking about this same subject a hundred years from now.

** The voice of the American public must be heard by industry, but what is even more important, the voice of the American public must be heard by its own Government.

Larry and Marie had talked for a long time about moving to Minnesota, where the air is cleaner, water is hopefully cleaner. but in Anoka, a city north of Minneapolis had the United States Army Arsenal in their midst also, and began having tainted wells any where near the arsenal. There seem to be a lot of work done quickly on the tainted wells, and people who had wells were hooked up to city water, and the whole incident seemed to die out. They bought a home in Maplewood Minnesota, a small suburb on the north end of St. Paul.

By September 1983 Larry had sold the trucking business he owned, operated and grew from one bobtail truck to fourteen semi tractor trailer rigs. He had built the business and the rewards were good. He owned the only contract carrier permit known for the entire state of Colorado, and he had a lot of people who would like to own it. He done it the only way that was fair, and put it up for bids. He said his good bye's to Erlich Sales for which he was a part of the Auction business for fourteen years. He says his hat off to Fritz Erlich, J. Lee Sears, and Henry Morimitsu, the dream team of Auctioneering. Larry wants them to know how much he enjoyed working with the best in the land, and being a part of.

The family had already left the last of August, Larry had to stay around to tie of all the loose ends. Marie had to get the kids ready for school, in a new setting. Then in October he loaded the pickup, hooked up a small trailer, and finished loading all the personables, and headed out toward Minnesota not knowing what he would be doing. He thought about getting started in Real Estate, Auctioneering, or Corrections.

Saying good-bye to Colorado his home state, he was saddened about what had happened there. He wants to share with everyone the Health Maladies him and his family suffered at the hands of the United States Government. SIGNS AND SYMPTOMS; headaches, shaky, dizziness, nausea, then all them plus loss of hair in patches like dime size, scruffy and tangled hair, blackened teeth, vomiting, sores in nose, mouth, bad rashes every where, loss of bladder control, Loss of bowel control, serious painful aching of the bones, shaking got worse became tremors, dizziness got to the point of loss of vertigo, burning-painful sensation in the face left or right side, little cold spots seem to appear spontaneously anywhere on the body suddenly, and the first thing your looking for is where the cold water is dripping from. Loss of teeth, rashes became bleeding sores, eye irritation with altered vision, sores in noses and mouth open and bleeding, sores in the throat, constant salivation feeling of tightchested, and getting tighter to the point it was difficult to breath. We had to think fast as to what the livestock and the people had in common. As we were nearing death, the livestock had already went through the above, and then became unable to stand would fall over and go into the final convulsions of Sarin Nerve Agent and others, with death soon to follow. The four legs would convulse in a rhythm with the eyes, and the head thrown way back.

Larry knows what he's told you above is like something out of a horror story, & it is as the Government stood by & done nothing to assist or even assist the Doctor's who were at a point that there were no tests for what they had been poisoned with, only the Army knew. The Tri County Health Department found in the blood of eight of the victims, a chemical known only at the Arsenal, "phenol" Used by the Army. (fingerprinted) to the Land wells and poisoned them and their livestock. Also Used in the manufacturing of Pesticides, & Nerve agents, at the very least they had the name of one toxic chemical. However the Land's got away from drinking the water or cooking with it in 1972.

This was a chemical they knew something about, and stated you are the walking dead, there is enough in each of your systems to be lethal, they called more Doctor's in, and finally called the Army in. The next thing we knew all medical records of the Lands were lost, no one had any idea of what happened they knew we were there but almost acted as if they knew nothing, there was no records. All though we had stopped drinking or cooking with the water, there was an improvement of the condition. The Doctor's didn't even want to talk of what the chemical was. They merely noted that they had stopped the use of the water for drinking or cooking and seemed much better.

The recovery was very slow for the longest time, Marie had a miscarriage, the same as the cattle, had about 75 head of cattle that were brought over from the Boxelder farm, within days of drinking the water they were about 7 to 8 months pregnant they all miscarried. The livestock that had been moved seemed to be recovering quicker than the humans, It was a little different at this point it was to figure out what the humans and livestock that got ill did not now have in common. (bathing). The pollution was absorbing through the skin. We stopped that and began to see dramatic recovery. As well as did the livestock recover very well however there was a failure to thrive problem. They were not growing, so Larry ended up with a lot of livestock that got no bigger. One that should have weighed 1500 pounds did not grow to weigh more than about half.

Larry was able to feel good about the family's recovery at least, and to look back and see how close they were to death. The signs and symptoms we suffered began to clear up immediately, then slowed down to a crawl. Most of them took between three years to ten years to disappear, and each time they reached peak of full recovery, it was a joyous occasion for all. The patches of hair steadily improved, and was over 5 years to reach full recovery, for others it was nearly ten years, some were more or became permanent.

"Belief is the knowledge that we can do something. It's the inner feeling that what we undertake, we can accomplish. For the most part, all of us have the ability to look at something and know whether or not we can do it. So, in belief there is power; our eyes are opened; our opportunities become plain; our visions become realities".

"STARK REALITY" by Steven Thomma; Washington Bureau

This is just another situation we all get into when using chemicals. Chemicals widely used on millions of lawns across America could cause cancer birth defects, gene mutations, or other maladies, according to a study just released.

To create the picture-perfect lawn, many lawn care companies rely heavily on chemical pesticides, & Herbicides, but these toxic chemicals may kill more than just weeds, Many have been found to cause serious adverse health effects as well. These chemicals are available at your local hardware store for the do it yourself gardeners, including such widely used brands as dursban, sevin, spectricide and Trimec/spectricide, these have already been banned on golf courses and sod farms, suspected of causing birth defects, nerve and liver damage. "Weiss" study a compilation of studies that is billed the first comprehensive examination of the 40 most widely used chemicals in the lawn care Industry. It was found "12" are suspected carcinogens. "21" have been shown to cause other long-term health effects including birth defects gene mutations, or damage to kidneys, livers and the nervous system. "20" have shown to cause damage to the Central Nervous System. 'CNS'

"36" have been shown to cause eye, skin, throat, and lung irritation in humans. "1" chemical 2,4-D a component of the Vietnam war defoliant agent orange is also now used in about 1,500 lawn care products. The national Cancer Institute, have it listed as a possible link between itself and a type of lymph cancer. "EPA stated that many suspected carcinogens are used in the lawn care industry. "This says a lot for EPA were caught lying in certain studies conducted by them, in a huge effort to cover the trails of horror left by Government Negligence, and by Corporate negligence, and greed".

"Let me spell out EPA. Environmental Protection Agency. They are there to protect the public and now take a look again as to what they just said, as they are trying to down play a decision by the National Cancer institute, EPA allows many lawn care products on the market with absolutely no protection for the public.

The EPA set what they called a safe standard on DIMP. a metabolite of sarin Nerve Agent at 600 ppb. Larry talked to the Colorado Health Department, with them using the same exact studies as was used by EPA, there safe standard was set at 8 ppb., and State Health officials were very upset when they heard that EPA had set a Standard. He said they cannot do that on the State of Colorado can set a standard, he said he would take care of that problem. EPA does not care about the people or whether or not they get hurt from chemicals. They seem to care about protecting Government and big Industry from the people who they are poisoning. "AMERICA THE POISONED" 'LEWIS REGENSTEIN'

Throughout the Country, and in every metro area, as the noose tightens on pollution, residents are learning the physical and social costs of a toxin in the neighborhoods. While to err on the side of caution, to do so is to sometimes harm people in a very real and immediate way. Everyone wants to trust their Government, and they really got little or no choice in the matter. But they must be made to take responsibility for what they are responsible for.

Larry arrived at his new home in Maplewood Minnesota, he was happy to be home and with his family. He decided just to kick back and work on the thing with the Arsenal. He has a lot of contact out of Washington D.C., and a lot of questions flooding in. We already had the new bill that is known as A Bill for the relief of the Land's et al.

The Cold War's toxic legacy lingers at the Hanford Site, where an estimated 40 billion gallons of radioactive liquid waste was discharged into the soil, eventually reaching groundwater. Plutonium-production reactors cooled their cores with river water and discarded it laden with cesium, chromium, and cobalt. The 50-billion-dollar clean-up may take 30 years.

Irony

Fifty years of bombmaking left the Hanford Site saturated with radioactive pollution.

JIM RICHARDSON, PHOTOGRAPHY

The Cold War's toxic legacy lingers at the Hanford Site, where an estimated 40 billion gallons of radioactive liquid waste was discharged into the soil, eventually reaching groundwater. Plutonium-production reactors cooled their cores with river water and discarded it laden with cesium, chromium, and cobalt. The 50-billion-dollar cleanup may take 30 years.

Irony

Fifty years of bomb making left the Hanford Site saturated with radioactive pollution.

CONFLICTING INFORMATION IS SO OFTEN GIVEN TO THE GOVERNMENTAL AGENCIES, BY OUR SO CALLED BIG INDUSTRY. THEN YOU HAVE PLACES LIKE THE HANFORD SITE, "JUST IN". THE GROSS, WILFUL, AND RECKLESS NEGLIGENCE OF IT'S PLANNING AND OPERATIONS. NOTHING TAKEN INTO CONSIDERATION OF KEEPING IT CLEAN AND POLLUTION FREE. THIS ISN'T EVEN THOUGHT OF, THEN THEY TRY TO FIGURE OUT WHERE THE FIFTY BILLION DOLLARS IS GOING TO COME FROM. (TAXPAYER). JUST TO ATTEMPT A CLEAN UP, PROBABLY A LOT OF COVER UP.

DEAN KRAKEL/Rocky Mountain News

Martin Marietta, which employs 12,000 workers at its plant in Jefferson County, uses hazardous chemicals in manufacturing and testing rockets and aerospace equipment.

NATIONAL GEOGRAPHICS
JIM RICHARDSON PHOTOGRAPHER
WRITER: BILL ALLEN
BELOW WRITTEN BY LARRY LAND

MOST PEOPLE ARE PROBABLY NOT AWARE OF THE GREAT COLUMBIA RIVER, IN THE PACIFIC NORTH WEST. HOME OF THE COLUMBIA BASIN SALMON. HERE WE HAVE HANFORD REACH, AND THE HANFORD SITE, SATURATED WITH RADIOACTIVE POLLUTION. THE CLEAN-UP MAY TAKE UP TO 30 YEARS, AND COST 50-BILLION DOLLARS. TAXPAYER'S WILL TAKE CARE OF THIS ALSO BECAUSE IT WAS OWNED AND OPERATED BY BIG BUSINESS AND THE GOVERNMENT, WHO HAD NO IDEA OF THE END RESULTS.

PACIFIC NEWS SERVICE
BY DENNIS BERSTEIN

PLUTONIUM, FOR THE ATOM BOMB DROPPED ON NAGASAKI WAS PRODUCED AT HANFORD. U. S. GOVERNMENT ONLY RECENTLY ADMITTED THAT NEARLY 500,000 RESIDENTS OF IDAHO, OREGON, & EASTERN WASHINGTON, HAVE BEEN SUBJECTED TO HIGH LEVELS OF RADIOACTIVITY IN THE AIR, GROUNDWATER, AND FOOD SOURCES FOR MORE THAN A QUARTER CENTURY. "IN SOME INSTANCES ACCORDING TO GOVERNMENT DOCUMENTS, THE EXPOSURES WERE NOT ACCIDENTAL."

BY LARRY D. LAND

THOSE WHO LIVED IN LAS VEGAS NEVADA, WERE ALSO SUBJECTED TO HIGH LEVELS OF RADIOACTIVITY IN THE AIR DURING THE 50'S AND 60'S. AGAIN ONLY RECENTLY, AFTER NEARLY 30 YEARES THE U. S. GOVERNMENT FINALLY ADMITTED THAT AT LEAST 500,000 RESIDENTS, HAD BEEN SUBJECTED TO HIGH LEVELS OF RADIO ACTIVITY, PLUTONIUM, CESIUM, AND COBALT. FROM TESTING AT JACK ASS FLATS NEVADA ONLY 50 MILES N. FROM THE CITY. "IN ALL INSTANCES THE EXPOSURES WERE NOT ACCIDENTAL"

[*This caption was placed at the side of the picture in the original.*]

No river in the United States has been more bent to the will of man than the Columbia in the Pacific Northwest, and no river stirs more controversy. Some 250 dams on its tributaries and 14 on its main stem, crowned by the massive Grand Coulee, have for decades provided affordable electricity to the region (and, in time of need, to California). They irrigate a vast agricultural economy. But the dams are also largely to blame for preventing 99 percent of Columbia Basin salmon from returning to their ancestral spawning streams. The very survival of these magnificent fish is in doubt. And so the debate: Should some of the dams come down?

There is no dam on at least one stretch of the Columbia, the Hanford Reach, where the river runs free. Free, but its bank tainted with radioactive tritium, strontium 90, and other contaminants from an old plutonium production site (above). Cleanup is under way, but long-term uncertainties remain.

Our in-depth report this month by Fen Montaigne and photographer Jim Richardson is an example of one of the things this magazine does best—bringing context to complex issues. ABC News thought highly enough of how we cover such a story to feature it on an upcoming *Nightline* program.

There are no easy choices along the Columbia River. Well-meaning people on all sides are frustrated, and so too, presumably, are the salmon. To paraphrase Scottish poet Robert Burns: The best laid plans of fish and men oft go astray.

[*This caption was placed below the picture in the original.*]

Shown is a 1998 aerial shot of the Rocky Flats Plant.

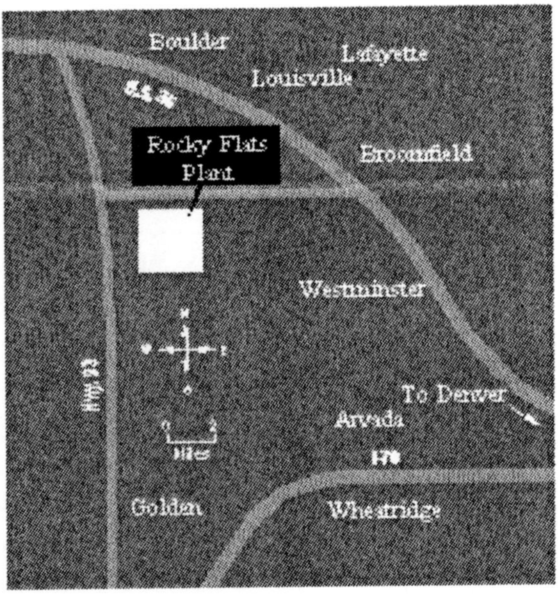

ROCKY FLATS, FORMERLY A NUCLEAR WEAPONS PRODUCTIONS FACILITY, THE ROCKY FLATS ENVIRONMENTAL TECHNOLOGY SITE, CONTAMINATED WITH PLUTONIUM AND OTHER RADIOACTIVE MATERIALS. IS ONE OF THE BIGGEST CHALLENGES FACED IN COLORADO'S SUPERFUND CLEAN UP PROGRAM.

NUCLEAR-WEAPONS PLANT PROBLEMS GREATER THAN AT FIRST THOUGHT, REPORT SAYS. HAZARDOUS

WASTES, AND CONTAMINATION ARE MORE SERIOUS THAN THE DEPARTMENT OF ENERGY REALIZED. IT WILL TAKE MUCH LONGER TO SOLVE THE PROBLEMS. OTA SAYS THERE ARE DEFICIENCIES IN NEARLY EVERY ASPECT OF THE DOE'S EFFORTS TO ADDRESS THE PROBLEMS. WHAT STARTED OUT COSTING ONLY A FEW MILLION IS NOW IN THE HUNDRED'S OF MILLIONS, EVEN BILLIONS.

LIBRARY FILE/Rocky Mountain News

FBI plane with infrared camera recorded illegal night burning of toxic wastes at Rocky Flats, shown in file photo.

ROCKY FLATS
Stapleton Airport
COLO.
Denver
50,000 neighbors versus Rocky Flats
Workers, neighbors band together to file lawsuits seeking millions in economic and property damages. Suits ask for independent study of health risks surrounding plant. Page 8

A LARGE PART OF WEST DENVER HAS BEEN CONTAMINATED WITH PLUTONIUM AND WHATEVER ELSE FROM THE ROCKY FLATS NUCLEAR WEAPONS FACILITY.

27 SQUARE MILE ARMY COMPLEX, ROCKY MOUNTAIN ARSENAL JUST NORTH OF DENVER
IT IS MORE CONTAMINATED AND POLLUTED MORE LAND THAN CHER-NOBLE, USSR

THE ARMY AND COLORADO OFFICIALS WERE BACK IN FEDERAL COURT LAST WEEK OVER PROCEDURES BEING FOLLOWED IN CLEANING UP THE CONTAMINATED ROCKY MAOUNTAIN ARSENAL. THE MAIN BONE OF CONTENTION: A FAILURE TO COMMUNICATE

LAST WEEK'S FEDERAL COURT ORDER DID ORDER THE ARMY TO COOPERATE WITH THE STATE OVER CLEANUP OF THE ROCKY MOUNTAIN ARSENAL.

THE STATE OF COLORADO HAS BATTLED IN AND OUT OF COURT TO GET THE ARMY TO FOLLOW STATE HAZARDOUS WASTE LAWS IN CLEANING UP THE MOST

POLLUTED PIECE OF LAND ON EARTH "ROCKY MOUNTAIN ARSENAL"

MAIN COMPLEX, ROCKY MOUNTAIN AERSENAL: Center of the death factory facilities that left fingerprints of death, it was not an accident but known well as the most contaminated piece of land on earth. It can never, ever be returned to it's once pristine state.

Larry registered in Pro-Source school in real-estate 1983, and 1984. Graduated from all four courses, took his State licensing exam in February 1984. Was offered a position with Edina Realty, as a market Auctioneer. At the same time had applied for a position with Washington County Sheriff Department. Still maintained contact in Washington D. C. They were suggesting he come to Washington D.C. and testify before the Judiciary Committee regarding the Rocky Mountain Arsenal.

The phone rang on Sunday morning the end of February, they said this is Washington County Sheriff Department, and would like to set you up for an interview. They said how soon can you be in, Larry how soon do you want me there. It was set for 9:00 A.M. the following morning. Larry knew they were really impressed as his Oral interview about his education and experience. He had several years as a Deputy Sheriff in Adams County Colorado, and then he completed the Denver Sheriff's Academy, and again several years as a Deputy Sheriff for the City and County of Denver Colorado. In Denver he was a specialist in Riot Control, and the training program, A team leader, A squad leader, A Field training Officer. He was a very driven person, what ever he done he would take it to the MAX.

The next morning 9:00 A.M., always very punctual he was waiting for his final interview. He was greeted by the Under Sheriff, and brought back to his office. Larry an him exchanged a few words, and he was asked if he could call the Denver Sheriff Department talk with them and see what they have to say. Larry said go right ahead, the number you have is for the Chief. He dialed the number and asked to speak with the Chief. They talked for a while with such questions asked as to how I handled my job. When they were done the phone hung up, the Chief Deputy turned to me, and said, are you thinking about going back to Denver, Larry said "NO" he then said would you go back if they offered you your job back, Larry said "NO" He said you know they want you back in the worst way, they just ask me to ask you to consider coming back. Larry said that makes me feel good about myself and them, what I was able to offer, but I'm staying in Minnesota, probably forever. He said if you want the job here you have got it. He then said when can you start? The answer was how soon do you want me to start? He got on the phone immediately called to find out when the physical and psychological could be completed. It was set up for Wednesday and Friday this week, he set Larry up to get his uniforms that very afternoon. Larry started to work the following Monday.

Washington County Minnesota Sheriff Department is probably the highest paying Department in the Country. Larry felt extremely fortunate in getting the position. He would now devote his time to working on the Washington D. C. issue, and his work on the environment, keeping the Rocky Mountain Arsenal at the forefront. He would have to hang his Real Estate License up at least for the time being. He wanted to use the remainder of his time to do research and development of several of his smashing ideas, to see if he could get any patents. and or copyrights, and of course writing is also in his thoughts, and besides that type work is good in tax right offs.

Everything was going real well for Larry and Marie, the children are all adjusting to new school's very well Larry has time now for the family is able to more of what he wants to do, and is putting together his first patent. On August 18, 1984 Larry received an unexpected phone call from one of his Attorney's.

"Larry, Congress is going to want your testimony before the House Judiciary Committee Washington D. C., in the very near future. They want you to bring everything you have, they will be calling back to verify the time and place. Your Congressman has a bill now pending before the House of Congress, regarding 1972."

The next few days were spent in preparation for the trip to Washington D. C. It was hard to believe that they had gotten this far. Finally the people who needed to hear about the horrors of the Rocky Mountain Arsenal, were willing to listen. They hurried to get ready. He made the arrangements ahead of time at his job so they would be prepared.

Larry knows, "The purpose of goals is to focus our attention. The mind will not reach toward achievement until it has clear objectives. The magic begins when we set goals. It is then that the switch is turned on, the current begins to flow, and the power to accomplish becomes a reality."

Unfortunately, the sometimes cruel hand of fate turned aside the pleasing breezes of Larry's milieu and placed his destiny on a different path. On the afternoon of August 24, 1984 Larry had decided to take a break and get some exercise and fresh air. It was beautiful outside, so he took his bicycle out of the garage and told Marie that he was going to a ride up toward the Church at highway 36, and Edgerton. Then Melissa his 8 year old daughter wanted to go with him, Larry said ok at least he would have good company. They started down their quiet street thinking about the days that lay before him. A soothing, Afternoon breeze ruffled his hair. They rode to County Road B, a beautiful pond on the corner, then down about a ¼ of a mile to the road leading to the church parking lot, The huge parking lot was empty, no cars. There is a pretty good slope going up toward the Church, then one can just kind of coast coming down back to County road B. After going up the slope twice, and about half way down Larry stopped to rest Melissa said I'll go ahead, and stop and meet you there. He thought that would be fine, and after resting for a few minutes, he then rode down as he was nearing the road leaving the parking lot he heard the roar of a big engine, a blue and white colored car was coming up the road toward him. It did not concern him as the road was wide and there was a great distance between the vehicles that traveled in opposite directions, but just to be sure he moved over several more feet.

** He heard the roar of the engine as the car accelerated toward him. Suddenly, at not more than two car lengths away Larry saw the driver veer towards him. Had he lost control? What was happening? He barely had time to think. The impact was absolutely incredible. He remembered those few seconds as though they were hours, traveling in painfully slow motion. The instant pain, as his head hit the corner post and windshield, he seemed to fly through the air, then the scraping of his flesh upon the asphalt, and then a sudden darkness. He came too, his head exploding with pain, the light burning his eyes, it seem extraordinarily bright. There was an incredible heaviness on his chest. As Larry fought to focus his eyes, he saw a brownish blond man of large build probably 240 pounds pot gut hanging over his belt, late 30's. I could feel his huge hands. The man had his knee buried in the middle of Larry's chest, and felt as though this man was forcing all the air from his lungs. His hands were around Larry's throat he was grinning. saying we finally got you, you little son-of-a-bitch. Larry finding it very difficult trying to defend himself, with a broken right clavicle his right arm was useless. Larry called on all the adrenalin, and police training he could muster at that moment, he delivered a very hard left handed blow to the man's face, The guy was half over side ways, his hands released from around his throat, and that is when Larry gave the final blow with his left knee right between his legs. Larry remembers yelling at him what the f____, have you been drinking, as the kick he received seemed to blow out all his breathe in Larry's face, as he was rolling over moaning in pain. The guy then standing over Larry holding his groin said F___ you. As he went for his car, Larry thought then it's over he looked for the man to get a gun, as he was struggling to get to his feet, he was surprised to see the man get in his car slammed the door, went squealing backwards in reverse. Larry thought then for a minute "OH GOD" he's gonna run over me, the dizziness was so great Larry fell to the ground, trying to lay real flat, and hope he misses with the tires. He then slammed it into a forward gear, and made his tires smoke, and howl as he sped away out to the street going toward freeway 36, turned on 36 and disappeared.

** Larry remembers a license number as the man left, LNT-459 the car was an antique 1961 Dodge Monaco, two door hard top, white on sky blue. There had been a very large antique car show in the Twin Cities that particular weekend. The license number came back to a late model Ford van. Two things, you could see the License plate had been removed, and replaced, The owner of the Van either owned or worked at a body shop in the town where it was from.

** Once again the darkness began to close in, Larry got to his feet picked up his bike, got astride the only thing he could think of at that point is to get help.

** Melissa had come back wondering where he was, saw him on his bike trying to make it go, as she rode up he ran into her bike throwing her off. She tried to wake Dad up but he was out. She ran over to a house and ask them for help. They came and looked then called 911, paramedics arrive taking him to the hospital.

** To this day Larry doesn't know what scared the man away, or what it was that made him leave when he did. but when he opened his eyes later, he was in the hospital. Larry found it hard to believe that he had actually survived the terrible ordeal. He had heard they wrote it up as if Larry had fallen from his bike. The Doctors had put Larry on a certain type steroid to help make sure the brain didn't swell. Larry kept going in and out of consciousness, and was delirious at times, suffered amnesia. The Doctor's at first thought it to be from the injuries. After about 5 days in the hospital they sent me home even though he was still having problems not knowing people, who he was, amnesia, and dementia. Marie called the Doctor's back and told them of the problems and he was again put in the hospital. The neurologist seen him the next day, and realized it was probably the steroid medication. They immediately took him off, and within hours improvement was focusing back in.

** Larry was able to talk with the police, and they come out took a statement, went to the scene, and could see the rubber marks on the pavement where it had happened. It was then re written as a hit and run, and reported to the State as such.

** Who ever sent the person, evidently knew that Larry was going to Washington D.C. to testify before Congress, and they were not taking any chances. Consequently, to the delight of some, and the sorrow of others they did not make it before Congress in 1984, and quite mysteriously it was put off again in 1985, 1986, and 1987. Although more information was placed in the hands of Congress.

** In March of 1986 the Army admitted, it was at least partially responsible for the pollution problem around the Southern Adam's County and agreed to build still another Water containment and treatment Center. Plans are also being considered for another Containment and treatment Center for the South plants and basin A area. The Army has now admitted polluting over 35,000 acres of Land in South Adams County that is off the Arsenal property.

** In 1986 litigation has now begun in the San Mateo County Superior Court in California, against Shell Oil Company for its part in this catastrophe between 1954-1983, and the future.

** The 250 some Insurance companies who represented Shell refused to pay, and in turn Shell Oil Company filed suit. In what is expected to the the longest and most costly case ever heard.

** One insurance representative from Lloyd's of London stated, "There is a fundamental principle of liability, Insurance companies pay for mistakes. What Shell Oil Company did over the period of thirty years at the Rocky Mountain Arsenal was no mistake. It was deliberate and intentional pollution to save money. Shell knew how to do it right but deliberately chose to do it wrong. ("what was said earlier, Shell is out for the almighty dollar"), and they don't give a damn at who's cost.

** This next bit of information is particularly close to Larry's heart because it follows a train of thought that he believes could be true, and it has been happening for many years. A thread, a common connection that is small, but noticeable, that links mental problems to toxic poisons that are in the everyday environment and water supply.

** New scientific findings have strengthened a controversial theory that Alzheimer's disease, and other brain disorders such as Dementia, Aphasia, Ataxia, affecting millions of Americans, may be caused by environmental toxins.

Brighton Standard, by Sharon Simms.

** A higher than normal bladder cancer rate was found in the Colorado Cancer Registry in the areas north and west of the Rocky Mountain Arsenal. "It appears to be greater than expected in that area." Said Dr. Theodora Tsongas, Colorado Department of Health. The 1981-85 rates reflected more than double the normal cases expected for both sexes.

Larry got the news again we had failed to get the bill before Congress, every two years if not heard a bill dies on the shelf, and must again be introduced. Our Congressman feeling stronger toward everything, said not to fear we'll rewrite a new one for the next Congress. Larry and his attorney's made several trips to Washington D.C. over the years, for various hearings, and to meet and talk with many Congressmen, and their aides.

In the mean time Larry's Attorney's called and let him know they had heard from some Attorney's in California involved with the Shell Law suit. They said we've got some documents from the Shell files in the Law suit, that also involve Mr. Land The attorney's said yes, they would make arrangements to be there in two days.

The Attorney's flew to California and went through some files, and discovered some very interesting documents. a particular one stating "LET THE SLEEPING DOG LAY". dated August 1984. What does that mean? Could it be it was Shell that had paid for a contract on Larry. It supposedly means their gonna get rid of a particular person, or they got rid of a particular person. Then to take the timing, of August 18, 1984, Shell Oil Company was made aware Larry was going to Congress to spill the beans. By August 24, 1984 they made sure it would not happen, and Larry's life was put in jeopardy, and out of the way. ("THEY COULD NOW SAY LET THE SLEEPING DOG LAY.")

The Attorney's also found many other documents, another in particular was some well test data material dated may 1975, These were test sheets done on the Land wells, and had each of the wells numbered and named and the chemicals that were found in each of those wells. Endrin- found at toxic levels, Deildrin found at very toxic levels, Dimp reported in each of the wells from 5 to 14 ppb. Chloride-237 ppm. Meaning that the Land wells contained at least 3.5 to 7 percent arsenal pollution in their wells.

This is the exact time the Colorado State Health Department had said they notified the Lands of Chemicals from the Arsenal was found in their wells. The only problem the Land's were never notified that the State or Tri County Health Departments never told the Land's contaminants were found in their well water's. It seems the State Health made sure all others who had as low as 2ppb Dimp in their water was immediately placed on bottled water. This leads Larry to believe the State Health Department "really hoped he would just die". was involved in the Cover up with the Army, and Shell Chemical Company, and are only say they notified the Land's, I haven't seen the letters yet to this day. Just recently in talking with the Health Department was told I must have my Attorney write a letter under the freedom of information act, in order to find out if there is in fact any letter's sent.

** Regarding the brain, and various brain disease, Alzheimer's, Dementia, Ataxia, Aphasia. The findings culminate more than twenty-five years of research on an extraordinarily high rate of brain disease among the Chamorro people of the island of Guam in the Marianas. Rate of the diseases, Parkinson's disease, Alzheimer's, and likened to Dementia, and Amyotrophic lateral sclerosis, were fifty to one-hundred times higher than in the United States.

** Now an international group of scientists have established that the diseases are caused by a "slow toxin," a toxic chemical that similarly produces disease years after ingestion or contact with the skin.

** Dr. Peter Spencer predicted that the findings will lead to the discovery of a whole new class of environmental chemicals that trigger the death of brain cells. Spencer, who is with the Institute of Neurotoxicology at Albert Einstein College of Medicine in New York, Headed the study group. Other participants were from Britain, West Germany and France.

** Proponents of the slow-toxin theory of brain diseases believe that initial exposure to such substances causes damage to specific regions of the Central Nervous System. The damage remains hidden, producing no obvious symptoms, for decades. During youth the brain is able to compensate for the damage.

** However, as time passes the slow toxins amplifies the effects of the gradual loss of brain cells that occur with advancing age.

** Where the search for slow toxin in the western countries will lead is unknown. In Guam, it led to what once was a staple item of the Chamorro diet, the seed of the false sago plant. The plant produces a seed that contains a starchy endosperm that Guamanians ground into flour for tortillas and other foodstuffs, and is also used as a folk medicine. Much of the plants use decreased with the adoption of a Western style diet by the residents of Guam.

** During food shortages resulting from Japanese occupation of Guam during World War II, many Guamanians lived almost entirely on flour produced from the plant. Researchers have suspected that the plant may contain a neurotoxin, a compound toxic to nerve cells, and could be responsible for Guam's high rate of brain disease.

** Larry believes there is an unspoken connection between the toxic chemicals saturating this country, the Earth, and the number of genetic brain disorders that occur in the general population. Some common thread must lie buried within these people, to account for their ever growing numbers. Frankly, the willful negligence, the reckless irresponsibility without any forethought as to the consequences when using or manufacturing, planning or taking responsibility for the end results of using or manufacturing Ultra Hazardous Materials. The trend of Our Government, large corporations & our military handling toxic pollutants, this seems a very distinct possibility.

** That seems to be the way corporate greed & Industry giants, big and little and the Governments Armed services have done over the years. They have. They've tossed, and toiled the laws to work in their favor. In doing this they have violated, and raped the people, everyone of us right here in the United States.

** "There is a fundamental principle of due care, that is owed to the people by the Government, and large Industry alike. The people have long paid for the mistakes of other's, and are paying the price today. What the Government and large Industry have done over the past sixty years through out the United States, and else where, was no mistake. It was deliberate and intentional pollution to save money. They'd do anything for the almighty dollar, and they don't give a damn at who's cost."

By Larry D. Land

** Just to name a few who didn't care at all. Shell Oil Company Shell Chemical Company, United States Army, and United States Government, the Pentagon, Rocky Mountain Arsenal, Industry, U.S. Army, the Government, and the Pentagon. Rocky Flats, again Industry, Government, and the pentagon. The Hanford Site on the great Columbia river in the Pacific North West, Government, Industry, and the Pentagon, Los Alamos, New Mexico, Government, The United States Army, and the Pentagon. Jack Ass Flats Nevada Nuclear Test Sight, U. S. Army, The Government, and the Pentagon.

** When Due Care attaches, the Law is there, not applied, if it is applied the Court throws it out. Accordingly the Courts can invoke the Federal Tort Claims Act, if Due Care is considered, or by law automatically attaches or they feel threatened in any way.

In August of 1987, two reports came to the house through a man who refused to give him his name. Larry realized how many people had approached him over the years with bits and pieces of information; those who were afraid to divulge their identities. Why should the public have to live in such fear for telling the truth? Why had our own Government instilled such fear in it's people

** The report told of an empty eight-thousand gallon storage tank at Buckley Air National Guard base just outside of Denver, that had caught fire, forcing the evacuation of seventeen-hundred people. According to the report there were also two other eight thousand gallon tanks at Buckley that had contained White Phosphorous, An Ultra-hazardous chemical that ignites when making contact with air. It had been created for the purpose of burning the enemy. Those tanks were to have been sealed and filled with inert gas. They would then be moved to the Rocky Mountain Arsenal. Another twelve supposedly empty tanks were stockpiled at the Arsenal. They were left over from its days as a major weapons plant.

** The Reason Larry said 'supposedly empty', is because no one really knows for sure what's been inside them. They were originally designed to hold White Phosphorus. Like many of the records that the Army was responsible for, they have long since disappeared. Yet, without knowing for sure, the Army felt that they were safe to travel on our highway.

** In September 1985, the Army supposedly emptied the last three tanks that held White Phosphorus. The tanks were then steam cleaned, rinsed and sent to Buckley in November. Buckley planned to use the tanks to store heating oil.

** At least one of the supposedly empty tanks still contained White Phosphorus, and it somehow became exposed to the air.

** Buckley personnel saw a small amount of smoke seeping from the empty tanks for a couple of hours, and done absolutely nothing before they finally realized what was happening.

** Even yet it seems surprisingly difficult to believe that an organization that has repeatedly bragged and stated how efficient and safety conscious its operation is, should be this careless.

** Then in a surprising new development that took place. A press release was issued by Shell Oil Company: Denver, Colorado (February 1, 1988). Shell Oil Company confirmed today that it has joined with the United States Army and four federal agencies in signing a consent decree settling litigation between Shell and the United States Government at the Rocky Mountain Arsenal, (RMA)

* R. G. Dillard, Shell Oil Public Affairs Vice President, represented the Company at a Denver press conference called by the United States Army to announce the signing.

* "Shell has always believed the remedial action at the RMS could be achieved more effectively by working cooperatively and forthrightly with the Army, Federal, and State Agencies, rather than through litigation," Dillard said. "We wish to emphasize that the consent decree does not mandate any particular remedial plan at the Arsenal. Future studies will determine the ultimate scope of the remedial plan, but the decree does provide for some very significant work to commence before the final plan is selected.

* Dillard went on to say, "Shell's policy has always been to conduct its business in a responsible manner, and we plan to play a constructive role at the Arsenal."

****** "Larry said he was glad to hear this after all they spent the last thirty (30) years playing a destructive roll at the Arsenal. The way their business was conducted for the last thirty (30) years has been willful, irresponsible, and negligent reckless conduct at the Rocky Mountain Arsenal, Denver Colorado."

*** In essence what this decree is saying is that Shell Oil Company and the United States Army agree to jointly finance the clean up of Ultra Hazardous waste contamination at the Rocky Mountain Arsenal. The Army estimated the cost of this clean up in the billions of dollars.

*** With a rudimentary knowledge of past Government spending the true cost will probably run into several billion dollars.

*** Of all the current superfund sites in the United States and its territories the Rocky Mountain Arsenal will be the largest and most expensive clean up of any Ultra hazardous waste site in the nation's history. Provided that it is done correctly.

* This cost will be shared by the Army and Shell under a formula contained in the decree. For the first $500 million, the Army and Shell will pay 50 percent; for any amount above $500 million to $700 million, the Army will pay 65 percent and Shell 35 percent; for any amount in excess the $700 million, the Army will pay 80 percent and Shell 20 percent.

*** By Larry Land: Shell Oil Company got away on this one too. The first Billion dollars spent will only cost Shell $380 million then after that it will cost them only $200 million for every billion dollars the Government spends. In essence the taxpayer's will end up paying a far greater amount because the Government bungled this also. The Government went into the negotiation session with Shell Oil Company knowing Shell was responsible for at least 85 percent of the contamination.

*** Shell got by with paying only 38 percent on the first billion, and 19 percent on $2 billion over all cost, if it were $3 billion less than 26 percent on the over all cost for Shell. That is either terrible negotiating on behalf of the United States, or damn good negotiating on behalf of Shell Oil Company. It should have been 1.5 billion to Shell instead of the taxpayers. That would have been 50 percent could have been $2.5 billion. for Shell Oil Company, Damn good negotiating on Shells part.

** Shell says this formula was based on a number of factors, including the relative amounts of contaminants disposed of or released by the Army and Shell, the degree of responsibility for handling of the wastes, and the toxicity of the wastes contributed by each of the parties.

** The consent decree provided that 13 separate projects, called interim response actions, will be carried out prior to development of the final clean up.

** The purpose of these interim measures is to deal promptly with the most pressing contamination problems.

** One of the priority projects will be the removal of liquids and other wastes from Basin F, the largest disposal basin at the Arsenal.

** In addition, the interim response efforts also include plans to build three new ground water treatment systems, including one on land directly north of the Arsenal's northern boundary.

** It might be of interest to note that this new ground water treatment system will be just one mile from Larry's farm.

*** "There are a few who thinks Larry is to critical, of their planning, and operations. such as the Hanford site on the Columbia river, Shell Oil Company, the Rocky Mountain Arsenal, Rocky Flats, Denver Colorado, Jack Ass flats Nevada, near Las Vegas Nevada, The Government, and the Pentagon. Hanford site affects five hundred thousand to one million people in at least a three state area. Shell and the Arsenal will probably be near ten million people, Rocky Flats near one million, Jack Ass Flats probably upwards of three to five million in at least a three state area. If you think your not involved guess again, you could be wrong. It's no longer something we don't have to worry about. This is just a short story, and my opinion from what I have heard and read, and am passing on. To impose stricter rules on the Death Factories from around the World. Some has been short term death, If you got long term death, then you first had to suffer long term "HELL" of about ten to forty years. Maybe some got an unexpected total surprise, a very short term hell and then death.

***** Larry's Mother and Father were on the long term program, they suffered the worst kind of hell for nearly 30 years before their untimely death, caused from low level poisoning with Ultra hazardous nerve agents, sarin, GB, and BZ including benzene, and pesticides from Shell Chemical Company. Larry and Marie's unborn son had about a month of pure hell before his birth, and untimely death, caused by pollution from the Rocky Mountain Arsenal, and Shell Chemical Company, and low level poisoning with GB, Sarin, and BZ nerve agents, and nemagon, deildrin, and Endrin, born short term death. The rest of the Land's are on the long term program

*** This is real, it's time to wake up, and I'm hoping you will consider this your wake up call. Talk to your senator's talk to your Congressperson express your concern both short term, and long term. Write them letters in your Home State, or even in Washington D.C. Tell them to buy the book "DEATH FACTORY U.S.A." it tells the stark truth. Tell a friend, Tell a relative, Tell them all to get Death Factory U.S.A. Their mostly awake now so tell them what you expect for your family, your relatives, and friends. Get all them to write also, and that is what has inspired me to let people know something about what happened, because we trusted our Government, and Industry in working with Ultra Hazardous Materials, and they have kept us in the dark. It's too late now for them to back up leaving no scars, but it's not too late for the people to ask our Government for better, and tighter controls and to get a handle the Armed Forces, Industry, and the Government as well, any one dealing with the Ultra Hazardous Chemicals. It may be a necessity but it's their responsibility. Next statement, "SHAMEFULL" no control even with the Government. Were talking of Deadly Ultra Hazardous Nerve agents, made by private companies, one out of Germany right here in the United States.

*** By the Los Angeles Times;

Washington- The U. S. Government is considering forcing two defiant chemical companies to sell the Pentagon a key ingredient for producing nerve gas, Pentagon officials said yesterday.

*** Occidental Chemical Corp., and Mobay Corp., say Company policies forbid sales that would contribute to the proliferation of chemical weapons.

*** Both have refused to fill Department of Defense orders for Thionyl chloride, a widely used industrial and agricultural chemical that is needed to make a lethal nerve agent.

*** Defense officials said the two companies are the only ones in the United States that commercially produce the chemical agent.

*** The Companies' unwillingness to sell, has brought the production of a new generation of U.S. chemical weapons, which began in 1987, to a halt.

*** The Army needs 160,000 pounds of the ingredient by June to proceed on schedule, the Pentagon said.

*** Government Officials said they can compel the Companies to sell the chemical under the Defense Production Act, a 1950 law designed to give the Pentagon first priority on war material.

*** The standoff between the chemical manufacturers and the U. S. Government is certain to put the Bush administration on the defensive as it works to stem the spiral of Third World chemical arms production and to negotiate a world wide ban.

*** Last December 1989, President Bush proposed to Soviet leader Mikhail Gorbachev that the super powers sign an accord at their summit this June 1990, that would call for the Destruction of 80 percent of their chemical weapons.

*** But the Bush administration has said it will continue production of chemical weapons until an accord sponsored by the United Nations banning the weapons goes into effect.

*** Though the Department of Commerce has the authority to enforce the Defense Production Act, spokesman Robert Kaylor said the matter is under review.

*** The Occidental Chemical Corp. is a subsidiary of Los Angeles based Occidental Petroleum Corp. The Mobay Corp. is subsidiary of the West German chemical giant Bayer AG.

Finally, after more than fifteen years, Larry appeared before Congress in the spring of 1988 to testify about widespread contamination on and around the Rocky Mountain Arsenal, and Shell Chemical Company. It had taken so long to get to this point and Larry was desperate to make it count. The Congress had to understand what was happening out there, and what happened to his Livestock and to his family.

Larry began with the painful memories. Memories of their private hell that his family had been through before they even knew what was causing it. The Cancer, the sores, the bleeding, the excruciating headaches, the dizziness, pain in the eyes unable to focus, the injury to the Central Nervous System, that left their sanity hanging by a thread. The humiliation, and the consequent guilt that they felt as their children suffered through the abrupt spells of nausea, vomiting, hurting eyes, sores in the mouths and nose, The horror of seeing their hair fall out in clumps. The unrelenting pain in their joints, the blackening and deformation of their teeth and gums. The private pain that they suffered when they lost their baby. The anguish of finding his wife sitting alone in the dark, lost in despair, and depression, he tried, but was unable to give her the solace that she so badly needed. She dealt with her wounds as they all ultimately had to, on their own.

Larry told them of the loss of his livelihood. He relived the depression he felt, that set in every time he had to hold his gun to the head of one of his cattle. The desperate fight to keep them alive. A fight that they seldom won. The red eyes matted shut, the vomiting and the convulsions that they fought day in and day out. Convulsions was the final blow, the deadly effect of Sarin, and GB nerve agents was death.

As Larry went into greater detail, with his attorney's and they began asking more questions, his emotions began to take over. He felt foolish, but he could not stop the tears as they fell from his eyes. It was as though the years were closing in on him. All the emotions of the past came over him like a flood. The loss of his parents came back as though it were only yesterday that they had died. At the root of it all was the indifference that the Army and Shell Chemical Company displayed, and had in regard to his family's pain. They didn't care and they made it abundantly clear. Their attitude drove the knife deeper into his soul and ultimately brought them too this point.

Everything Larry had up to that point was turned over to his Attorney's and Congressman Hank Brown, including the Infrared photographs of the Arsenal, and the list giving the number of people who had died from cancer in that area. We would be going before the House Judiciary Subcommittee on the following day.

Larry's attorney said to Congressman Brown, The United States Army is standing before the Congress of the United States of America holding the smoking gun, a gun they cannot discard. The Army will never be able to cover up the lies, the fraud and deceit, willful negligence in handling Ultra Hazardous Chemicals, a failure to notify, and due care attaches, damages, and the contamination & pollution they caused at the Rocky Mountain Arsenal. Now known as the most contaminated piece of land on earth.

While all this was going on in Washington, D. C., the draining and excavating of Basin F had begun, sending noxious fumes over the country side. People getting sick, nauseated, vomiting, shaking, passing out.

Then in the strangest turn of events to take place since Larry had become involved with the Arsenal. It was released to the newspapers by the Army, that this almost pristine acreage, with all its buried bombs, artillery shells, grenades, and hundreds of thousands of buried fifty-five gallon drums of toxic and carcinogenic waste in some fifty huge pits, along with hundreds of acres taken up by what is called basins A, B, C D, E, and basin F 240 million gallon chemical waste lake, several other burn pits where they have burned some of the chemicals, and all the storage pallets, boxes, trash, that were contaminated. would be a beautiful urban wildlife refuge open to the public. He could not believe what he was reading.

This 27 square miles, is considered by many leading authorities to be the most contaminated piece of land in the world. Ground that could never be purged of contamination and made safe enough for humans was going to be opened to the public. By comparison, this makes the Watergate burglary look like a Sunday school picnic. To think President Nixon lost his job over the Water gate and said he never told a lie. The Army has told a lot of lies, never tells the truth, and hides behind the United States Government, and the Pentagon. The army has sealed off many of the huge pits, with clay caps, because they have no records as to what is buried in them, and their not going to find out. Just like all the contaminated soil, the buildings, and any thing else is going to be pushed into large pits and covered up. The sewer lines will be left in place and sealed in with a clay cap.

"The opportunity for an urban wildlife refuge this close to downtown Denver with the skylines and the mountains in the background can't be matched anywhere," said United States Fish and Wildlife Biologist Peter Gober.

A number of conservationists and wildlife enthusiasts viewed the agreement as an indication that the Army would like to see the Arsenal turned into a wildlife preserve after clean-up was completed. The next thirteen pages are the transcripts of Congressman Hank Brown, and Larry's tow Attorney's Sam McClaren, and J. Stephen McGuire testimony before the subcommittee.

STATEMENT OF THE HONORABLE HANK BROWN, MEMBER, UNITED STATES HOUSE OF REPRESENTATIVES

STATE OF MR. BROWN

Mr. BROWN. Mr. Chairman, I have a statement that I would like to submit for the record and I will just briefly summarize it.

Mr. FRANK. Without objection, it is so ordered.

Mr. BROWN. Thank you. Mr. Chairman, what we are talking about is a piece of land called the Rocky Mountain Arsenal. It is in the Fourth Congressional District in Colorado. It has been described as perhaps one of the most polluted sites of land anywhere in the world.

It is designated by the Department of Defense Environmental Restoration Program as a number one clean-up priority that the Department has. Set up on 27 square miles in 1942, it was used to manufacture chemical weapons, including nerve gas. Obviously, that was a secret for much of the period of the production at the Rocky Mountain Arsenal. Obviously, everyone understands the reason for keeping it secret during some of that period.

In addition to the manufacture of nerve gas and other chemicals, it was leased to a private company or a portion of the ground was leased to a private company to manufacture pesticides. So, you have an enormous production and center of toxic chemicals that have been produced there. What happened is that we have had some incidents throughout the time period where you have had some of these chemicals get into the groundwater.

The Land case is simply one where an adjacent landowner had livestock and members of his family poisoned, in effect, by some of those chemicals. What they are asking for is very simple—an opportunity to bring their case before the Court of Claims that would only establish an amount recommend to them that Congress would later have to consider.

So, basically, all they are asking for is an opportunity to make their case. The Army has not covered themselves in glory with this. The reason I say that is frankly this. The incident developed; the Army flatly denied that there was anything that could have contaminated the groundwater. Period. It turns out, several years before, there was evidence of contamination that was filed with the Health Department.

In other words, the Army's own statements made after the incident took place were wide diversity, not only with the truth, but with their own statements. These folks, I think, deserve an opportunity to make their case. There is no question there is contamination on the Arsenal. There is no question that it has gotten into the groundwater.

From going from a position where the Army said there was no reason to worry about the groundwater, we now have an enormous barrier that is built across the fence line at the Arsenal, with a huge concrete abutment with pumps and filters to filter the groundwater. We have had a settlement with the Water District that is adjacent where they have proved there is contamination.

The Army themselves have admitted that they are partially responsible for contamination of the drinking water in the adjacent area. My hope is that the Committee will simply allow these people to make their case. There is no question they have had damage. There is no question the Army has produced toxic chemicals there. There is no question but some of that has gotten into the groundwater.

I hope they will have an opportunity to at least make their case before the Court of Claims so that the Congress can later consider whether they want to award any damages here. One critical point I might mention—the question has arisen why these folks did not bring their action under the Tort Claims Act.

My understanding is the Tort Claims Act has a number of exceptions in that Act. The feeling of the parties involved here is that their case does fit under the exceptions. In other words, the Tort Claims Act exempts them or provides exceptions for their ability to obtain recovery under that Act.

Mr. Chairman, thank you. I would be happy to respond to questions

[The Statement of Hank Brown follows:]

****************STATEMENT FOLLOWS********************

NAME: HJU133020

Mr. FRANK. Did you want to present anyone else? You have done a perfect job. Maybe you want to have someone join you for questions.

Mr. BROWN. I would like, if the Committee is willing to consider the testimony from Mr. Land and his attorneys.

Mr. FRANK. Yes, we will put that in there. Do you want us to hear them? You have to go, but let me just say that I appreciate this. As you know, as a former very distinguished member of this Subcommittee who we miss, this whole area of the Tort Claims Act and the discretionary function exemption, et cetera, is one that needs a lot of work. So you undoubtedly have an area here that we think is one of the most unsatisfactory areas of the law as far as we are concerned and we have had a number of cuts at it and we will be looking at this and we will also look at a more general kind of approach.

It does not mean that your decision ought to have to wait for a general approach to this thing, but I do say this as something we do know. The Subcommittee and the full Committee have dealt with this on several cases. Thank you. I know you have other business.

Mr. BROWN. Frankly, while I am delighted to have left the Committee, I appreciate your hard work and endeavors. I do think I would add my voice to the need for reform here. What we have run into is the Army has simply flatly refused to make an offer to try and settle the thing.

Mr. FRANK. We will, of course, tell Mr. Rodino and Mr. Fish what you said. But we will not put that in the record.

Mr. COBLE. Mr. Chairman?

Mr. FRANK. Yes, Mr. Coble.

Mr. COBLE. Very briefly, Mr. Brown, and I am not trying to be a fly in your ointment, but just for my information, the dumping of these agents or pesticides—could that person whom you represent—could that not have been apparent through diligent search? I would like to hear from you on that.

Mr. BROWN. Well, what has happened here is the Army simply flatly denied that there could have been any causal relationship. The tragedy of it is, they had admitted to the Colorado Health Department a couple of years prior to that that they had a problem and there were incidents prior to that. Since then, it has come out, in a number of instances, that they have contaminated the groundwater water in the surrounding area, so what you have—

I do not think anybody can come before you with a straight face and say they Army did not impact the groundwater around there. They had and frankly, they simply stonewalled it.

Mr. FRANK. Maybe that is why they sent a written statement and did not come here. It is easier for a statement to keep a straight face.

Mr. COBLE. I guess, Hank, what I—and I will be brief about this, Mr. Chairman—Hank, I guess the course from which I am coming or the angle from which I am coming is, should he have known this prior to the purchase of the land? That was my point.

Mr. BROWN. Oh, I am sorry. Yes, he bought the land in 1971. The incident occurred several years thereafter. The statements of the Army were in 1978. Should he have known that there was groundwater pollution or the potential for groundwater pollution?

Mr. COBLE. Yes

Mr. Brown. I do not know. I do know—I mean, I assume the point is that he—

Mr. FRANK. He purchased it prior—he was the owner at the time the incident took place? This was not a subsequent purchase?

Mr. BROWN. Yes.

Mr. COBLE. Okay.

Mr. FRANK. I think that was what Mr. Coble was asking.

Mr. COBLE. And then your answer to that, Hank, is that the Army was not candid and truthful, I presume.

Mr. BROWN. Yes. The problem is, obviously, that polluting the groundwater is a violation of Colorado law and Federal law and whether you had known they were producing something toxic over there, you do have a right to expect that they will not poison your groundwater.

Mr. COBLE. Thank you, Mr. Brown. Thank you, Mr. Chairman.

Mr. FRANK. Thank you. Mr. McGuire, why don't you come forward?

Mr. BROWN. Thank you, Mr. Chairman.

Mr. FRANK. Mr. McGuire? Is that who is representing Mr. Land? Whoever you want to present this. We just ask you not to repeat facts that Mr. Brown just mentioned.

NAME: HJU133020

STATEMENT OF MR. J. STEPHEN McGUIRE, REPRESENTING MR. LARRY LAND, ACCOMPANIED BY MR. SAMUEL J. McCLAREN

STATEMENT OF MR. McGUIRE

Mr. MCGUIRE: I will not do that, Mr. Chairman. Thank you. My name is Steve McGuire, an attorney from Denver, Colorado. With me is Mr. Samuel McClaren, an attorney from Denver, Colorado, whose expertise is in geology and chemistry and then Mr. Larry Land himself.

I do not want to expound on any facts beyond what Mr. Hank Brown has stated. I would like to just briefly get the map up here and I could just show the Chairman and the Subcommittee. This map depicts where the Rocky Mountain Arsenal is in Denver. The dark outlines and the Stapleton Airport are right down at the bottom.

Larry Land's farm is still small rectangle up above there that I am pointing to. For years, any effluence from the Arsenal from the basins and all the chemicals stored here were generally to the west and northwest. The studies made by Dr. Konoco, et cetera, later in the 1970's, are depicted here which show that definitely the underground effluence to go to the north and that the rise in the water table could definitely have brought remnants of nerve gas, et cetera, to Mr. Land's farm in the early 1970's when this occurred.

So, you asked earlier about the direction of the flow of effluence or whether he should have known before hand. Well, Mr. Land, when he purchased the property—he purchased it from his parents who had lived there since the 1930's. There was no problem at all. But the studies show and the evidence shows that this northern boundary has a channel underground with the aquifer to where definitely it could happen and Mr. Land was the first one to really get hit with it.

The Army, several years later, put in, up along this northern boundary of the Arsenal, a deep underground filtration system to stop all groundwater from going north. That tells me a lot right there. We feel that the Lands were poisoned by the remnants of nerve gas destruction and other chemicals that flowed through the underground aquifer to the north, up towards their land in 1972 and the family was really devastated.

They lost over 200 cattle and the family has had severe medical effects ever since this happened. There is no question that the remnants of the hydrolysis destruction of nerve gas do not occur in nature, yet they were found in the well at that time and we are talking about water that is going underground, segments of underground water in an aquifer and it hit the Lands in 1972.

NJU133020

[The statement of J. Stephen McGuire and Larry Land follows:]

***********************STATEMENT FOLLOWS***********************

NJU133020

[The statement of Department of the Army follows:]

***********************STATEMENT FOLLOWS***********************

NJU133020

[The statement of Department of Justice follows:]

***********************STATEMENT FOLLOWS***********************

NJU133020

Mr. FRANK. Mr. Land, do you want to add anything to this?

Mr. LAND. Pardon?

Mr. FRANK. Did you want to add anything?

Mr. MCCLAREN. If I could expand a little bit upon Congressman Coble's question on the thing?

Mr. FRANK. Sure.

Mr. MCCLAREN. As Steve indicated, the Land family had been out there since 1930, prior to the establishment of the Arsenal. So therefore, they really did not know what was going to happen some 14 or 15 years later. I became involved in this in 1974 and as a graduate petroleum engineer, it took me three years of digging to figure out what happened in this case.

This information was brought to the attention of the Army, to Shell Oil Company and to this day, although they have acted upon the information, they have never admitted that this is the situation. In anticipation of your next question, I would also point out that this is a very—on the question possibly, are we impacting other people on t his by setting a precedent—this is a very low density populated area.

In addition to the Lands, there are probably ten other people there in the area that has the same geologic conditions as the Larry Land property does. In 1974, we did an extensive survey to determine if there were anybody else who would be interested in presenting claims against the United States Government and there were none. So, we are dealing here with a situation as to whether or not this Committee feels that a reference should be made to allow the Lands to make their claim.

Mr. FRANK. Let me ask—there is a list of people up here Is this your extended family? Is this all from the same piece of land?

Mr. MCCLAREN. That is Mr. Land's extended family, with the exception of the Phinney's, who were living on the property at the time and—

Mr. FRANK. But it all comes from the same piece of property?

Mr. MCCLAREN. It all comes from the same piece of property. Yes, sir.

Mr. FRANK. I have no questions. We want to be clear because we will put in the record the statements from the Army and Justice Department. What we are being asked by Mr. Brown to do is not to take any position on the merits, but simply refer this so that the Court of Claims could then reach it on the merits?

Mr. MCGUIRE. That is correct, Mr. Chairman.

Mr. FRANK. Mr. Coble?

Mr. COBLE: No questions, Mr. Chairman. Thank you.

Mr. FRANK. Thank you. We will be back to you.

******"News just in on chemical weapons, The United States* Army, the Pentagon, fully intends mass production of new and more sinister, and efficient Ultra Hazardous Nerve Agents, and we have a Government that condones, (Corruption in politics), their actions. In total violation of the treaty on proliferation of Chemical weapons. It appears we have an Army, and the Pentagon who has secreted it's deadly past and now will secret it's deadly and sinister future. How can we continue to to allow them to call what they do an accident, when it goes so monstrously wrong. The citizens pay with their lives by the Millions, and pay with taxpayer money by the $ Billions to clean up that which was no mistake. I'm sure much like Shell Oil Company, their would be no Insurance Company to condone or that would Insure anything the United States is doing." Information and thoughts of Larry.

Then again in 1988, when the City and County of Denver was trying to sell the region on building a new airport, instead of expanding the existing Stapleton International Airport onto the Arsenal land, they were quick to point out that the Arsenal was much too polluted to build an Airport on. They had a great plan just to cover it with reenforced concrete.

** This is 18,000 acres that was once worth $40,000.00 per acre to the City of Denver, and or Adams County who wanted the Airport to expand onto the Arsenal, which sits in Adams County. They were trying to strike a deal with Denver, and had many Plans for new highways, and Hotel's and Warehouses pushing toward Brighton which is the County Seat. Even though Adams County still got the site for the new Airport by Bennet, Colorado, I'm not sure of the relationship between Denver County, and Adams County regarding the new Stapleton International Airport.

** The Arsenal property that was once worth upwards to one billion dollars, is now worth nothing, and the only hope now is maybe a wildlife reserve. remember that's for animals that live on the surface only will have a chance. Anything living under ground doesn't have a prayer. Even those on the surface and in the air have been found contaminated, or found dead.

Animal concentration camp 6 and 7 with chain sence to keep animals in.

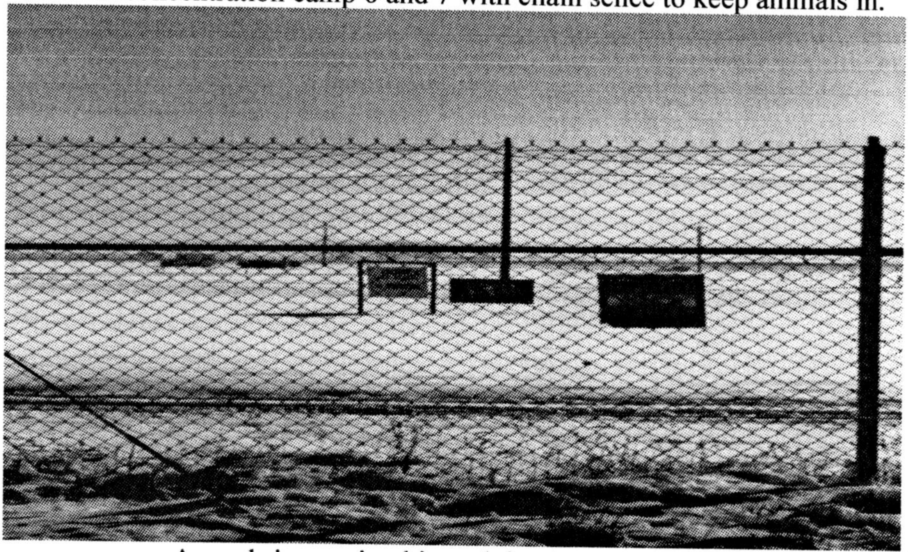

Army brings animal in and they cannot get out.

* Larry has wondered, and tried to figure out why the eight to ten foot fences, of the chain link type were put up around the entire Arsenal in to 1980's. Then he started to remember how much the wildlife preserve was raved about by Government officials. Now he thinks he knows, It was the only way the Government, and the wildlife Department could keep the animals from running off.

* The beaver's the rabbits, the Coyotes, the kangaroo rats, and the prairie dogs, they were just exterminated by the Army's own elite exterminating team.

* The fence was good enough for over forty (40) years. Larry as a child in the 40's growing up around the Arsenal, often played on the Arsenal, catching many animals and birds' taking them home and then turning them loose, didn't remember seeing many deer, and if he did they weren't sticking around. He remembers seeing, huge numbers of birds, ducks, geese, hawks, and numerous small birds. He also remembers' seeing many Jack rabbits, Cotton tails, rodent life, such as the kangaroo rat, Prairie dogs, Coyotes, and beaver's.

* By the time the 60's rolled around, rabbits, Coyotes, birds, beaver's kangaroo rats, deer, hawks, were nearly non existent. The Prairie dogs, if the contaminated ground didn't kill them, the Army exterminator's did, because they were the food of the Bald Eagles the Government was using as a prop, Eagles, and deer were believed brought in to replace the dead and really deter the thought of what they had done to make it look as a pristine wildlife refuge just outside Denver Colorado. what we really have here is a wildlife concentration camp, and the most contaminated piece of land on earth. The Army here has lost all public trust, where the arsenal is concerned. There have been far too many misstated facts (Lies), in their attempts to calm public concerns, about the Rocky Mountain Arsenal and Shell Oil Company. Public relations seems to be a greater priority than public safety, for the Army.

* At one time Larry remembers speaking with a person who's job was to each day ride around the Arsenal picking up dead animals, dead birds, if any were looking sickly they had to be killed and taken out so they would not spoil the beauty of what was really a deadly trap.

Hundreds of walls all a part of the Water Boundary Treatment System
The Deadly Trap

* He was also told of how huge flocks of geese or ducks would circle in on any one of the basins, and especially basin "F". Almost the minute they hit the water, the birds would shoot almost straight up, and fall down dead. It was that immediate. He was also told of another time when a couple of rabbits came up to basin "F" before the fence was put around it. They seemed to sniff at the water, and drank immediately turning around, running stumbling and falling, then fell over and the legs would be jerking in the air, as the convulsions turned to death.

*** This is when the basin F was fenced off and they installed what Larry understood are large machines that go off like cannons', on a timer so this continues every few minutes, scaring any animal or bird that may be nearing basin F.

*** The Army and Shell Oil have continually refused to provide documentation to citizen advisory board members, to support their claims the arsenal is safe for visitors. These concerns were well founded, but this public input has been discarded by the Army and Shell Oil Company. We will wait for the reports as to the full extent of the hazards.

** The Army has lost all credibility, and public trust in their all out effort to make the public believe they are cleaning up what will be a Federal wildlife Reserve. Yet the EPA. takes 15 years to ban pesticides that endanger the drinking water, and it took them 40 years to do tests on Sarin, and GB nerve agents at the arsenal. It had been secreted, and classified by the Army, until 1975. There was no way to get medical treatment, there were no tests for it, The Colorado state Department of Health was kept totally in the dark. They didn't know what it was even when they seen it first hand as it was happening. The real affects of low level poisoning of GB, Sarin, and BZ nerve agents, along with lethal levels of deildrin, as was finally found by the State Health Department, but they didn't even know the effects of the levels they found, and ignored the true health maladies, signs and symptoms, they had seen first hand.

*** Shortly after the public was convinced of the idea of a new airport several miles to the east, the public relations machine of the Army swung into a stance of damage control and began selling the public on turning the Arsenal into the Rocky Mountain Wildlife Refuge. They began publishing slick brochures adorned with scenes of eagles soaring, and deer frolicking on the land. No doubt they hoped that the public would be lulled into a sense of security as to the safety of this place. Many times Larry wondered how the same piece of land could be far too polluted to pour concrete runways, and to just cap the rest of the area with cement. yet they try to convince the public it is safe enough for the public to roam across with their picnic baskets, and binoculars, on many lazy, kick back, summer afternoons.

***** You can quickly see above how the citizens advisory board was treated by the Army, and Shell Oil Company. The advisory board wanted the Arsenal, and Shell Oil to support their claims, showing the Arsenal is safe. These well founded concerns were discarded by the Army and Shell Oil Company.

Why is it that he could not find out, from anyone, what was buried out there. He was able to point out 42 huge dump sites, "HOT SPOTS" requiring more information and showing very clear on infrared photo's, it was either classified, or Internal Information only, really what they meant to say is there are no records of what was buried out their. The fact is that the Army plans a cosmetic clean-up of the soil and water, capping all the buried pits with a clay cap, and a covering of top soil, for prairie grasses or what ever. and to leave everything else 'as is'.

***** I hope that somebody from our Government, meaning Senator Wayne Allard, Congress woman Diana DeGette, Congressman Bob Schaffer, Congressman Scott McGuinnes, Senator BEN NIGHTHORSE CAMPBELL, Governor Bill Owens, pays attention to what the Army is planning on leaving for Colorado, and the future. All the offices have been supplied with information that spells out very clearly the Army, has lost all public trust, and they have no credibility left in their bag of tricks, all that is left are Deceit, Fraud, Misuse of Due care, Negligence, Wrong doing, Willful and Reckless misuse of Ultra Hazardous Materials. Trespassing, nuisance. The Government and the Courts have lied to Congress over and over.

*** Larry envisions a Fish and Wildlife Service Biologist, squinting through a spotting scope at some immature bald eagles squatting on ice covered Lower Derby Lake, ignoring the fact that it was once used as a cooling pond in the manufacture of pesticides and Herbicides.

*** Supposedly the Arsenal was to become a sort of "nursery" for juvenile eagles, a place where they could roost close together.

** With five thousand acres of prairie dogs on the property, it would seem that there was an unlimited food source for the Hawks. However the prairie dogs had lived too long in this area and were themselves badly contaminated by the Toxic materials which they had ingested. As a result, they had to be exterminated, and a new supply was imported from the Buckley Air National Guard base which had more than an abundant supply of the animals. In moving them they were replaced in area's on the Arsenal that were hopefully less contaminated. This was another way to save the hawks, and still have an abundant supply of food.

* Gober and his partner, Mike Lockhart, are in the second year of a three-year study on the endangered bald eagles wintering on the heavily polluted Arsenal.

* Their $350,000.00 project is funded by the Army, Department of Highways, and the new airport authorities, who want to know how disruptive clean up activities at the Arsenal and construction at the airport will be on the eagles.

* The concern is that there is an abundance of nesting habitat in Canada for bald eagles, but the birds lack wintering grounds in the lower states. Wintering on the Arsenal have contaminated the bald

eagle population causing them to die, and led to birth defects, eggs with a thin covering no shell, cannot be used for reproduction of the birds. The True reason the bald eagle went almost extinct was the deadly dilemma they faced with chemicals they much like those who lived around the arsenal Ate, Drank, Bathed, and breathed pesticides, Herbicides, yes probably the various types of nerve agents that were produced and dismantled here at the Rocky Mountain Arsenal by the United States Army, and Shell Chemical Company, and others alike.

** "You can crank out a lot of youngsters, but if they don't survive, it could become a problem for the species," Gober said.

** The big attraction for eagle's at the Arsenal are the prairie dogs, but rabbits had also made up part of their diet. One can not help but wonder what the chances are that the eagles are ingesting contaminated animals even with the Army's attempt to refurbish the supply with an untainted population. In turn what will happen as the food chain works its way through its various stages. Unfortunately, it would appear the eagles are simply serving their purpose as an integral part of the Arsenal's public relations program. By inviting the public to ride their quaint double decker English bus through the Arsenal land to a sight where they can "view the Eagles", the Arsenal continues on its quest to reassure all of us that everything is OK. That there is no need to worry about anything., "or is there? Once again the Army and Shell have refused the citizens advisory board any information to support their claims the Arsenal is safe for visitors. All public input has been discarded by Shell Oil Company, and the Army.

*** On December 20, 1988, San Bruno, California, a jury ruled that Shell Oil Company was responsible for cleaning up three decades worth of pesticide pollution at the Rocky Mountain Arsenal. A project that insurance Companies estimate could cost up to two-billion dollars.

*** The state court verdict was a major victory for the two hundred and fifty insurance companies, which argued that Shell knew of the pollution and therefore was not covered by its eight hundred different policies any more than a homeowner who set fire to his own home. What they done was no accident.

Shell officials said they were 'extremely disappointed' and that the company would appeal the decision.

*** In the meantime, in Colorado, residents of communities which are with in close proximity to the arsenal, began to hear rumors that the Army was discharging DIMP, a Nerve Gas by product still containing about 3 percent GB, and Sarin, also IMPA, a by product of BZ nerve agent into First Creek. The Creek in turn, flows into Barr Lake which recharges nearby private domestic wells as well as those of neighboring Brighton, a city of 15,000 residents and again the farmer's in that area irrigating with the water.

**** Later it was disclosed that a plume of at least five chemicals had been detected flowing into the South Platte River.

*** The contamination, which reaches the river in a rural area near Brighton Road and east 120th avenue includes: nerve gas, nerve gas By-products, Industrial Solvents, and two pesticides banned in the United States because of their link to cancer.

** This announcement marked the first time that Federal environmental officials publicly admitted that toxic wastes dumped by thee Army and Shell Oil Company had reached the South Platte River. However, the residents of that area had known about the contamination for a long time.

** The South Platte River supplies water for drinking and irrigation to thousands or rural high plains residents, all the way across Nebraska, and into Iowa.

*** Although the plume contained many pollutants, the EPA said they were most concerned about five; DIMP, a Nerve Gas By-product that was dumped by the Army; PCE, an industrial solvent; Chloroform; Deildrin a Shall product banned in 1974 because of the links to cancer; and DBPC, another Shell pesticide banned for most uses in 1979 because it caused sterility.

** Does the intrim plans at the Arsenal, for Shell Oil Company include the purchasing of all homes, and properties on or near 96th avenue. They are very rapidly purchasing the homes, moving the people out and bulldozing everything into a large hole, almost as if no one was ever there. This all started after finding chicken eggs on one nearby farm were found to be polluted with unsafe levels of a banned pesticide once made by Shell. The U.S. Environmental Protection Agency advised one neighboring farmer to stop eating his eggs contaminated with deildrin, a Shell manufactured pesticide that was outlawed in 1974 because of links to cancer.

** It is believed that Shell was buying the homes to preclude the property owners from filing costly environmental lawsuits against Shell. The Oil giant polluted the Arsenal northeast of Denver and surrounding neighborhoods with tons of toxic waste between 1952 and 1982, when it made pesticides at the federal facility. But one landowner said he suspected Shell is trying to ward off any new lawsuits. They paid upwards of $21,000.00 per acre.

*** January 7, 1989, by "Janet Day Rocky Mountain News" Nationally known Environmental attorneys yesterday took up the cause of 15 families living in a mobile-home park near the Rocky Mountain Arsenal. These attorneys participated in cases involving the evacuation of residents near the Love Canal, N.Y., and Times Beach, MO., toxic waste sites and the Three Mile Island nuclear power plant. They asked a federal judge to order the Irondale Trailer Park evacuated saying residents' health is endangered by pollution from arsenal cleanup efforts.

*** Governor Roy Romer has refused to evacuate the residents. Short term medical problems at the time do exist, the long term effects are not known. Residents have been complaining for months that excavation of Basin F, one of the Arsenal foulest waste sites, sent chemical fumes into their homes. Besides choking, from the odors residents claim that the chemicals have made them dizzy, sick, irritable, and itchy.

*** This policy of trying out cleanup methods to see if they cause health problems, rather than confirming in advance that they are safe, amounts to little more than trial and error. "Of course we all know that is exactly how the Army, the Government, and large Industry like Shell, are not by accident". "It's called by the seat of their pants". If they get caught, they hide behind the cloak of the Courts, Lie, use fraud and deceit, put people down, try very hard to discredit, or destroy their credibility.

***** Although the EPA acknowledged the pollution for the first time, it said that by the time this poison reaches the river it is so diluted that is poses no immediate health risks.

***** What does 'no immediate health risk' mean? Does it mean that in the future there is a health risk, but right now it's all right? Are we to be comfortable with the presence of DIMP and the other chemicals in our children's Kool-Aide?

** We need to think, we have to keep in mind that we are listening to a government agency. Just imagine, how many gallons of contaminants have past into the waters of this river before it posed no immediate health problems. One very credible estimate puts the amount at seventy five thousand gallons a day.

** In their endless wisdom the United States Army and Shell Oil Company decided to construct an underground barrier, containment system along the Arsenals north and northwest boundary lines designed to cleanse and treat the ground water flowing outside the facility.

** The problem was the barriers leaked and more contaminants left the Arsenal by simply passing around the edges of the barrier system. Now somewhere along this Countries great history, someone forgot to tell the Army that two Plus two still has to equal four when you build something.

** It was also at this time that federal officials, while draining toxic waste from the ninety-three acre pond called Basin F, came to what they thought was the bottom. Until one of the vehicles began to sink through this bottom. When they started to dig they discovered and additional ten millions gallons of crystalline compounds was a mixture of chemical weapons, and pesticides, that had escaped below the bottom of the basin.

*** Basin "F" has been considered for years to be the most dangerously polluted site at the Arsenal. Officials for years thought the basin held somewhere in the area of a few million gallons. They were stunned when they learned it was much higher than they was led to believe. They were almost overwhelmed when they learned it held over 240 million gallons of chemical waste.

* Why did the Army have no clear idea of how many gallons the basin held when they are the ones that built it?. Surely even in the United states Army, they have heard of mathematics: a process that is used for determining such volume. Also, which report is accurate? One report stated that Basin F held an estimated two-hundred forty million gallons. Another said only eleven million gallons could be contained. This leave us with a huge discrepancy. Plus the ten million gallons they found under the basin.

* Is it possible, that all during its thirty + years of existence, Basin F was losing that many gallons and the Army took one of their now famous guesses? You have to admit it does make a person think.

* Draining and dredging the basin was suppose to be a temporary step to stop chemicals from leaking into the soil then into the groundwater.

** When the additional waste was discovered, Officials decided to pump it into a nearby unlined pond for temporary storage. This pond was designed to hold rain and snow runoff that would be used to decontaminate equipment.

** Some State officials were concerned that this temporary use of the pond would end up being a long term solution, and that its use may violate a Federal Law on disposal of hazardous liquid waste. Much like Basin "F" was designed as a temporary basin with a life span of 10 years. The Army violating every law on the books for "DUE CARE" working with Ultra Hazardous materials, they used it for 30 years knowing full well it was leaking, and polluting the underground water off the Arsenal.

** During the draining and dredging of basin F, huge bluish and sometimes yellowish clouds of toxic, emitting a strong offensive, nauseating and sickening odor that rose into the air over Adams County neighborhood.

*** Residents west of the facility closed their windows and complained to health officials who in turn talked to the Army. In the trailer park a resident said, "During the cleanup of Basin F, we had real trouble breathing. It seems to effect the lungs, your eyes are constantly irritated, shaking, gettiness, headaches, so bad, so intense, they bordered on madness." "The clean up of Basin F, literally gassed the residents of Irondale. They repeatedly complained to the Governor, and Health Officials Congressman and Senator's who seemed to do nothing. Reporting over and over again of nausea, headaches, vomiting, dizziness, gettiness uncontrolled shaking, loss of memory, numbness, tingling feelings in the extremities, and irregular menstrual cycles. entire body in pain.

III Officially, the Army was already calling the basin F clean up a 'success story.' And was still insisting that all it caused was a bad smell. Yet, in the process, gases were being released from the soil that haven't been exposed to the air for more that thirty years. They weren't really sure what it was, probably unknown by-products from the chemicals that had been mixed, used and put into basin F. such as GB Nerve agent, BZ Nerve agent, Sarin gas, Nerve agent, DIMP, DCPD, IMPA, Mustard agents, Lewisite, Deildrin, Endrin, Aldrin 2,4-D, Agent Orange, and then the heavy metals mixed with the 2100 chemicals, that made the witches brew. Mercury, Silver, Lead, Sodium, Sodium Floride, Copper, Strontium, Boron, Magnesium, Calcium, Chloride, Titanium, Scandium, Chlorine, Sulfur, Cesium, Barium, Iodine, Lanthanum, Cerium, Praseodymium, Indium, Niobium, Molybdenum, Zirconium, Cobalt, Manganese, Chromium, Nickel, Gallium, Arsenic, Selenium, Rubidium, Yttrium, Bromine, Lithium, Boron, Magnesium, Aluminum. These are just some the the chemicals, that are known to be in Basin F. along with Ion Exchange Resins, we have reason to believe there were some radio active materials also. These materials were found using the same equipment that was available to the State Health, The Army, and Shell.

***** The methods and technology's used, and the uncaring attitude of the Army was well planned, knowing there were gonna be a lot of sick people, the plan they used was much a plan of force and see what happens, "It was no accident".

Larry said I imagine me naming most of the heavy metals used by them and disposed of in F will be shocking proof of how the people have been lied to, and how many of the items listed and tested for by the air monitor's. I have the actual test data naming every item and more. That they have been very careful not to mention. Well now it's mentioned. Some being major components, much more than were even hearing about.

*** However, the Irondale residents were frightened and the public outcry was growing, The media was on sight daily to interview those who lived in the community. Finally even Colorado realized that something had to be done. Governor Romer made a visit to talk to the people. He reassured them that steps were being taken to assure their safety. The bottom line came when an Irondale resident invited the Governor Mr. Roy Romer to move his family into her home. Her feeling was that if it was safe for her family, it must certainly be safe for his. The Governor declined "OF COURSE".

*** Again the Army's right in every detail; however, honest, hard working, loyal Americans were wrong. How can the American people hold their heads high and defend their government when they were being deceived constantly and their neighbors are dropping like flies all around them?

** The Army would argue, but 1988 was not a very good year for certain residents of Colorado. Enacting Governor Romer's promises to serve and protect the health of residents near the polluted Arsenal has been difficult, Frustrating and, at times down right embarrassing. By January of 1989 things were not much better.

*** Romer, in late December, promised about eighty residents of the Irondale Trailer Park, on 88th Avenue; that round-the-clock monitoring of air pollution would occur; free medical tests would identify pollution related health problems; and contamination would stop.

**** They monitored the air, they just didn't know exactly what to look for. Months later, none of these promises had been kept. Leaky or faulty equipment, equipment that was not properly calibrated, or not calibrated at all. Internal disputes, lawsuits, and insufficient staff, untrained techs, or lack of money, had complicated efforts to determine how much of a danger existed and what to do about it.

**** "There's a credibility crisis. its been a real nightmare," said Ellen Mangione, who led the Colorado Department of Health's effort to evaluate the residents health.

** The Department's official position, based on soil, water and air samples, is that no "imminent health risk" existed for mobile home residents. 1 mile west of the Arsenal's Ultra hazardous toxic waste site Basin "F".

***** Larry says the biggest problem with the Army, Government, and Industry. There isn't such a thing as better safe than sorry. They should first learn to make it safe, and safely, and then to safely take these chemicals apart there for having a safety valve for the public. They need to stop dumping these deadly chemicals on their people. It's as if the Army, Government, and Industry giants are running blind waging a chemical war on its own people.

*** Yet several Health Department employees, who asked not to be identified, said such an assessment was premature and the there could well be a serious long term risk to those people. Comparing it to the Hanford plant, 30 to 40 years later those people are dying terrible deaths, and nobody seems to care, especially the Government and Industry they admit what was done there was no accident. Years from now we'll probably hear that same story over and over again. Larry thinks what was done at Irondale was no accident. They had no idea of short term affects until it happened right in front of them, and they sure as hell don't know what the long term affects will be.

*** The Health Department was to have been responsible for overseeing the health of the general populace. Instead, It seemed as though they were sitting in judgment over the very people that they were supposed to protect. One person took it upon herself, Ellen Mangione, Colorado Department of Health sounds as if she took it upon herself to ask the question who is credible. Insinuations made was these people were all ok and had nothing else to do but complain. Acting as if they had no health problems She didn't know what affects the materials would even have on people yet she is making a decision that the people are not credible. Those who didn't live out there, were not credible, and should not have been put in charge of something they didn't know.

***** There is no way in hell for Ellen Mangione or any one else in the State Health Department to make a statement their decision; "Long-term health risk is not high."

***** The Colorado Department of Health had a chief Chemist by the name of William Dunn, who kept the people in the dark as well as the Health Department was concerned. There were tests available to him to know whether Water was contaminated by the arsenal, It was him that allowed those tests to be discontinued, when they were sure chemicals were polluting the environment, the water. He made sure they would stay discontinued through most of the sixties, and into the seventies, and stated "There is no way in Gods green earth that chemicals could be leaking from the Arsenal, and into your wells". He had the power in his hand to do or not to do and he chose not to do in favor of the Army, and Shell Oil Company. It covered up and concealed years of vicious polluting, not with just toxic waste, but Ultra toxic waste. This man was responsible for telling everyone that it was ok.

*** I'm disgusted, and I'm scared," said Irondale resident Jeannette Buschman. "I'm ready to move into my camper and live at a

KOA campground somewhere." The people were sick and it was the Colorado Department of Health's responsibility to find out why.

*** The claims and reoccurring technical problems, worried the residents of the Irondale Trailer Park. When the draining and dredging of Basin F started, clouds drifted over the trailer park. Resident complaints and demands for evacuation led to the December meeting with Romer and his subsequent promises.

** That week, excavation stopped and what remained in Basin F was covered with ("Larry says really covered up"). a three foot layer of clay and soil, but residents say the odors and illnesses continue.

** Looking at this last statement gives one an uneasy feeling. You would have to assume that by saying 'what remained in Basin F', that this basin was not completely cleaned up, and the Army, with full knowledge of the EPA and Health Department, covered this basin with clay, in hopes that the people would quiet down. Is this just another part of the gigantic cover up created by the Army and their followers? Or did in fact the Health Department make medical discoveries they were unable to talk about, and was possibly kept hush-hush.

***** Soon after this incident occurred Army Officials found a leak in a tank holding 1.3 million gallons of Ultra Toxic liquid. Then health officials discovered a manufacturer had miscalibrated air monitoring equipment which was leading to erroneous readings, and readings not sent in. Therefore erroneous information given to everybody.

***** Again the American public was right and the Army, EPA, and the Colorado Health Department were wrong. The residents of the trailer park suffered long and hard while their so-called credibility was again being established.

*** By 1989 it was decided that burning, or incinerating the toxic waste around the –clock for a year and a half would be the cheapest and safest way to get rid of the 8.5 million gallons of dangerous chemicals stored at the Arsenal.

*** A study of disposal plans, ranked 'incineration' at the top of the list of possibilities based on safety, environmental soundness, technical feasibility and cost.

** The corrosive salt, and metal-laden chemicals, are being stored in three tanks, and a smaller, newly built pond just north of Stapleton

Airport. Come to think of it, who heard of the new pond, was it lined, was it sealed, "OH WELL" if you can't trust your own Government, who can you trust. I wonder if the State Health Department was aware of this. As you noticed Larry had given the names of many of the heavy metal products, that will have to be monitored very closely. Heavy metal poisoning will show signs and symptoms that may differ from chemicals. Incinerating the waste at the Arsenal will leave heavy metal emissions, needs to be monitored very close for each one.

*** Arsenal representatives say emissions from incineration would be minimal because of the air pollution equipment and air monitoring equipment available. The main pollutants expected to be released are Carbon Monoxide and nitrogen oxides, or so they say. During a test incineration, conducted at a Pennsylvania company, using one-thousand gallons of waste brought from the Arsenal, traces of suspected cancer causing chemicals were found in the emissions. What other emission equipment do you suppose they had in mind to cover that aspect?.

**** For years the Army and Shell Oil Company said that the amount of toxic and carcinogenic chemicals that were contaminating the water, and the air were at minimal levels. Even as the citizens of Adam's County were dropping dead with cancer, they claimed no fault on their part. per capita, around the Arsenal has the highest rate of cancer deaths anywhere.

**** They are not crying wolf, or saying the sky is falling, or the end of the world is at hand. What they are saying is that they have to pay the price, and in many cases the ultimate price, for not watching, for assuming that large corporations and the military have our best interests at heart.

** In the final moments, as much of us stands before he who judges, what excuse can be given for what we have left our children?

*** Diane Gessler has spent all her adult life looking at, and worrying about the Rocky Mountain Arsenal. Her Commerce City home is located across from the new water treatment plant built by the United States Army, that is suppose to eliminate one more threat to public health. Gessler doesn't want to have to worry about another potential threat to her family when the Army begins to burn the 8.5 million gallons of toxic waste taken from basin F. The same waste that contaminated her drinking water.

***** Diane's feelings reflect those of thousands of Americans throughout the country who have had enough. "We've lost a lot of faith in our bureaucracy," she said at a recent public hearing on the incineration plan. "I don't want my children to die long before their time from the Ultra Hazardous chemicals, they already know leave a trail of carcinogens for everyone. They have lied to all of us, and until they get caught up in the lies." They denied injecting the chemicals under basin F into the underground aquifers for nearly 30 years, they were aware of what they were doing. It was not by accident as they would like you to believe. They said they weren't putting it in the air also but what they done was no accident, and now the Army wants to put more into the air without even addressing the problems they know exist when tested back east, and were notified the carcinogens it was emitting into the air. Now the Army expects us to believe them just because they say it will be all right, "I don't think so!" What about all the heavy metals, have they even give it a thought.

***** Army Officials assured the residents that the emissions from the incinerator would be carefully monitored and that the pollution would be no worse than that coming from fifty cars.

***** Environmentalist's called the Army's statement misleading, because it does not address the Ultra Hazardous Chemicals, and Toxic metal pollution that may escape from the incinerator, nor did it discuss the known carcinogens.

***** Larry says "It sounds as if they want the people to accept their long term death plan, at first it will only make you sick, and may be a nauseous odor, difficulty breathing, "OH" and you will have to watch out for long term birth defects, The mutations you will see. The problems with learning disabilities, and Things that you will no longer be able to do, You could become multi chemical sensitive to everything, including carbon monoxide. But don't worry you'll probably be here today and gone tomorrow well before your time."

*** While neither the workers nor the public was threatened by the leak, Environmental officials remain concerned about the possibility of a larger, more dangerous leak in the future.

*** Representatives of the EPA and the Colorado Department of Health inspected the leak area. Neither agency was previously aware of the canister-testing program. The canisters are in the southeast quadrant of the plant which is about a mile west of its eastern border, and one and a half miles from the closest home.

**** "I'm not really concerned about this accident, but what may happen in the future. I'm concerned about how they will deal with an incident if it happens again in greater intensity," said Jeff Edson with the Colorado Department of Health.

*** Lewisite is a clear, nearly odorless gas that was mixed in Mustard gas for use as a chemical weapon during World War II. The systemic poison can be fatal within ten minutes if inhaled in high concentrations. Exposure at lower levels produce an immediate stinging sensation and reddening of the skin within thirty minutes. Severe blistering appears about thirteen hours later.

***** The experience the Kurd's in northern Iraq, and Iran were virtually sprayed by plane with this type material during the 80's, sprayed over entire towns and cities, Leaving them dying, killed and maimed, and left laying dead in the streets. They were sprayed like spraying insecticides on crops for insects. Larry say's "The word is out that this was a purchase from the United States of America by Iraq to be used on the Iranians, during the Iraq, Iran war. Mustard agents, Lewisite, Dimp, Impa, and GB or Sarin nerve agents. Larry said this was probably a part of the Weapons for Iraq Controversy during the 80's."

***** The other one called BZ nerve agent, "the gas that makes you happy and laugh as your dying." Larry says "I heard about it when and where it was produced and how hard they have tried to cover it up. They have lied to the Congress of the United States of America regarding BZ nerve agents." "He Also believes it was involved in the weapons deal with Iraq in the early 1980's."

***** "Larry spotted in the news right after the Gulf war stored in the chemical weapons storage depot in Iraq, the very same thousand, and two thousand pound containers that held the mustard, Lewisite, GB/Sarin, BZ agents, the DIMP, and IMPA when they were stored at the Rocky Mountain Arsenal." Larry's first thought was "it is Almost as if those canisters just changed hands from the United States to Iraq. Some biological and bacterial warfare agents, and lots of other technology, it is understood also came from the United States of America, France and Germany. It is believed it was to have been used on the Iranians. I don't believe it was expected that Iraq would then turn on Ku-wait, and Saudi-Arabia, and the Kurds in northern Iraq. Their towns and homes Literally sprayed like insects in a field, people left laying dead in the streets clutching their babies.

***** Then the Americans in the Gulf war incident are believed to have been poisoned by low levels of the very same agents. Has it ever been thought when the wells were set on fire, at least some may have been set up to release nerve agents or biological agents at the same time catching everyone unsuspectingly as they were caught in the smoke. It may have been thinned too much carried into the air and dissipated somewhat, so everyone was getting low levels of the toxic materials being released. What ever it may have been Iraq had probably everything known in their arsenal, and still does.

*** The testing, was conducted by the Army as part of a program to identify canisters that still contained lewisite. Contractors from the Tennessee Valley Authority opened the containers and inserted probes to test the air inside.

*** About one-hundred seventy-five containers have been tested so far, with sixteen registering positive readings for Lewisite.

*** Every activity concerning hazardous waste at the Rocky Mountain Arsenal must be cleared by state and federal agencies to avoid potentially dangerous leaks. That is exactly what they said during the 60's, 70's and early 80's. How do we trust them now".

**** At the time of the testing neither the EPA, nor the Colorado Department of Health were aware of the canister testing, even though all hazardous waste programs at the Arsenal are governed by the United States Superfund Law.

***** Larry said the law is very clear "Every activity concerning hazardous waste at the Rocky Mountain Arsenal must be cleared by State and Federal agencies to avoid potentially dangerous leaks." Yet the Army continues their belligerent attitude, and avoid doing what the law tells them to do. The Army at least needs a different command inside the Pentagon. So they cannot deceive Congress, and conceal to the public what they've done by stamping it classified material, exactly what they did in 1969 when they knew full well the basin "F" was leaking and was unsafe. The Pentagon was totally aware of what was going on at the Arsenal for 30 years.

** I figured out one of the Army's problems, If you say NO, they do it their own way.

*** The Superfund Law was designed to clean up the most dangerous toxic waste sites in the United States. They Rocky Mountain Arsenal is number one on the list, and requires notification of activities, risk assessments and hearings.

*** Evaluating whether there is waste or not in the containers is something that needs to be done within the process that applies to every Superfund site. This process requires EPA approval, public involvement, identification of standards, assessment of the risk, or in other words, the equivalent of getting a permit, to make sure it is done right.

**** Two weeks after this incident happened, Army contractors at the Arsenal resumed testing of the large containers, even though EPA and Colorado Department of Health approval had not been given, and said "NO" not until it is approved. As Larry said you say no and they do it anyway.

*** The purpose of the Arsenal was the manufacturing of chemical weapons. With the closing of the Arsenal that eliminated one of the Army's major sources, but the Army wasn't worried because it had other sources all over the country. Yet, According to the politicians, and the media, the federal government was dedicated to eliminating all chemical weapons throughout the free World; especially in the United States.

**** Unfortunately there was two men who wanted, and was going to change all that in the 80's and 90's. No two Presidents had spoken more forcefully for eliminating all chemical weapons than Ronald Reagan, and George Bush Sr. "If I'm remembered for anything," said candidate Bush in 1988, "It would be for this: a complete and total ban on all chemical weapons."

*** When George Bush was president his administration had gotten itself into a dream-world position on chemical weapons. President Bush made far-reaching proposals to ban chemical weapons. The United States pressured West Germany to stop its firms from aiding the chemical warfare efforts of terrorist nations such as Libya, Iraq, and Iran.

*** The Bush administration quietly considered taking legal action to force one German firm, and a U. S. firm to supply the government with the chemicals necessary for the Army's manufacturing of chemical artillery shells. These firms, the only two in the country that make or knew the ingredients, had refused to sell Thionyl Chloride to the Army because they wanted no part in the production of chemical weapons. It was totally against the Free Worlds Treaty on the proliferation of chemical weapons, and the United States had signed and were a part of that treaty.

*** The German firm had actually prohibited its subsidiaries from dealing in chemical weapons, a policy it initiated after the United States pushed the West German's to curb any chemical exports to the third world.

*** The Bush officials were in a bizarre position. If the two companies couldn't be coerced into selling its poison gas ingredient directly to the United States the officials would ask the West German firms to export to us the very chemical we have been pressuring them to stop exporting to others.

This had become terribly embarrassing for the president.

*** Administration officials conceded that they had put themselves in a most awkward situation on the one weapons program where Bush had been determined, to be seen by all the world, as the "Hombre" in the white hat.

*** This whole episode could not have come at a worse time for President Bush. In June of 1990, Mikhail Gorbachev came to a summit that had as its centerpiece a bilateral agreement to push for the elimination of all chemical weapons.

*** The two countries could agree to destroy old stockpiles of deadly unitary (one stage) chemical weapons. The United States would agree to destroy eighty per cent of its stockpile, and the soviets would destroy the same. In the end, it was hoped that both countries would have reduced their stockpiles to the same size.

*** Bush's commitment made a diplomatic virtue out of a legislative necessity. Congress had required that ninety per cent of the existing United States Chemical stockpiles be destroyed by the 1997.

*** Still, the United States had another trick or two left up its sleeve. It had a new category of chemical weapons that were to be left untouched. They were binary, or two stage chemical weapons. The United States began producing them in 1987 and the Soviets weren't producing them at all. So much for American honesty in the political arena.

*** These binary weapons, unlike the old unitary weapons were supposedly safe for stockpiling. They are made up of two chemicals that are kept in separate containers and are assembled on the battlefield. Still we must ask, do we need such a weapon if our strongest antagonist doesn't have them.

*** The Army was especially anxious to obtain the Chloride quickly. If it didn't get its chemicals by June, it would lose forty-seven million dollars in appropriations. Congress had appropriated the money because the pentagon convinced law-makers that binary weapons were needed as a deterrent against the threat posed to Europe by the chemical Arsenals of the Warsaw Pact? Suddenly that threat had disintegrated. Seeing that peace had neutralized the Warsaw Pact, the pentagon now offered the following rationale: Chemical weapons are needed to deter the use of chemical weapons by the third world countries.

*** Every citizen must examine this explanation carefully, not only with and open mind but, with his or her heart as well. Did our chemical weapons program deter Iraq, Libya, or Iran from seeking these weapons, or deter Iraq's willingness to use them? Our conventional military might turn out to be our strongest deterrent.

CHAPTER EIGHT

The people have a right to the truth as they have a right to life, liberty, and the pursuit of happiness. It is not right that they be exploited and deceived with false views of life, false characters, false emotions and false notions of self sacrifice.

By mid June of 1990, the pending Soviet tour of the Rocky Mountain Arsenal caused the Army to rush clean-up efforts, cut corners on safety practices and completely ignore hazardous waste regulations, that were witnessed by State and Federal officials.

Former nerve gas production building in South Plants area which must be torn down and disposed of along with all former manufacturing buildings at Rocky Mountain Arsenal. "impossible" contamination site.

[*This caption was placed below the picture in the original.*]

The Army was to begin decontaminating a former Nerve Gas plant within the Arsenal in preparation for the Soviet visit. On the afore page you will see one of the original chemical weapons buildings, and in the foreground of that building at the South plants area you can see part of the expanded wildlife program at the Arsenal. They intend introducing elk, antelope, turkeys, sage hens, and grouse to this cesspool of Ultra Hazardous toxic chemical wasteland, before the clean up is even half over. Connally Mears of the Environmental Protection Agency. "His primary concern would be whether theirs anything in those pond sediments that would be spread further by some sort of expansion."

"It's not that it's an impossible thing to do, or an inappropriate thing in the long term, but there is the potential for some problems with contaminants," said David Shelton of the Colorado Department of Health.

The Army gave State and Federal environmental officials only days to review the plan for workers, public and environmental safety. Then four workers were injured while working inside the Arsenal.

In a disclosure to Army officials the state said, "although the state appreciates the importance of the Soviet visit, it does not believe that the health of its citizens and the environment should be placed in jeopardy in order to attain this goal. The state simply has not received enough information on this plan to positively affect human health and the environment."

The Army responded by saying, the program would not endanger public or worker health and that it would have been undertaken without the Soviet visit. "The tour definitely is a consideration in making doubly sure the place is squeaky clean, to make sure it is clean when they come later this summer."

Still not satisfied with the Army's assurance, state officials sought a federal court order stopping the decontamination program, until their concerns were resolved.

The Army contractors were scheduled to resume the program by flushing a caustic solution, much like Drano, through the same pipes that carried nerve gas thirty-three years earlier. This was how the four workers were injured earlier, when this caustic solution spewed out on a pipe pressure valve and burned their backs. Since then the Army, EPA, and the Colorado Department of Health had been arguing over worker and environmental safety.

"The Army is not about to resolve any of our concerns and is prepared to begin decontamination activities once again," said an official from the Department of Health. "We want it stopped until our concerns are addressed and we feel this clean-up is not being rushed." Larry say's "This is why the Army has injured and killed so many people around the country, they do it without following good safety procedures, then count the injured and killed."

***** Larry says. "What the Army has done, where ever they have been, they have shown to quickly set up shop, irresponsibly not following the rules of law with willful negligence, reckless endangerment, careless wrong doing, Hap-hazardly, in every thing they do, very plain and simple they don't stop and think, They don't stop and listen, they don't keep safety in mind." Larry thought, "We have an army much like that of Adolph Hitler's that finally just took over by force, if you did or didn't agree you were dead. It wasn't just the Jewish people, if you didn't fit the profile you were dead." "Larry said he takes his hat off, and salutes the Jewish, Polish, Bulgarians, Russians, Yugoslavians, French, Prussians, Bohemians, Austrians, Hungarians, Belgiums, English, Romanians, and to all other Europeans, and to all who fought the war, and gave their lives, and to all those I failed to mention."

***** Then we also take our hat off to the Chinese who were so brutally tortured and killed, at the hands of the Japanese." "I SALUTE YOU ALL."

***** "Larry's head bowed in shame for what Hitler and his henchmen did in Europe.

***** Again his head bowed in shame for what Japan done to the People of China.

***** Then his head still bowed in shame for the chemical warfare the U.S. Army, the Pentagon, Industry and the U.S. Government is waging on it's own people right at home." Larry says the Army treats people as if they are expendable, ten, twenty, thirty, or forty years of pain and suffering and then finally death. Today or tomorrow, or in the future, there are people dying from what the Industry giants and the Army has done in the past, and today are still exposing them to Ultra Hazardous chemicals. To see what effect it will have on humans." "What they have been doing was no accident." "If people don't like it they just force them, as they did the Irondale

Trailer Court in Colorado., and the very same they are now doing to EPA, and Colorado State Health, who by law are in charge of the clean up. The Army will not follow the rules. They depend on the Government, the pentagon, and the Courts to skirt the issues the law requires, leaving it to the Army's very own irresponsibility, as the past has shown.

*** Larry said "We didn't have to face Hitler or his Army in Europe, but like so many other's right here in the United States, we have had a warfare with large Industry, and our very own United States Army, and the Pentagon because they tend to handle the same death agents here Irresponsibly. The very same ones and more, that Hitler was involved with in Germany, 1936-1945. It was Adolph Hitler who said at a staff meeting, we've got to think of something else, were wasting too much ammunition shooting people, that is when they started using pesticide to see how well it worked then by 1938 they had developed their own death gas. "Nerve agents."

*** "The Army here at the Rocky Mountain Arsenal has made a judgment that the national security governs this particular clean-up action, and we at the State Health Department, do not think they've followed the appropriate procedures to make that type of judgment."

*** Two days later, United States District Judge Jim Carrigan rejected the State's request for a temporary restraining order that would halt the clean-up for at least two weeks. Carrigan said the state hadn't proved that irreparable harm would be done to workers or the environment if the decontamination plan continued, at the Rocky Mountain Arsenal, Denver Colorado.

*** Colorado officials were disappointed, to say the least that they didn't win the temporary halt, which would have allowed a safer decontamination process, but were happy they had won one legal victory and that was when the federal judge agreed to allow the state to use the courts in the future to protest other Army clean-up actions.

*** Sitting at his patio table is the spot where Larry would do his best thinking, and this particular night was a typical Minnesota Summer evening. A warm southerly breeze, drifting across the land strange and was so wonderful Larry slipped back in time sitting on the front steps of his parents home. He imagined the scents of sage and cedar blended with the evening air, along with the chirping of the birds, back to the time he was a boy, Larry remembered his father sitting with him out on the front steps one night talking over all the boyhood problems and encounters while growing up. He remembered the two of them looking up to the stars and him hearing his father say that if man could only talk to one another and learn from those talks, all questions that man could think of, would be answered. He remembered the night his father taught him how to blow through his hands, using a single blade of grass, making a typeshrill whistling noise, different tones with different widths of grass.

*** Then Larry began to remember back when he first went back to Colorado in 1969, a very proud and successful business man. He would stop over to see his mom and dad. Them just being in their early 60's they got around like people in their 90's he thought. Mom had to be helped up and down any steps. his father could no longer work. It never dawned on him that his parents were being slowly poisoned from the Arsenal. All the fruit trees were gone, in fact most of the trees were gone, Larry never giving it a thought where did they go. The parents being so old so young. Old way before their times.

***** He can remember his mother and father just wasting away in their 60's, This made Larry Think "OH GOD" I don't ever want to be 60 if this is what life is all about. He remembers his grand parents living into their 90's and in a whole lot better health. He thought at times could it be what the Army had already done. not realizing they were still doing even worse, more irresponsibly.

** Then Larry began to remember in 1972 when he bought the home place where he was born and raised. This was like a dream of a life time coming true. Owning the very place he was born and raised. He built fences, barns, corrals, had set up to handle up to one thousand head of livestock. He had his very own type breeding program for breeding Holstein diary heifers that would be sold across the country. He was already buying and selling heifer's or young cows in California, Arizona, Nevada, Idaho, New Mexico, Texas, Colorado, Kansas, and Nebraska. He was considered and expert as an entrepreneur in animal husbandry and sales, as a buyer, and Auctioneer. Larry already owned two hundred head of heifer's.

*** Then March 29, 1972 Larry his family and the first two hundred head of livestock arrived from his Boxelder Creek ranch 75 head bred heifers, and 125 mostly yearlings. Larry had hooked up the stock tanks, and his own place up to the parents well His was going to be drilled on about April. 10, 1972. Then on March 31, 1972 the first shipment of Holstein calves came in, very active and perfect condition, they come in special trucks with special ventilation, and well strawed, The shipment was less than 12 hours, they were inspected by state Health to see if their were any signs of sickness or shipping fever, they had been cleared, checked in Colorado and approved. all calves were eight to sixteen weeks old, weaned at seven to eight weeks of age. They were all given nasal-gen vaccine, was to eliminate any chances of upper respiratory illness such as pneumonia. Within two or three days there were a couple of very sick calves. Larry called Dr. Scott the veterinarian asking why could it possibly be the nasal-gen, he examined the two found blister's in the nose, and mouth, and were getting difficulty in standing. With in a couple of days the two had died. Then Larry began noticing others with ears drooping, lapping at the water, drooling, blister in the nose and mouth, some were experiencing jerky movements.

*** Then by the fourteenth of April they completed drilling the new well, and the water was running almost red in color. They said pump it, it was iron and it would clear up, the second day it had cleared a little so Larry Had the Health Department come out and check water was ok, it's only iron and won't hurt the animals. The water was hooked up. Little did Larry know all they had checked for was bacteria in the water. With pesticide and nerve agents in the water their would be no bacteria. Then arrived an additional one hundred head of heifer's twelve to eighteen weeks old. Larry was buying up all the local calves he could get for his program.

***** With in days of hooking up the water calves began staggering and going into convulsions. In order to allow the State and Federal officials to see first hand at what was happening, as well as the calves in convulsions. Dr. Scott was convinced by then it was the water, sick humans, and sick livestock, the only thing they have in common would be the water. Then the pregnant heifer's began aborting their calves, most were about eight months pregnant, within three days all calves were aborted, and dead, some heifers died in birthing. This whole operation had a definite plan to be at one thousand by the end of the year. It all seemed to get the younger calves first.

***** A bad sign the cattle's teeth were turning black, and falling out of the gums, they were losing hair in large round spots, then came the vomiting, the diarrhea, lapping at the water, drooling, sores in nose and mouth, staggering, ears drooping a little further each day, unable to stand, falling over and going into convulsions. other cattle brought in and within days the same process begun. Dr. Scott done autopsy's on a couple and noted that there was no "rumen" in the stomachs. Rumen is cattle's digestive system bacteria, the stomachs were void of the rumen. The liver, heart, kidney's lung's, Thimus gland were destroyed Dr. Scott said then and there it is the water, let's get the livestock left to a different place. Of course the health department was calling it hemoragic septicemia, which is contagious pneumonia. To put them with another heard would pass this on to the other herd. Dr. Scott was so sure it was not A contagious pneumonia, that it was the water, talked another big dairy man a few miles away into loaning Larry some of the pens for these calves. Well Dr. Scott and the other man was somewhat worried what if it is contagious pneumonia, however they said move-em and we did moved nearly 278 into the new corrals, some were lifted on and off the trucks. Within a few days the difference could be seen by all, and their was absolutely no contagious disease passed on to the other herd, which only had a barbed wire fence between them. The Colorado State Health Department, and the Tri County Health Departments still trying to say it was pneumonia, and not something in the water, yet 1972, 1973, 1974, and to the middle of 1975 was Larry's water ever checked for anything except bacteria. Larry would say check the water for something from the Arsenal, all he would ever get is the Chief Chemist Bill Dunn, saying there is no way in Gods green earth your wells are polluted by the Arsenal.

***** The cattle improved all the bad signs and symptoms slowly disappeared, about five of them moved didn't make it by we did have 273 head left, and back in what we thought was good health. So Larry kept them all for another year, and began noticing they were not growing, they were no bigger than they were a year ago. It had been determined what ever happened, stopped any further growth. Out of 635 head of livestock, Larry lost 362 head, then discovered he had actually lost the other 273 head for a total loss of 635 head of dairy animals, around $1800.00 per animal,

**** During the 1970's the Army and Shell denied any responsibility yet at a meeting in 1973 between the United States Geologically survey professor, the Army, and Shell Chemical had agreed that it was Arsenal pollution in the Land wells.

***** In May and June of 1975 the Tri County and State Health Departments supposedly had written letters to the Land's their water was contaminated by Arsenal pollution. "Yet these letter's were never delivered to the Land's. The parents were both dead not knowing what went so wrong for them. It was not discovered until 1996 at the Trial that such a statement was made by the Health Department. Let Larry tell you why he don't think they ever intended to notify the Land's. The Land's lived at the same place until mid 1976 without any notification at all. The Health Department was delivering water to hundreds of other houses who had tested positive for DIMP. 2ppb for over two years, none to the Lands, their water tested positive for DIMP 13ppb, also found was Endrin a Shell Pesticide, and Deildrin found at lethal levels. We were probably pretty lucky that we had quit drinking or cooking with the water, however we were still bathing in the water, and wondering why our health was not clearing up. Well we found all this out the hard way, and not the health department.

***** The Tri County Health Department Doctor's had done blood tests on eight of us, and with in a day called us back into the clinic to tell us that we had in our blood a material called Phenol, believed used by the Army and by Shell Oil Company at the Arsenal. This worried them because the tests showed well above any safe standard for phenol, and could be lethal. The next day they called in more Doctor's all agreeing this was dangerous. So then they called the Army's attention to this, and all records disappeared from the files. No body knew anything, even though some knew the records were there and disappeared.

***** Larry was so good at buying livestock, he thought well I found a new and different way to make a living by buying one day selling the next. However he was soon stopped by letter from the U. S. Marshall's office Washington D. C. stating your right to a bond to buy and sell livestock has been revoked. From this day forward you will be fined up to $5,000.00 dollars and six months in jail for each head of livestock purchased for resale. This is the area of time Larry was investigating and finding how badly the Army had lied to everyone including Congress. The first cease and desist order had been delivered to them. The Justice Department being the ones who will defend the Army for their mistakes, again tried putting a stop to his way of making a living. To get their attention every week Larry would release some more information to the news about the Army. Then through out most of the 70's he had several attempts made to run him off the road the same way they did Karen Silkwood in Oklahoma.

*** By 1979, Larry had been able to pay off all his debts, and the loans on the livestock. He worked as an Auctioneer, Deputy Sheriff for the City and County of Denver, Colorado, and operated his own trucking business which had grown fast enough that it had paid the debts completely by 1979. By 1983 Larry and Marie decided to leave Colorado and move to St. Paul, Minnesota where her family was from. At least the air and water were cleaner. Every two years researching more, and 1983 was the Year Shell Oil Company, and the Army were done. Shell was forced to move, and the Army forced to clean up the worst disaster in American history, 27 square miles of the most polluted land on earth.

**** Finally by 1988 Congressman Hank Brown brought before the House Judiciary Sub-Committee his request for House Resolution 61 for House Bill 816 for the relief of the Land's. By the fall of 1988 it was going before the full House of Congress for approval. It cleared the House unanimously. Then sent to the Court of Claims a fact finding commission to investigate and return to Congress their complete lists of all the facts and findings, everything the fact finding commission discovered. This was a period of time between 1988, and the time there would be a trial for damages. The first complaint was turned in by late 1988, by 1989 the second complaint was necessary for the Court. The Court made a decision that it would take two different trials. He ordered the trials be bifurcated, one for causation, and the second trial for damages.

** Hearing the door open, Larry turned to see Marie coming outside to sit with him. He did not think she would admit that really she was very happy our new place in Minnesota since it was the very same place she had grown up. They had purchased her parents home, and her children were able to live in the same house and go to school where she had. She was at last happy, Larry was happy for her. In many ways that even today he is unable to explain, he feels like she is almost a physical part of him. They held their heads high through hell and high water "together", or they probably would not have made it.

*** Marie asked, "what are you thinking about you've been out here nearly two hours. Marie sat down with him.

*** Larry said, "Colorado, my parents, our livestock, and what it did to us, and the kids and where were at at this point.

** Larry laid his head in Marie's arms, then lifting his head, Larry suddenly felt a strong desire to speak to her. Yet he was hesitant to speak or to move for fear that the slightest sound or movement would shatter the moment's spell and leave him with nothing. As long as they were both silent, the intangible communion between them existed, and she remained for him the stuff dreams are made of.

** In the night's vast quiet there was between them, an invisible link, perhaps forged by some mysterious bond of stars and stillness. They were drawn together by the silence, the night sound of crickets, and frogs, and the shadows of the evening itself. Was she, too, sensing that his moment was special? That here, for the moment at least, each of them, even more so now, belonged to each other in their silence.

**** That winter, the Arsenal, and the clean up of basin F, again came into the public eye.

**** Once the excavation began, local residents complained that, the odors and chemical fumes being stirred up by the heavy earth moving equipment and truck traffic, were making them sick.

*** Fumes became so bad that Ebasco twice evacuated all employees not wearing protective clothing, once on December 5, and again on December 12, but no one bothered to inform the nearby residents of the Irondale Trailer Park, just across the street.

*** The Army twice evacuated its employees from Arsenal offices because of the fumes. This bluish green haze was so bad that eventually it shut down the entire job site. again residents of Irondale were forgotten.

*** The same contractor responsible for the clean up of Basin F, was responsible for monitoring any air pollution caused by the clean up.

*** The test samples were sent to a laboratory for full analyses, but only certain samples were analyzed. Records show that no laboratory analyses were conducted for what was perhaps the two worst days, December 5, and again on December 12.

*** Employees said that at times Ebasco deceived both the Colorado Department of Health and United States Environmental Protection Agency.

*** One Employee said, "The whole operation was a joke, we knew in advance when the EPA. or the State Health was coming out. Drivers were told to drive slowly and not to stir up any dust when EPA or the State Health was present. Otherwise there were just ton's and ton's of the chemical pollution in the air. This soil is heavily contaminated with Ultra Hazardous, toxic and carcinogenic chemicals."

*** Ebasco said the roadways were constantly watered to reduce the amount of dust, and Ebaso denied timing the watering with the EPA, and State Health visits.

*** Ebasco, in September of 1987, won the twenty-two million dollar contract to remove liquid waste from Basin F. They were also given a twenty-eight million dollar contract to study pollution problems at the Arsenal. (That's interesting, I understood that's what the Superfund program was for, "The study and removal of hazardous materials.")

*** The Occupational Safety and Health Administration Citation lists violations ranging from improper emergency training, to a lack of toilet facilities for the dozens of workers excavating Basin F.

*** Ebasco constructors, a branch of Ebasco Services Inc. of Lakewood Colorado, has contested the huge amount of violations. Osha area director Gerard Ryan classified the violations as "Serious" (only serious when death or injury has already taken place.) In his seventeen page citation. Serious denotes a situation that if there was an accident or illness, it would result in permanent, or irreparable harm to the employees.

*** At least one clean up worker reportedly had contracted chemical pneumonitis, an inflammation of the lungs from exposure to pollution. Others have complained of respiratory problems and skin irritations. The pesticides, Nerve agents, and mustard agents, and their by-products, chemical's, solvents, and a variety of other toxic chemicals that workers worked in, had combined to form new compounds that have not yet been identified.

*** The citation, which included a possible fine, stems for Osha inspections when Ebasco was dredging Basin F. Employee complaints prompted the inspections. Violations include insufficient worker health protection and employee training for the personal handling of Ultra Hazardous Materials.

*** The citation also listed Ebasco's alleged failure to analyze the hazards of Basin F liquids for workers, and provide workers with proper respiratory equipment. Other alleged violations also include the company's failure to give workers the results of medical tests, or to provide them with drinking water.

*** Employees, were not afforded the opportunity to use the toilet facilities when desired, due to supervisory pressures. Meanwhile, the Army awarded Ebasco seven-hundred thousand dollars for finding ways to cut costs.

*** By August of 1989, Shell Oil Company had for the sixth time, agreed to buy out and move a person living in the path of the pollution created at the Arsenal, bulldozing all property.

***** On September 8, the Colorado Department of Health fined the United States Army $1.5 million for violations of the State toxic waste law.

**** Army lawyers dismissed the charges as "ridiculous," saying the Federal Government would refuse to pay. It is the biggest environmental fine in the state's history.

**** At issue was the thirty-five million dollar clean up of a ninety-three acre dumping pond that once held over 240 million gallons of a conglomerate of toxic chemicals and pesticides. The Ultra Hazardous fluids, or at least part of them now are being stored in three stainless steel tanks and another dumping pond protected by two plastic liners. State Officials said the Army is violating Colorado law because it can't prove that those toxic holding facilities, as well as a giant underground cell holding basin F soil, aren't leaking.

*** Federal Officials disagreed.

**** "There is absolutely no indication or evidence to believe that there has been any release into the environment," said John Moscato of the United States Justice Department.

*** Still some fluids were believed to be seeping through the first protective liner in a newly constructed dumping pond.

*** In the compliance order that was filed, state officials also said the Army violated Colorado Toxic waste law by failing to file a Basin F Clean up plan in October 1986. *** In September, United States Army Officials said the penalty would only divert scarce resources from cleaning up the Arsenal.

*** In their first comments since being slapped with the penalty, the Army said Colorado Public Health Officials had no reason to levy the fines and may have overstepped, or even abused, their authority.

*** It is unfortunate that we now appear to be resuming active litigation instead of attempting to cooperatively address clean up problems," said Col. Daniel Voss, the Army's program manager.

***** Larry said "this is the path of a psychopath, doing every thing his way, and to hell with the Law. This Army reminds me so much of Hitler's Army. The more they got away with the more that was done, they had only one way and that was the wrong way. The United States Army seems to be going right down that path with rockets and nerve gas in hand."

*** State Health Officials defended the size of the fine, saying they cited the Army for more than thirty violations at the Arsenal. It's like Larry said "and that's only the ones they got caught with, where you see one rat there is a hundred."

***** Where is the common sense which God gave to us all, yet seems to have escaped our leaders. "They thought he said No sense at all so they stayed away, and they had no common sense, so really that left them with no sense at all."

*** In September, Federal Inspectors cited Ebasco Construction inc. of Commerce City, with six serious violations of worker safety regulations during clean up operations. These are uncaring and needless incidents of willful negligence, and probably has sentenced each one of these people with life of misery, and premature death. Have you ever ask your self how many are suffering unnecessary misery, and premature death, because of another person's willful negligence??.

***** Why was a medical waste incinerator allowed to be built right next door to an elementary school in Omaha with the consent of the Governor? A hazardous waste incinerator to clean up a contaminated site in Texas was built by Monsanto butting up against a housing development and again one block away from an elementary school. How many of these children will suffer a life of misery, and premature death, again because of someone else's willful Negligence? Could this be your child, or your sister or brother's small child, a friends child? I only remind you that it is out there, and it could happen to you so watch out at least for the sake of our children.

***** Larry said, "This reminds me of a time, I believe back in the 1950's When the Army at the Rocky Mountain Arsenal had made a fungus type material to destroy the wheat in Russia. They evidently wanted to see if it would have any affect on the American wheat crops. So they sprayed by plane millions of acres of wheat in the what was known as the north central wheat belt, Colorado, Nebraska, Kansas, Iowa, and Minnesota, unbeknowenced to any one. what are the effects, I have heard it causes lung damage, is highly carcinogenic, and possibly a mutant. Well no body will ever know for what ever reason it was discontinued, and like so many other projects that went wrong it got buried and no one will talk about it. Then the Army also done many chemical tests on the people in Minneapolis Minnesota, right down town and near and at least two elementary schools. They were using the children and adults as guinea pigs, so they would know first hand the affects of each of the chemicals. They didn't stop at just Minneapolis, San Francisco, Los Angeles, Denver, Cheyenne, Dallas, Fort Worth, Atlanta, New York, Chicago, Boston, Pittsburgh. Topeka, Kansas City, Tampa,"

*** The nuclear regulatory commission was at one time, planning to allow nuclear waste from nuclear power plants to be made into furniture, Kitchen utensils and so on. These types of issues are going on everyday and can be prevented if the public is made aware and takes a stand.

*** In Schuylkill County Pennsylvania there are fifty-four toxic waste sites within a six mile radius. Five of these are listed as Superfund sites.

*** This is not a unique situation, there are thousands of toxic catastrophes all over this country. Is the Government using our tax dollars to help resolve these problems, or is our Government helping the companies avoid responsibility and letting these cases be prolonged. Lets not forget the Courts are playing a bigger role than ever, and so many incidents where the Government was wrong the court will go as far a invoking the FTCA, to stop all proceedings where the Government cannot be found wrong.

*** At the Rocky Mountain Arsenal the Government is letting the responsible parties, those who created the problem, do the technical analysis to determine the magnitude of the problem. In other words, the Government is trusting the rustlers to count how many steers they have taken and to stand guard over themselves at the gate. In other words they've left the Fox guarding the Hen house.

*** Along with all the problems we as a nation are facing, Larry has been told that there are roughly fifteen nuclear weapons plants in the United States, and all are leaking, and will cost more that one-hundred billion dollars to repair. We do not have the money or the technology to do either of these jobs correctly. We had better have an understanding of what is lurking in our water supply, in our air, and in the atmosphere of the beloved planet we call home. (Remember the Hanford site). or (Rocky Flats).

*** In Minneapolis they have in operation a garbage incinerator right in the downtown area. There is no question that if you mass burn garbage, including plastic's and chlorinated material, that it will produce a very wide range of highly toxic, and carcinogenic chemicals. Particularly the chemical known as dioxin, which is a highly toxic chemical contaminant that has been used in chemical warfare, specifically Agent Orange, and Furan, which is a colorless liquid that is used as a solvent for resins, plastics, etc., or as a tanning agent. It is grossly and recklessly irresponsible for selling this bill of goods to the public should be brought up on criminal charges.

*** One final word on incinerators. If a problem arises in Japan with one of their incinerators, they have three hours to correct the problem or they are forced to shut down completely. In Minneapolis they have three days. Have you any idea on the amount of damage that can be done in three days?

*** We will get rid of more garbage by going back to returnable bottles and by subsidizing the recycling of paper and lawn debris, and it will be cheaper.

**** On September 22, Justice Department attorneys sue the Colorado Department of Health to block the $1.5 million fine against the Army for hazardous waste violations at the Arsenal.

**** In the United States District Court lawsuit, Justice Department lawyers argued that the fine is illegal and unnecessary.

*** State officials had won the right to apply state law to the clean up efforts, but Federal Officials are appealing that decision.

*** Justice Department lawyers argued that following state laws would force the Army to violate the Federal Superfund law that governs the clean up of the nation's most dangerous toxic waste sites, and that it would cost too much money and siphon money from other projects. They also claimed the Federal Government is immune to state fines and prosecution. Justice Department representatives characterized the lawsuit as, "an act of frustration".

*** The State agreed to the clean up measure when it was started, but later raised concerns about work methods and fined the Army. The Arsenal clean up is also the subject of three other Federal lawsuits.

*** "This lawsuit, makes it clear that the Army does not know what it is doing," said Mike Hope, Colorado Deputy Attorney General. "The state's real goal is to force the Army to substantially comply with the same law everyone else must," Hope said. "It doesn't matter if the Fed's or the state is doing the clean up, just so the health standards are met."

***** "The Army wants to do it their way, & that's the wrong way."

*** In Washington by mid October, two of Colorado's leading Democrats, Senator Tim Wirth, and Representative Pat Schroeder, had reached an angry impasse over how to clean up the Arsenal, with each side claiming to have the State's best interests at heart.

*** The controversy surrounds an amendment by Schroeder to next year's defense budget, aimed at speeding up the clean up of the Arsenal. The amendment would preserve the twenty-seven square mile Arsenal for wildlife, Which means the Arsenal would only have to be cleaned to a level that would support wildlife and prevent any spread of contamination.

***** Larry Says, "That is not what was agreed to, and I'm shocked at Schroeder's back up in anger to allow the Army off the hook of cleaning it up as was first agreed. (How ever we can possibly understand why, 'since she was on a committee for the Armed forces in Washington D. C. The Army made this Ultra Hazardous problem, let them undo what they did in the first place. he said also he was glad to see Senator Tim Wirth to stand his ground, and see to it the Army must clean up the Ultra Hazardous disaster they made, let them undo, and clean up the disastrous and most toxic piece of land on earth. All over the United States the Army has destroyed, Ultra Hazardously polluted any thing they touch then give it back to the State, all until Colorado at the Rocky Mountain Arsenal, at least at this particular site they must be forced to clean up the land they polluted without and regard to human life."

***** "Larry wants to show how anything can hit right at home, no matter where you are. "Be on the alert" "The Arsenal devastation happened to Larry when he lived only one mile from the Rocky Mountain Arsenal. This was his home place where he was born and raised, and for him to continue a tradition like his parents did when they purchased the place in the 1930's, Larry was born there in 1941. Then came back home in 1972 purchased the home place and was planning his life. He was an Auctioneer, a cattle buyer, His wells became contaminated with low levels of GB, Sarin, and BZ nerve agents, low levels of DIMP, IMPA, DCPD, Nemagon, Endrin, Hydrazine, and Aerozine rocket fuels, and Lethal levels of Deildrin a Shell Pesticide. All these finished products were found in his wells, along with an array of heavy metals, such as lead, mercury, ytrium, boron, silver, all this plus many more at toxic levels. Livestock were killed and destroyed within days. Larry lost one son, his mother and father, also a young man who worked for him Bill Phinney, leaving his wife and 3 children penniless and very, very, sick."

***** You've already seen all the signs and symptoms, most of health maladies from what happened there by Arsenal poisons The devastating effects over a eight to ten year period, had mostly reversed, healed and somewhat back to normal once again. he says he can truly say never give up. By 1983 Larry passed a physical and psychological evaluation with no problems he shared all information with the Doctor's. Then just six months after he began his job he was ran over in a hit and run incident, just four days after he was called and told he will be going to Congress to testify on the Rocky Mountain Arsenal, and Shell Oil Company. On August 24, 1984, the automobile incident, Please note he calls it an incident, because like he says "it was no accident." He had a broken right clavicle, a brain concussion, and was unconscious for a short period, was placed on a Steroid to keep the brain from swelling. That had an affect of leaving Larry suffering amnesia, not remembering who he was or who his own family was, lasting for almost seven days. Once that problem was under control, he was released to go back to work on October 10, 1984, and by December 10, 1984 he was given a complete, physical, psychological recovery from his Doctor's. All injuries were reversed and he reached 100% medical improvements.

***** How ever his employed Washington County Sheriff Department, facilities Department head Lyle Doher, and his assistant Ed Kapler, were the two who approved the purchase of the paint to be used in painting the interior of our workplace inside the jail. This started on or about February 1985, Inside a place like this there was no windows, no doors, no ventilation for anything like this. There was like a cloud hovering in every area, everyone was getting sick and again all the signs and symptoms of being poisoned, vomiting, nausea, headache, sweating, shaking, everyone turning white, dizziness, worse as time went on for nearly 3 weeks, this paint was being sprayed with air sprayers, Larry repeatedly ask the supervision to stop, since they were all in a different part of the building it didn't bother them. They would not listen, he called OSHA, the man had him get the serial number off a can they were using. He had to wait and watch because they had been bringing the paint in unmarked containers. he finally got a can they were using in the back room got the name serial number, and noted the can was black 5 gallon can of paint, with an 18" x 18" label telling how dangerous this paint was. It had in each upper corner of the white label a red skull and cross bones. It stated industrial bridge paint, high rating for lead. That afternoon Larry got home and called OSHA, gave him the information, within twenty minutes he called back really stirred up, and said that paint will kill you. It is Ultra Hazardous, It is a high level lead based paint it is just totally poisonous. He said don't go back into the area. and he explained it was his job. He said then please call me if they are still using it tomorrow, please have your supervisor call me, I will be here by the phone.

***** Larry went back to work the following day, would not allow the painters to start until it was cleared with the supervisor. Well they went on ahead into the area to be painted, and Larry

Locked them in so they could not get the rest of the equipment or plug in the equipment. This paint was not only deadly, it was also very flammable. He told them they were under arrest for violating both State and Federal laws. The reason he locked the doors behind them they were not going to stop when asked.

***** The Supervisor came in within a half hour, he was given the phone number to OSHA and that particular person. He took the number and went outside to his office and made the call. Within ten minutes he came back into the jail, he was very pale, told Larry to get the painters out of here now, tell them to take all their equipment, and not to come back the painting was over. His Supervisor turned to him after the place was clear, and said I apologize to you, and the sheriff did also, and it will not happen again. The Paint being used was Ultra Hazardous lead based bridge paint that had been banned in the United States since 1967 that required an airline oxygenated respirator when used on bridges, and the painter was to have lead levels check at least once a week, normal blood levels are .06 ug Larry's blood was elevated to over .21 ug over a long period.

***** All the signs and symptoms once again disappeared, The headaches with medication, and the pain in the joints with medication. The headaches were being caused by Hypertension, What causes hypertension (Lead Poisoning). The pain in the joints took almost four years before the Doctor's were able to diagnose it as Severe Arthralgia. What causes Arthralgia (Lead Poisoning). Lead poisoning can also cause premature death. Larry had a permanent disability from 1972 when poisoned by the Arsenal, causing the bones to ache Larry often calls it the long bone aches and pains. What was discovered here is rare, and was caused by nerve agents at the Arsenal. It is called calcification of the ligamentus attachments to the bones. every place the ligament is attached to the bone parts of the ligament were calcified, causing continual pain even with the slightest of movement. The only way Larry can get complete relief is lying down without the slightest of movement, other wise for every thing he does or thinks is painful.

***** Larry wants everyone to realize, something as ignorant as above, this is willful negligence on the part of several Ed Kapler, and Lyle Doher Facility heads, also on the part of Sheriff Trudeau, and Sergeant Kimble for not stopping this on the very first day. They were made aware immediately, yet they allowed it to become injurious to several.

***** Larry says to Duane Spoors, the under sheriff who we tried to explain to him what it was doing, He could smell, and was told it will get-cha, it he was right their also, hacking, coughing and choking, Larry's telling him there's something wrong here with this paint It's killing us. By the way he was an investigator. His statement was to another jailer, and he was overheard by Larry stating, (I hear we have a G__ Damn jailer here who can't take a little bit of paint fumes. If he can't take that he better go find another job). Larry says "Duane in case you read this I hope your getting along with your emphysema, probably the lead based bridge paint you allowed to be used in painting the jail, I noticed each time I seen you were always hacking and coughing during and after the paint incident, I see it got cha.

***** Then in 1990 they did the same exact thing with a different paint, stating it was latex paint, and gave the health department data sheets for latex paint. Again Ed Kapler facilities, and also the head of the Washington County Safety Committee. Purchased a Mercury based paint two part epoxy paint where xylene added begins some type isometric changes in the radical compounds. This paint had been banned in the United States of America, and was never intended to be used in an interior, or to put anything painted to go into an interior. The radical changes from this paint continues to emit toxic fumes for years after. Sheriff Trudeau was ask by letter from Larry and his Doctor's to please stop the painting it was killing them. He made no effort to stop the painters. Captain Richard Becker, the jail supervisor, was asked over and over again, and was written letters asking them to stop the paint situation at least until it can be investigated. He refused, a personal statement when Larry was begging for help from him Captain Becker was "where the F____ do you think your going with this." Larry went then to the County Health Department, with a label from one of the cans showing it was not latex paint in fact was a petroleum based paint for exterior's only. Again in the jail painting it with no windows, no doors, no ventilation and with a very flammable paint. The health department official Lowell Johnson. was shocked he said they told me that was latex paint. He immediately called Ed Kapler's office and ask for the data sheet on the paint they were using.

***** The true paint data sheets were locked in facilities safe. The facilities Department are the ones who purposefully issued the wrong data sheets to the health department. They also were to be displayed in the area being painted and there was nothing. The proper data sheets were delivered to the Health Department Lowell Johnson who had made copies to be hung in the work area. The paint was Mautz line 8900 petroleum based Industrial Enamel. Containing toxic levels of Mercury, and Xylene if not properly used. Lowell Johnson was unable to get them stopped even though the paint was very toxic. However he did issue Lupp Paint Contractor's a warning letter asking them not to put any more of the toxic and flammable paint material and cans in the County's compactor. Larry didn't realize that the Washington County Health Department had so little power. Larry even failed to be heard before the County commissioners, He was allowed to explain what was going on in his continued effort to get it stopped before it caused injury, The Washington County Board stated they were powerless, since he had filed a suit with workers comp.

***** Larry then called OSHA, at that time he had over forty other employees that had been made ill or injured from the paint. all though everyone was worried about their job, would not file workers comp. And would not sign complaint. Even though it also became overwhelming health maladies for others, since the did not file for workers comp at the time they were not allowed too later. Larry was worried about his job also, but he was also worried about his life and the life of his coworkers. He did file workers comp. And was the only one allowed workers comp. This had been going on nearly 30 days for 10 hour each day. The Sheriff, chief Deputy, all the investigators who had offices even close to the jail, Captain Becker, would come in pick up mail or papers and were gone, One investigator who was coming into the jail stopped short just outside the jail door, and stated the odor was so pungent, he ended up with an overwhelming headache, and dizziness, Larry decided what he had to do was not that important. He had tried everything under the sun to get them to stop but they would not, even take the time to listen. It was as if to say we are going to force it on you. Larry understood, Other employees were told if they join Land on this their jobs was gone.

***** There was a standing joke amongst the Washington County Minnesota investigators and the Deputies, "Sheriff Trudeau was trying to get the County Commissioners to fund a new jail since they were having at times severe overcrowding." "Everyone was saying the Sheriff solved his overcrowding problem he just gassed all the prisoners at the jail". 'ha, ha'.

***** Larry called OSHA, signed a complaint, and they did an inspection of the paint scene. Washington County was issued Severe Citations, the Painters Lupp Inc. were Issued severe Citations. Captain Becker was issued his supervisory citation for not doing anything. Washington County tried their best to get out from under the bad business practice of using a toxic paint and Causing injuries' or caused illnesses that led to disabilities or death. The Teamster's Union, and the AFSME union attorney's joined Larry at this time in the Court actions. After six hearings the County lost. were given citations to post in the work areas, and they were also fined nearly $2,000.00.

***** Larry suffering many medical health maladies, he tried to get help but most of the injuries were beyond help, and there would be more he was told this can also be progressive, and more things may go wrong as time goes by. Mercury enters the body via the nerves where ever contact is made. therefore doing a great deal of injury to the Central Nervous System. Larry was made aware by Doctor's that some of the minor symptoms now may become major as time goes by. Unlike the signs and symptoms, and health maladies from the Rocky Mountain Arsenal which most eventually reversed themselves, The Hit and run incident the injuries reversed themselves. The 1985 Paint incident with Lead Based Paint was very dangerous, leaving him with two disabilities for life. Hypertension, and a painful Arthralgic condition.

***** The Injuries in 1990 from the Mercury/xylene poisoning done damage to the Nervous System, because of the intensity of Thirty, ten hour days, and the entrance into the system is through the nerves, there was no way to stop it, and that is why the intensity of the injury in the Central Nervous System. His Mercury levels were still above safe standards nearly a year after the incident. Larry will talk later of the disabilities that he was left with and how he dealt with them, and how they affected his job, and his life. The injury to the Central Nervous System is rated at 40% under workers comp. The Chemical over exposure in 1972 at the Arsenal had left Larry with what he called an allergy when near chemicals. After the over exposure in 1985 by the Lead based paint, that allergy grew to a hyper allergenic health malady for him, Leaving him very susceptible to all chemicals, he was very affected from the very first day. Then after the 1990 over exposure to Mercury, Xylene, and Epoxy isomization he was diagnosed (MCSS) Multi-Chemical-sensitivity-Syndrome. This is again a very serious disability causing respiratory shut down, or death. He must carry medication with him at all times, and has a pick up with special ventilation when driving, and Emergency service immediately available at all times.

***** By December 18, 1990 Larry was still having some great problems getting accustomed to his disabilities, and also his employer. One afternoon at work Sergeant Tim Adams ask him to get a large box of books from an upper shelf in the back room, he went with Larry and showed him what box. Larry told the Sergeant at that time he has a real hard time looking up trying to do something, he has a disability losing his balance because the loss of vertigo during the last paint issue, and is one of his permanent disabilities. Adams said go ahead I can't since I injured my back, but I'll help you. As Larry was standing on his tip toes, and brought the box out from the shelf on his fingertips. When the box came out from the shelf, his fingertips went down, his top toes went down as he lost his balance to the right. He ended up with a severe compression injury to four lower vertebra, and herniated disks at four different levels. The box was not light it weighted over one hundred pounds. work comp rating at 24.5 total disability.

*** Larry was off work for nearly three months, in traction and therapy, before it reached it's maximum medical improvement. and was able to return work with restrictions, even though it will always remain just another disability, making it more difficult to do my job, now looking at three on the job injuries that all occurred because of negligence on the part of the Employer, mercury, Xylene, and lead are all listed as Ultra hazardous, and most toxic chemicals. Mautz paint company is large industry. what he's trying to help you understand it all runs downhill, you need to remember all of us with all our children are at the bottom of that hill. It's up to all of us to work toward making it safe at the top of the hill. The Government, Industry, Army,

Pentagon, our employers in order to make it safe for all of us at the bottom.

*** At this point we'll go back a little to 1989, and the Court of Claims had Larry the Plaintiff file an amended complaint. this was completed, and rapidly filed with the Court. Then the Justice Department filed an answer back to the complaint in 1989. Then what's called the Interrogatories, (written Question from both sides each side sworn to tell the truth and be notarized.) second is the production of documents, (Any documents that may be used as evidence are to be copied, or arrangements made to view, and copy. Third comes the depositions, where one appears with their attorney, and answers questions asked by the Defendants Attorney, you are sworn to tell the truth and nothing but the truth, all information is stenographed for the Court, the defendant, and the plaintiff. It also works the other way around when the plaintiff can ask question of the Defendant, and their Attorney. This information is prepared by both sides in a effort of what is called discovery. This went on for three to four years at a point motions can be made, motions to dismiss, motion for summary judgment. Discovery was still going on when the Defendant served a motion for Summary judgment.

**** Summary Judgment (victory) "for one side in a lawsuit, or in one part of a lawsuit". Without trial when the Judge finds, based on pleadings, depositions, and affidavits etc. that there are no genuine factual issues in the Law suit. (He would rule for the defendant and dismiss the complaint.)

***** However in Larry's case the Judge ruled there was sufficient genuine factual issues to support the Plaintiffs to trial. A hearing was held by Judge Robinson to decide whether or not the trial would be bifurcated. Bifuraction would mean two separate and distinct trials or just one Trial for all issues.

***** On September 13, 1990 Judge Robinson ordered that the two issues here for trial would be bifurcated. Their would be two very distinct Trials. One for Causation, "That which produces an effect). (Motive or Reason). (Just Cause). (Proximate Cause). (Theory for which a lawsuit is based).

***** Filed October 13, 1993, Motion for summary judgment by the defendants, Scope of an equitable claim,; and Scope of wrongful act.

***** The plaintiff challenges the motion. Urging that the record does reflect genuine issues of material fact precluding summary judgment. For the reasons stated below, defendant's motion was denied.

***** In 1942, the defendant through the United States Army, established the Rocky Mountain Arsenal (RMA), located near Denver Colorado. From 1942 thru 1982, It was to manufacture toxic chemicals, and incendiary munitions. later part of that time was spent in demilitarization. Portions of the main arsenal were leased to Julius Hyman Company, who sold to Shell Chemical Company that produced some of the deadliest forms of Pesticides. They were also believed in manufacturing deadly herbicides, and dioxins. The handling of and disposal of chemical waste was done by the Army.

***** The USGS and University of Colorado each concluded in their studies that the primary contaminants were sodium and chlorides, which were carried off-post by an underground water plume in a northwesterly direction. Basin A was identified as the likely source of the contamination. As a result of the studies the Army constructed Basin F, located northwest of Basin A. this basin consisted of approximately ninety-six acres and was constructed of clay, and soil lined with a 3/8 inch thick membrane of blown asphalt, meant to water proof it. this basin was temporary, and designed with a life expectancy of 10 years, it would hold up to 180 million gallons of liquid waste. From early October 1955, through, 1975 all of RMA's chemical wastes were pumped to Basin F, and Shell Oil Company's waste was pumped up until 1983.

***** The Army used that basin for nearly 30 years, 20 years past it's life expectancy. They not only used it for an additional 20 years, they used it know it was leaking, and using it on the premise it was contaminating underground water off the Arsenal. The U.S. Department of Health, and the U. S. Army Hygiene Department both in 1969 advised the Army to stop using Basin F immediately it was leaking, and to continue using it would be knowingly using it while it was leaking.

***** The RMA. halted monitoring in 1960. It refused to resume testing until 1974-75, despite the fact that a report on groundwater contamination in the relevant area, by the U. S. Department of Health, Education, and Welfare (HEW), completed in December 1965, recommended that the well monitoring programs of the Tri County Health Department, and the Colorado State Department of Health, and RMA itself be continued.

***** In 1965, the Army Environmental Hygiene Agency recommended that steps be taken "TO ELIMINATE LAKE F', as soon as possible, and thereby remove much of the present environmental hazard of exposed surface storage of toxic wastes." By 1969 there was evidence that Basin f was leaking: indeed, a physical inspection reflected that sections of the protective membrane were absent. Thus by the spring of 1970, the Army understood full well, and without any doubt, that it was operating Basin F on the premise that it was leaking.

@!@!@ This warning was after an inspection of the basin from two Army Generals from the Pentagon. Who had discovered the liner of the basin had vanished.

@!@!@ They then continued filling Basin F with Ultra Hazardous Chemicals, from both the Army and Shell Oil Company. The Basin when built was given a life time expectancy of 10 years, They began using it in 1954, and in 1965 one could have understood had they responsibly been taking care of business, and would have stopped it use., However they didn't they used it for nearly twenty more years after the life time, and knowing it was leaking. The peak capacity of the Basin was suppose to have been about 180 million gallons, however when they knew it was leaking and for nearly 20 more years they had bumped the capacity to over full, at 250 million gallons.

@!@!@ When Shell Oil Company got it's first cease and desist order from manufacturing Ultra Toxic Chemicals. Any ordinary business would have been expected to stop. If you would not have stopped you would be in jail and closed down. but not the Army and not Shell Oil Company, by 1976-77 They doubled the production. They received another cease and desist order, as well as the Army for handling the waste for them. By 78-79 Shell tripled the production of Ultra deadly chemicals. These were chemicals that had been banned in the United States of America. The Army was aware, the Pentagon was aware, So Colorado again issued cease and desist orders on both the Army and Shell Oil Company. By 1980 Shell had then quadrupled the production for the Ultra deadly, and banned chemicals at the

Rocky Mountain Arsenal. That is why the deadly pollution placed in Basin F increased from about 180 million gallons to 250 million gallons overfilling the basin. causing like a toilet ball effect to the North and Northwest of the Arsenal, it being filled to that extent it was by now polluting underground water wells where it had not before with a controlled level in the basin.

***** Plaintiffs rely on a report by Dr. Frederick W. Oehme, D.V.M., Ph.D., which opines that the cause of the death or eventual destruction of all but three of the calves was the presence of the Land's well water of a man-made compound DIMP, and it breakdown product IMPA.

**** Larry has received information now that the product IMPA, and the Army's claim that it was a by-product of DIMP was a lie. There is now evidence that will show that IMPA was a metabolite of still another nerve agent called "BZ".

**** The also rely on a recent report by Dr. Daniel T. Teitelbaum M.D., concluding that the cause of the health problems suffered by plaintiffs and Larry Land's calves was DIMP, and IMPA contamination of the Land well water and the air.

***** Review of the records which were supplied to me establishes unequivocally that the Land family's groundwater was contaminated with materials from the Rocky Mountain Arsenal. In particular, the finding of DIMP... establishes beyond any question that the source of the material in the groundwater under the Land's property and in their drinking water and hygiene water wells was the Rocky Mountain Arsenal. No other source of this material is conceivable in that area. This very-well Finger-printed chemical establishes without any question that the toxic material which reached the Land family had, as its source, the Rocky Mountain Arsenal GB manufacturing facility. In addition, the finding of IMPA, a metabolite of DIMP, further identifies the material as having a source in the Rocky Mountain Arsenal.

***** RMA and Ft. Detrick established programs to determine the effect of DIMP on wheat growth.? Plain water with DIMP in it had little or no effect on the wheat. But Arsenal water, containing DIMP, would have an immediately killing effect on the wheat. James Land, Larry's brother had a wheat field just across Hwy 96 at highway two, where the water was contaminated with DIMP. The tests they ran showed conclusively their were other chemicals not yet detected in the water, and we know their were, but the Army had the only key to unlocking the real truth to what it was.

***** Larry thought, One could almost imagine after seeing this that the Army, and Shell were standing holding the smoking gun, and were unable to discard it.?

***** In December 1974, the Colorado Department of Health detected DIMP in a well near the city of Brighton, located seven miles north of the RMA and six miles north of the Land ranch. Although the quantity of DIMP detected was small, it indicated that ground water traveled in a northerly direction. This direction of contamination of the ground water resulted in the State of Colorado Department of Health issuing three cease and desist orders on April 7, 1975, against Shell Chemical Company, and Rocky Mountain Arsenal. The orders required that:

1. SCC and RMA immediately stop off-post discharge (both surface and subsurface) of DIMP and DCPD;
2. Take action to preclude future off-post discharge (both surface and subsurface) of DIMP and DCPD;
3. Provide written notice of compliance with item (1);
4. Submit a proposed plan to meet the requirements of item (2); and
5. Develop and institute a surveillance plan to verify compliance with items (1) and (2).

*** Andy Feinstein, an aide to Schroeder, said the purpose of the amendment is to end the six-year legal battle between the State and the Army over who will control the massive clean up of the Arsenal, so that the process can move forward.

***** Feinstein, must not have read the rules of the Superfund clean up procedures that was worked out by Congress. It states the EPA, and the State of Colorado will together control the massive clean up at the Arsenal. This is the damn stubborn way the Army throws a tantrum to get the pentagon and Congress to change the rules because they don't like being told how to do it the right way. They want to do it the Army way, (haphazardly).

*** Some analysis estimate that the clean up may cost taxpayers two-billion to four-billion dollars If a lengthy legal battle continues, there may not be Federal money available to clean it up, Feinstein said.

*** "We as a nation could not clean up all the contaminated military facilities in the country in the next ten years if we spent the entire defense budget to do it," Feinstein said.

***** Larry says "Does this alone tell us how willfully negligent the Army has raped, and ravaged this Country. This should come out of the Army's budget." "I have never in my life heard of any of the other Armed forces having been so destructive like the ARMY." "The Army has failed to clean up behind themselves." "They have completely failed good management, and they have lost all credibility everywhere." "I think Feinstein has told the truth on the Army, and it's about time they bite the bullet." The legal battle that is always waged because the Army will not follow the law, or practice good business.

***** On the other side is Wirth, several national environmental organizations and an array of Colorado Officials including Governor Roy Romer, Attorney General Duane Woodard and the Adam's County Government.

***** Jim Martin, spokesman for Wirth, said the concern of opponents to Schroeder's amendment is that it doesn't require the Arsenal to be cleaned up to the highest level possible, which means making it fit for commercial or residential use.

***** Wirth supports the goal of preserving the land for wildlife, but he still wants it cleaned up to the maximum extent possible, and for the state to have approval power over the clean up plan.

** Martin acknowledged the Arsenal clean up will be significantly more expensive if Wirth and State Officials succeed.

** Wirth's and Schroeder's staffs met, but couldn't agree on a compromise.

** "Their strategy is to force the federal government to give the state more leverage and more money in the process. We do not think their strategy is going to work," said Feinstein.

** Feinstein said the state wants to "soak" the Army, but in the mean time all the money is going toward legal cost.

** "My answer to that is, we're clearly not doing that and neither is the state," Martin said. "Our highest and first priority is to make sure this place is completely cleaned up so the twenty years down the road we don't discover that one of the Aquifers under the Arsenal is contaminated."

** Meanwhile both sides say they are willing to continue discussion, but don't hold out much hope for a compromise.

***** Larry said, "What I think they meant to say is, not to hold your breath"., "Or close your eyes." Make the Army give back to the State of Colorado that 27 square miles of land the same way it was when it was confiscated by the United States Army, in 1940. The United States Army deprived many farmers of their prized farmland by confiscation. Tore down the homes, moved em or bulldozed them, and set their property in the street.

* By January of 1990, it was discovered that the Defense Department had cut that year's budget for environmental clean up at the Arsenal by ten-million dollars, prompting an angry letter from Denver Congresswoman Pat Schroeder.

***** Larry said, "This reminds in 1972 when we were first poisoned by the water, cattle dying, we were dying a slow death, and ask Congresswoman Schroeder for help. She wrote the Army asking them for medical intervention for the Land's. no reply. She wrote the Army again, and told them they were about as slow as molasses, as she demanded medical intervention for the Lands. they made promises they were going to set up some type of medical monitoring program. Well it's been thirty years now and they never answered that letter or any other letter's.

*** Schroeder sent the letter to Defense Secretary Dick Cheney demanding an explanation for the budget cut.

***** "I am at a loss to puzzle out the logic," Schroeder wrote. "The Army has a terrible reputation for environmental concerns in Colorado. There are still many in the state who attack the Army's efforts and question the Army's motives. A cut in environmental funding at the Rocky Mountain Arsenal will strip away the minimum levels of trust which are just starting to be built again."

*** The reduction was made even though Congress appropriated $600.8 million for that year's budget to clean up pollution at military facilities, which is $83 million more than president Bush had requested.

*** Schroeder said in her letter that she was told, the money was transferred to naval programs.

*** "We do not intend to delay major milestones at the Arsenal, nor change existing provisions in the clean up agreement signed by the Army, EPA, the State of Colorado, and Shell Oil Company."

*** Two months later, nervous residents patiently waited in hopes of seeing for themselves, the findings of a recent study that said burning of toxic waste will not endanger the public.

*** Concerned residents and environmentalists had hoped the draft of a public health risk assessment would have been included among incineration documents open for review the previous month.

*** It wasn't. Primarily because it was completed late in the public-comment period and only in draft form. It will be open for public scrutiny as part of the incinerator design process, an Army spokes-woman said, but not until this summer.

*** The Army, last year, proposed incinerating 8.5 million gallons of toxic chemicals that drained from Basin F.

*** "They asked us to accept their statement that this was the best choice for protecting the environment and the public, but there was no documentation, no proof," said Clara Lou Humphrey, a member of the public technical advisory group studying the proposal.

*** "The public should have been able to go through the risk assessment, to see if it is as good as the Army thinks it is," she said. "This may answer our questions, address our concerns."

*** Researchers looked only at possible airborne contamination, not at skin exposure, contamination of livestock and garden produce, or ground water and surface water contamination.

*** Earlier state health officials supported the incineration proposal, but only if the pollution-monitoring plan is modified and independent oversight of the operation is allowed.

*** It is my opinion, that while most businesses comply responsibly with regulations governing hazardous waste disposal, they need more assistance to actually reduce the amount of hazardous waste produced.

*** More than ninety-nine percent of the governmental spending done for pollution control, is devoted to controlling the pollution after waste is generated. Less than one percent of the spending is devoted to the reduction of toxic waste itself.

*** Approximately two-thousand eight hundred Colorado businesses are registered with the state as generators of hazardous waste. These businesses cover a wide range of specialties that include automotive repair shops, circuit board manufacturers and paint shops.

** Such businesses can reduce hazardous waste output and increase profitability at the same time. In a recent study, conducted by the Colorado State University's Minimization Assessment Center, experts evaluated the waste output of ten Colorado businesses. Funded by the EPA, the center provides technical assistance, free of charge, to companies to help reduce waste generation. If the businesses follow the center's recommendations, experts predicted and annual overall combined reduction for the ten companies of 9.6 million gallons of hazardous liquid waste.

** The estimated one-time cost per company, to make suggested changes to reduce hazardous waste, was thirteen thousand dollars. in return, companies, on average, could expect to save thirty-three thousand dollars a year in hazardous waste treatment and disposal charges.

** While many firms have plenty of engineering expertise they lack professionals trained in hazardous waste reduction techniques. A Colorado manufacturer of aluminum cans produced twenty tons of hazardous solid waste every ten days. The company paid six-thousand dollars a truckload to properly dispose of sludge that contained traces of a toxic chemical

** Experts from the center informed the company about a different chemical can washer that does not contain toxic chemicals.

** The company began to use the new solution, and effectively eliminated all solid hazardous waste output.

** Pollution prevention pays, not only in Colorado but in other states and for individuals as well. The less hazardous waste generated, and stored throughout the United States, the less danger we face of serious contamination to our environment. Such as the crisis we are all facing at the Rocky Mountain Arsenal.

** Businesses do not have a difficult choice. They can, and should, take the steps necessary to reduce environmental pollution for at the same time they will be increasing the company's profitability.

** By June the Army was attempting to ship toxic waste from the Rocky Mountain Arsenal, and to conduct on-site pollution tests, without studying the potential public health risks as legally required by law.

**** Larry Says "The old saying is you can't teach an old dog new tricks. I think that is the problem with the Army, They have for so many years done it their way, without any concern for public health, or the environment, they won't follow laws or rules that are made for them to follow. Here they are trying to sneak one in on the Colorado Department of Health, and EPA, what that is really called deceit and fraud but their use to that too. I believe it's time to order a new army or let one of the other branch's take over their duties and get rid of them. I know they got problems but that what life is all about, fixem when they happen. The United States Army creates their problems by being just plain Reckless, and willful negligence. The Army don't give a damn about the public, if anyone says something they just blow up a smoke screen".

** In their attempt, the Army tried to ship barrels of the highly toxic fluid polychlorinated biphenyl's (PCB's) without EPA approval. The agency halted the shipment. Another shipment of toxic waste was sent from the Arsenal to an Arkansas incinerator.

***** Larry said, "Now we know it was probably shipped by train, can you imagine a train derailment near your town, your home, your children's school. If it were to happen in the wrong place at the wrong time, their could be a lot of victims. "WHY?" Because the Army didn't want to do it the safe way, or the right way, or under the law way, they just improvised and done it the Army way.

** The army's efforts to decontaminate pipes at the Arsenal, will add thousands of barrels of corrosive liquids to the plants already large inventory of toxic waste.

****** Larry said, "It sounds like the Army needs a lesson in pollution prevention, like all the private firms that the Government is working with to prevent pollution. It sounds like the only thing the Army knows is how to create it."

** Representatives of the Tennessee Valley Authority, under contract to the Army, spent six weeks using caustic solutions in an attempt to flush Nerve Gas vapors and residue from the pipes in the old production area known as the "North Plant".

***** Just a few pages back Larry explained. "How several workers for the Tennessee Valley Authority, were seriously injured as they were trying to flush the nerve gas vapors and residue from the pipes with this very caustic solution."

** This decontamination work preceded a visit by a Soviet delegation that was to verify that chemicals were no longer being produced at the Rocky Mountain Arsenal.

***** Also in June another new twist was added to this toxic nightmare.

***** Particles called colloids apparently can attach themselves to, or encase, bits of toxic elements and allow them to flow for long distances in water undetected, researchers said at the Los Alamos National Laboratory.

***** Study's suggest that Colloids could confound predictions of how far pollutants travel in ground water. "That means some hazards might have been underestimated. Colloids probably play a more important role than people expected in the past," said Wilfred Polzer, one of the study's co-author's.

*** In August of 1990, Nikita Petrovich Smidovich, chief of chemical weapons issues for the Soviet Union's Ministry of Foreign Affairs, led a fifteen member Soviet Delegation on a three day tour of the Arsenal. Their objective: To see for the first time how the United States produced Nerve gas, and Mustard agents, and to verify that the Arsenal is no longer in operation.

***The group accompanied by the U.S. delegation, toured old Nerve Gas and Mustard Gas facilities within the Arsenal. At the end of the tour both delegations went to the chemical destruction facilities at the Pueblo Army Depot.

*** "We've already learned some very important lessons on this visit that will be reflected in the negotiations," Smidovich said. "It's better to see the U. S. facilities once then discuss them for years."

*** Members of both groups then met again in Geneva to continue talks on destroying chemical weapons and verifying their destruction.

** Here's some information on an event that took place in September of 1990. In itself it's not very amusing, but it does confirm some information which I was later given.

** Officers at Buckley Air National Guard Base had reached an uneasy truce with the enemy. Airmen would move prairie dogs from the Air base, to the Rocky Mountain Arsenal instead of gassing them. This is to replenish the prairie dog families into new locations on the Arsenal, since the thousands that once made the Rocky mountain

Arsenal their home were so chemically poisoned The United States Army exterminated them all so they could save the hawks on the reserve. This is the Army's continued effort to keep the wildlife on the Arsenal replenished and alive.

** "A spokesman for the Air Force said" "We're looking for good, effective, passive measures to control them," "We decided not to use poisons or gases. That is not an effective way of managing them." "Moving them to the Rocky Mountain wildlife Reserve needed to be replenished. "Hope we can find a good home."

***** Larry says, "The Air Guard said poisons, and gases are not an effective way of managing them. I wonder what will happen to them on the most polluted piece of land on earth? Will that be a better and more effective way of managing them?."

** Chris Bishop, vice president of Prairie Dog Rescue, reacted to the news with wary optimism. "It's Hard to believe, because in the past they have always relied on gassing them for control," replied Bishop. "But I would certainly be happy if they decide to deal with them in a more humane manner."

** The three-thousand acre Buckley Air Base, on the east side of Aurora, Colorado is riddled with seventy-thousand prairie dog burrows. Officers are particularly concerned about their proximity to buildings and runways.

*** The only thing we'll do this fall is respond to a request from the Arsenal for prairie dogs," said a spokesman. "Their prairie dogs died and they will have eagles coming down from up north that depend on them for food."

***** "The prairie dogs at the Rocky Mountain Arsenal were so contaminated, with various toxins, as they lie on the ground dying or dead the eagles would swoop in for a meal. They then become contaminated and died passing on the contamination trail to the young and in many cases what was suppose to be the young. The egg Shells were so thin, even though the eagles tried hatching what they had the hatchling's didn't make it. Driving the bald eagle to near extinction.

**** Sometime earlier, I had been shown a report that stated prairie dogs on the Arsenal had been killed by cyanide gas because they were severely contaminated. Prairie dogs remain through out the year on the Arsenal, eagles are there for only a few months. When the eagles arrived and started eating the prairie dogs, they too would become sick and suffer chemical death, the Army did not want that because it would ruin the image of the pristine wildlife refuge, which it is definitely not. (The Army is blowing a smoke screen).

***** If, after getting this far into the book, and nothing has shocked, or made you realize just how bad the military and large chemical corporations have defiled this land we call home, maybe this will.

*** The following are military bases located in Colorado that the National Toxic Campaign Fund calls "potential contamination sites."

ARMY Number of toxic sites identified

Aurora Army Reserve Center No. 1	....	1
Aurora Army Reserve Center No. 2	....	2
Boulder Armed Forces Center	....	6
Denver Army Reserve Center	....	3
Fitzsimmons Medical Army Center	....	1
Fort Carson, Army Base	....	49
Fort Carson Armed Forces Reserve Center	....	1
Fort Carson Army Center	....	9
Fort Collins Army Reserve Center	....	11
Pueblo Army Depot Activity	....	35
Rocky Mountain Arsenal	....	155
Subtotal of sites for United States Army	....	273

Air Force Number of Toxic sites

- -

Air Force Academy	....	11
Air Force Plant PJKS-Martin Marietta	....	22
Buckley Air National Guard	....	9
North American Aerospace Defense Command	....	1
Lowry Air Force Base	....	11
Peterson Air Force Base	....	9
Air Force Subtotal	....	63
Air Force and Army toxic Chemical sites in Colorado	....	346

*** Looking at these figures, you can plainly see the number of contaminated sites for just one state, is awesome. If we multiple just 20% of that total by the number of states, we can get a pretty rough idea of the type of environment we are living in.

*** There are four more items I would like to discuss before ending this chapter, I would like to share some medical information That I find disturbing, and it touches us all.

***** Steep increases in mortality from a mysterious affliction called Motor Neuron Disease have occurred in the United States since the 1960's.

*** Motor Neuron Disease, or (MND), refers to a group of neurological conditions that damage nerve cells in the brain and spinal cord, causing muscle weakness, wasting and other symptoms perhaps the best known condition is Amyotrophic Lateral Sclerosis, or Lou Gehrig's disease.

*** "A steep rise in the incidence of MND suggests that the population has experienced greater exposure to causal factors at some time in the past few decades," said Dr. David Lillenfeld.

*** He cites the possibility that the same factors may be involved in Alzheimer's disease and other degenerative nerve diseases.

*** Lillenfeld, who headed a new study, said deaths from MND increased in all population groups over age forty, with the steepest increases occurring among older people. He is with the division of environmental and occupational medicine at Mount Sinai School of Medicine in New York.

*** The study, which was reported in the Lancet, a British medical journal, examined changes in MND mortality over the last twenty-three years. It found that the largest increases have occurred among older people.

*** MND has increased by 19% among men and 46% among women age's 55-59; 33% among men and 45% among women age's 60-64; 46% men and 86% among women age's 65-69; and 5% among men and 117% among women age's 70-74.

*** Lillenfeld noted that MND's predilection for older people suggests that it will become an increasingly common medical problem as the American population ages.

*** MND mortality for women age's 50-54 is alarming it has increased by 50%, but for men of the same age, it has remained essentially unchanged.

*** Lillenfeld and his associates,, in contrast, argue that the increases are real, and cite statistical indications that the environmental factors are playing a very important role.

By Sandra Blakeslee; The New York Times.

PESTICIDE PRODUCES PARKINSON'S SYMPTOMS

***** An organic pesticide widely used on homegrown fruits and vegetables and for killing unwanted fish in the nation's lakes and rivers produced all the classic symptoms of Parkinson's disease in rats that receive steady amounts of the chemical in their bloodstreams scientist said Saturday.

**** While it is much to soon to say that the pesticide, rotenone, causes or contributes to Parkinson's disease in humans, the scientists said the finding was the best evidence thus far that chemicals in the environment may be factors in this devastating disease.

**** Their study, the first to implicate rotenone in Parkinson's disease, was described here on Saturday at a workshop on the neuro-biology of disease, held in conjunction with the annual meeting of the Society for Neuroscience, the nations largest gathering of brain researchers.

**** The workshop involved work carried out by Dr. Timothy Greenamyre and colleagues at Emory University in Atlanta. The results of the study will be published in the December issue of the journal Nature Neuroscience.

**** "This is a very important new study," said Dr. William Langston president of the Parkinson's Institute, a leading center for research and treatment of the disease in Sunnyvale, Calif. "It is the next major step in Parkinson's disease research."

**** Dr. John Trojanowski, an expert on neuro degenerative diseases at the University of Pennsylvania and the moderator of the workshop, said, "This is the best model we have ever had for this disease being associated with an environmental agent."

**** But Trojanowski cautioned that the findings "may not represent what anyone would experience in the real world." For one thing, the rats in the study were exposed to the chemical through their jugular veins, so it was not broken down or metabolized in the digestive tract.

*** Rotenone is extracted from the dried roots, seeds and leaves of various tropical plants, including the jewel vine derris, custard apple and hoary pea.

*** Rotenone is found in 680 compounds marketed as organic garden pesticides and flea powders.

*** Unlike many artificial pesticides. which linger in the environment, rotenone breaks down within days in the sunlight.

*** Rotenone is also widely used in liquid form by fishery managers to destroy pest species.

***** Parkinson's disease is one of the most common neuro-degenerative diseases, affecting nearly 1 million Americans over the age of 50. The disease is caused by the steady loss of cells, in a tiny region of the brain called the substantia-nigra, that produces a chemical, dopamine, which is crucial for movement and cognition. Patients develop jerky, tremulous movements that the worse with time.

***** What happened in the Tokyo Japan subway, is something that could happen right here, whether it be a subway, a skyscraper, the water supply

"By Nicholas D. Kristof New York Times News. and Eric Talmadge Associated Press.

*** On March 20, 1995 Terrorism suspected 6 killed, 600 injured apparently by nerve gas, during the rush hour early today in what police called "a guerrilla attack" on the Tokyo subway system.

*** No group immediately has claimed any responsibility, police said that they thought that a known terrorist organization had set off the attack as an act of sabotage. Police said it's "highly possible" the gas was Sarin Nerve Agent. An Ultra toxic, and volatile nerve agent developed by Nazi scientists in the late 1930's.

*** "This is a case of organized and indiscriminate murder," said Masahiro Terao, head of the Metropolitan Police First Investigative Division at a hastily called news conference.

*** Ambulance sirens wailed in Tokyo as police and rescue forces rushed to the affected subway stations on three main lines that traverse central parts of Tokyo. Among the stations where people were gassed were those serving the national parliament and the Foreign Ministry.

**** As trains pulled into the stations, passengers staggered out onto the platforms and collapsed. Emergency workers set up tents outside subway stations, and passengers were rushed out on stretchers and lay on the ground with bubbles coming from their mouths. In some cases, blood poured from their noses.

**** No explosions were reported, and some passengers reported that a liquid appeared to come from lunch boxes wrapped in newspapers. Police said that these boxes may have been the source of the gas. They warned passengers not to touch any such box and to report it immediately.

**** "I saw no gas, but I saw a transparent liquid spreading on the floor, and people falling on the ground one by one," A young woman told Japanese television. She was not hurt.

**** Prime Minister Tomiichi Murayama ordered an all-out rescue effort, and the government ordered increased security at all public railways, airports and ports.

**** One reason why police think terrorists were responsible is that the gas came from perhaps as many as 15 different points, all in subway cars. The subway stations affected were on the Hibiya, Marunouchi and Chiyoda lines.

**** In one case, passengers reported that a man in sun glasses-unusual for the subway-apparently left a package behind when he got off the subway at the Ebisu station on the Hibiya line. The package was blamed for one of the gas outbreaks.

**** A similar but much more minor incident was reported on a subway on March 5, in Yokohama, a major port city adjacent to Tokyo. Eleven passengers in a subway car were hospitalized after they complained of Dizziness and eye pain but police said this morning that they never found the source of the gas or made any arrests.

**** Last June seven people died in a mysterious case of gas poisoning in Matsumoto, a city in Nagano prefecture in central Japan. They were in their homes at the time of the poisoning, and it was unclear just how they had been gassed. Sarin was also blamed for that outbreak, but police never made any arrests or determined a motive.

**** Sarin is not available in Japan. It is made by mixing chemicals containing fluorines, and organic phosphorous, and it causes breathing problems and seizures, it attacks the central nervous system. Authorities said the 0.5 milligrams of sarin is enough to kill an ordinary-size person.

**** The Nazis did not use the poison gas they had developed, but Japan in close cooperation with the Nazis developed poison gases itself. Japan used poison gases during World War II in fighting the Chinese troops——though not in battles with Americans or Europeans—and large quantities of poison gases were left behind in China at the end of the war.

**** Many foreigners live along the Hibiya line, near such stations as Hiroo and Roppongi, but no foreigners are known to have been among the dead or injured. The Hibiya line was closed.

**** Other subway stations affected were Kamiyacho, Kokkai-Gijidomae, Hibiya, Hacchobori, Kodenmacho and Nakano-Sakaue.

Kokkai-Gijidomae is the station that serves the national parliament, while Hibiya is a central business district near the Imperial Palace.

**** Passengers reported dizziness, blurry vision and nausea. "I was losing my sight," a subway employee told Japanese television, "I couldn't see, and I was feeling dizzy."

**** At one station, subway employees tried to pick up the boxes from which the fumes were coming and carry them out. Then they fainted.

**** "I saw people coughing, and I thought they were sick," a male passenger told the television. "but then it started all over."

**** There have been periodic incidents of terrorism in Japan over the years, but Monday's incident does not fit the pattern of past events. The leftist Red Army Faction was active in the 1970's but has been quiet lately. Another leftist faction, Chu-kaku ha, or Central Core Faction, has set off rocket attacks apparently in an aim to embarrass a government that it regards as illegitimate.

**** The Chu-kaku ha is supporting several groups that oppose an expansion of Narita airport, the main airport serving Tokyo. While the dispute over Narita has been blamed for several bomb attacks, subways have not been a target and in any case there has recently been a move toward conciliation of the Narita dispute.

**** Right-wring groups have attacked some individuals but have not been blamed for general terrorism against the public.

***** Mass subway poisoning fills up Tokyo hospital. Subway passenger's are carried out of Tsukiji station this morning in Tokyo as medical personnel wait to take them to the nearby ST. Luke's Hospital.

**** Hiromi Oka stood at a public telephone in the hospital lobby, an intravenous drip hanging from a wheeled rack behind him, a needle in his arm.

**** "No, I'm all right." he said weakly into the phone. "But I never expected anything like this to happen on my way to work."

**** Neither did the staff of St. Luke's International Hospital in downtown Tokyo, which after today's suspected mass subway poisoning was filled to overflowing.

**** On the first floor of the hospital dozens of people crowded around as nurses taped up the latest list of patients. In two hours, the poster-sized lists had swollen from just two to five, each one bearing the names of about 30 people.

**** "I just wanted to go to work," said Akio Masahata, 21, as he held an ice pack over one eye and sat up on his gurney.

**** Masahata, who looked pale and tired, said he stayed on the train for few moments after it stopped and the passengers had been told there was an explosion.

**** "Then I started to smell it," he said. "It hurt to breathe, I could feel it in my nostrils. When I realized it must be gas, people were starting to collapse around me."

**** "It's terrible," said Dr. Ayumizo Mikami.

**** "We've never had anything like this before."

***** What you've read is real, Terrorists has warned the world, "We'll getcha." "in the subway's, skyscraper's, plane's, buses, or train's". This is real & could happen you, be alert for signs of terrorism anywhere anytime. "So they can never say we gotcha."

UNION CARABIDE TO PAY $470 MILLION IN BHOPAL LEAK:

"Rocky Mountain News" February 15, 1989.

***** "Around the world in Bhopal India, right near home for thousands, a devastating gas leak from Union Carbide in 1984 left more than 20,000 people suffering from the effects of the gas, and victims die at the rate of at least one a day, according to Indian officials.

***** "This is a betrayal of the gas victims," said Babulal Gaur, A State legislator from the opposition Bharatiya Janata Party. He called the settlement negligible.

***** "The government has once again come under the pressure from the lobby of multi-nationals," said Abdul Jabbar Khan, who mobilizes demonstrations of gas victims.

***** Many of the victims of the disaster were unaware of the settlement because they are too poor to own radios.

**** Slack says it is still Union Carbide's contention that the gas leak was the result of sabotage by a disgruntled employee. The Indian government contends that the disaster was the result of negligence.

**** NEW DELHI,, INDIA Union Carbide Corp. agreed yesterday to pay the Indian government $470 million in a court-ordered settlement resulting from the 1984 gas leak at Bhopal that killed more than 3,300 people in the World's worst industrial disaster.

***** Activists in Bhopal denounced the settlement as a betrayal of the 20,000 victims who still suffer from exposure to the deadly gas that escaped from a pesticides plant on December 3, 1984 The government had sought $3 billion in damages.

**** Chief Justice R. S. Pathak interrupted a government prosecutor's routine argument when the court reconvened after lunch and order the U. S. based multinational Company to pay the damages by March 31.

***** Attorneys for the government and Union Carbide promptly agreed.

***** "It was apparent that there was an out-of court agreement between Union Carbide and the government." said a court official who spoke on condition of anonymity. "for such an order, there should have been excitement, but there was no murmur even." "It wasn't entirely out of the blue," said another source, also speaking anonymously.

***** Pathak, citing "the enormity of human suffering," said a settlement was needed to "provide immediate and substantial relief."
*****More than 2,000 people were killed almost immediately when the white vapor of methyl isocyanate seeped from a storage tank at the plant operated by Union Carbide's Indian subsidiary and drifted over nearby shanty-towns and into Bhopal. *****The leak occurred shortly after midnight, and some victims died in their sleep. Others, blinded by tears and gasping for breath, tried to flee but collapsed in death.

***** More than 20,000 people still suffer from exposure to the gas and victims continue to die at a rate of at least one a day according to government gas relief board. It says the death toll has reached 3,329.

***** Pathak, speaking for a five-judge supreme Court panel, ordered Union Carbide to pay $470 million to the Indian government "in full and final settlement of all claims, rights and liabilities related to and arising out of the Bhopal gas disaster."

***** He also ordered all civil proceedings transferred to the Supreme Court and quashed all criminal charges, including one of homicide filed in 1987 against Union Carbide ex-chairman Warren Anderson.

***** Pathak gave no details of how the money should be paid to the victims, but he directed government prosecutors and attorneys for the Danbury, Conn. Based company to submit a detailed agreement today.

***** "This is a fair and just settlement," Gopal Subramanium, told a reporter.

**** In New York, Union Carbide spokesman Earl Slack said Pathak's order "was based on its review of all pleadings in India and the U. S., applicable law and facts and the enormity of human suffering that requires substantial and immediate aid."

**** James Wilbur, a vice president with Smith Barney,, Harris Upham and Co. Inc. in New York, said Union Carbide had set aside $200 million and had $250 million in insurance in Bhopal, a city of nearly a million people.

"WELL" Settlement boosts stock of carbide:

***** DANBURG, Conn. Wall street was delighted yesterday by Union Carbide Corp's $470 million settlement over the 1984 Bhopal gas disaster that killed more than 3,300 people, but an attorney who represented some victims said the chemical giant got off to cheaply.

***** Union Carbide's stock price soared $2,121/2 to $31.25 a share in New York Stock Exchange trading after news of the settlement with the Indian government and the company's estimate that it expected a relatively mild impact on it earnings.

***** Melvin Belli, the attorney who represented the interests of about 2,000 plaintiffs before the cases were transferred to Indian courts from U. S. courts, called the settlement "much, much too little." He said two to three times as much would have been fairer.

***** "They got off criminally cheap," Belli said from San Francisco. "We had that amount of money two years ago in New York, and it wasn't enough, and it isn't enough."

***** The Supreme Court of India, sitting in New Delhi, ordered the $470 million payment by March 31 to settle all claims resulting from the leak of deadly methyl isocyanate at the Bhopal pesticide plant, which was operated by Union Carbide's Indian subsidiary.

***** More than 3,300 people have died as a result of the December 2, 1984 leak, the world's worst industrial disaster. The plant has been closed since the incident.

***** Leslie Ravita, a research director for Solomon Brothers Inc. in New York, said of the settlement: "Psychologically, it's terrific. Financially, it's reasonable....This relieves the pressure on Union Carbide and the stigma."

***** The company said it expected an earnings impact of no more than 50 cents a share from the settlement.

**** In the year that ended Dec. 31 Union Carbide earned $20 million, or $5.31 a share, on revenue of $8.32 billion. That compared with 1987 earnings of $231 million, or 1.76 a share on revenue of $6.91 billion.

**** I think it's a decision that was pretty much anticipated," he said.

**** Union Carbide had offered to pay the Indian victims $350 million as early as 1986. The settlement was accepted by U. S. Courts but rejected in India.

***** WE'LL GO BACK TO FACTUAL DOCUMENTS AND REPORTS, THAT WAS BROUGHT, AND PLACED INTO EVIDENCE WITH THE COURT OF CLAIMS BY LARRY LAND THE PLAINTIFF, AND HIS ATTORNEY'S 1989, 90, 91, 92, AND 93. THIS IS THE REPORT AT THE TIME OF THE HEARING OFFICER WILKES C. ROBINSON JUDGE. THIS IS FACTUAL EVIDENCE THAT WAS NEVER REPORTED BACK TO THE CONGRESS OF THE UNITED STATES. THE COURT OF CLAIMS IS A FACT FINDING COMMISSION WHO WAS TO REPORT THEIR FINDINGS BACK TO THE U. S. CONGRESS. SIGNED BY HIM AND FILED WITH THE COURT ON OCTOBER 31, 1993.

*** The plaintiff urged the Court that the record does reflect genuine issues of material fact precluding summary judgment. For the reasons stated below, Army's, (defendant's) Motion is denied.

*** Judge Robinson "heard and was placed into evidence there were 520 head of calves 250 were 10 to 12 weeks old with 270 head at 16 weeks old to yearlings. Land had purchased in 1972, and the remainder 135 head of 2 year olds, was purchased and raised in 1970, and 71, at his Boxelder ranch. Approximately 80 were already with calf. Bill of sales was supplied to the Court for evidence on all livestock."

*** Judge Robinson's order recorded as evidence "Plaintiffs rely on a report by Dr. Frederick W. Oehme, D. V. M., Ph.D., which opines that the cause of the death or eventual destruction of all but three of the calves was the presence in the well water of the man-made compound DIMP and it's breakdown product IMPA."

*** Judge Robinson's order placed into evidence. "They also rely on a recent report by Dr. Daniel T. Teitelbaum, M.D. concluding that the cause of the health problems suffered by plaintiffs and Larry Land's calves was DIMP and IMPA."

*** Judge Robinson's order, was placed into evidence, "The RMA. halted monitoring in 1960. It refused to resume testing until 1974-75, despite the fact that a report on groundwater contamination in the relevant area by the U. S. Department of Health Education and Welfare (HEW), completed in December 1965." Showing 25,000 acres off the arsenal chemically polluted.

*** Judge Robinson's order, and was placed into evidence, "Recommended that the well-monitoring programs of the Tri-County Health Department, the Colorado State Department of Health, and RMA. itself be continued. HEW states:"

***** "The United States Army, was totally aware in 1965 that contamination from the Arsenal was polluting wells off the Arsenal, and were advised by the Army Hygiene, and the U.S. Department of Health they were in fact using basin F on the premise it is leaking." By Larry Land

*** Judge Robinson's evidence order stated "Contamination of the Aquifer is likely to persist for many decades if the trend of relatively small decreases in chloride concentrations continues."

*** Judge Robinson noted in his order and placed into evidence. "In May, and June of 1975 the Colorado Department of Health had written a letter to Mr. Land noting their drinking, and hygiene water was contaminated with pollution from the Rocky Mountain Arsenal. The Tri County Health Department also wrote a letter to Mr. Land explaining to him that Arsenal Pollution contaminated his drinking water and hygiene water and that for the livestock. These letters were placed into evidence at that time."

***** Review of records and reports supplied to Judge Robinson, to enter into evidence was, "Dr. Daniel T. Teitelbaum, M.D. review of the records which were supplied to me establishes unequivocally that the Land family's groundwater was contaminated with materials from the Rocky Mountain Arsenal. In particular, the finding of DIMP....establishes beyond any question that the source of the material in the groundwater under the Land's property and in their drinking water, and hygiene water wells was the Rocky Mountain Arsenal. No other source of this material is conceivable in that area. This finger-printed chemical establishes without any question that the toxic material which reached the Land family had, as its source, the Rocky Mountain Arsenal GB manufacturing facility. In addition, the finding of IMPA, a metabolite of DIMP, further identified the material as having a source in the Rocky Mountain Arsenal.

***** This was acknowledged by Judge Robinson was part off his order, and placed into evidence in this case. In 1965, the Army Environmental Hygiene Agency recommended that steps be taken to eliminate Lake "F" as soon as possible, and thereby remove much of the present environmental hazard of exposed surface storage to toxic wastes. By 1969, there was evidence that Basin "F" was leaking; indeed, a physical inspection reflected that sections of the protective membrane were absent, Thus by the spring of 1970, the Army understood that it was operating Basin "F" on the premise it was leaking.

***** Presented to Judge Robinson, and in his order and placed into evidence. "December 1974, the Colorado Department of Health detected DIMP in a well near the City of Brighton Colorado. Was located 6 miles straight north of the Land Farm. 7 miles north of the Arsenal. proving that the pollution from the Arsenal traveled in a northerly direction, under the Lands property. The State of Colorado Health Department issued Cease and Desist orders April 7, 1975, against SCC. and RMA. The orders required to immediately stop of post discharge both surface and subsurface., and to develop and institute a surveillance plan to verify compliance."

***** Judge Robinson's order and in evidence., "Activities on the Arsenal have resulted in one of the worst hazardous waste pollution sites in the country. The Army estimated at this time at least 120 contamination sites." 'Larry says has now grown to near 160 sites.'

***** Judge Robinson's order, and in evidence. "The House of representatives referred this case to the Court of claims, September 22, 1988. Plaintiffs filed an amended complaint on October 15, 1990, requesting legal and equitable relief. Plaintiffs allege 5 counts. (1) negligence; (2) strict liability, ultra hazardous activity and inherently dangerous activity; (3) trespass; (4) nuisance; and (5) deceit based upon concealment."

**** Judge Robinson's order placed in evidence. "H. R. 816 states A BILL; Be it enacted by the Senate and House of Representatives of the United States of America in Congress assembled, That the Secretary of the Treasury is authorized and directed to pay, out of any money in the Treasury not otherwise appropriated, the sum of $__________________ to Larry Land, and Marie Land et al. This sum shall be in full and complete satisfaction of all their claims against the United States based upon health problems and other related injuries resulting from the operations and activities at the Rocky Mountain Arsenal, Denver Colorado, conducted under the auspices and control of the United States Army

**** Judge Robinson's order placed in evidence "H. Res. 61 states; RESOLVED, That H.R. 816 entitled "A bill for the relief of Larry Land. Marie Land, and others," together with all the accompanying papers, is hereby referred to the Chief Commissioner of the Court of Claims pursuant to section 1492 and 2509 of title 28, United States Code, for further proceedings in accordance with applicable law."

***** Judge Robinson order placed in evidence: "This case is before the court pursuant to 28 U. S. C.: 1492 (1988) which provides that "[A]ny bill, except a bill for a pension, may be referred by either House of Congress to the Chief Judge of the United States [Court of Federal Claims] for a report in conformity with section 2509 of this title. "Under section 2509, the hearing officer is charged with proceeding "in accordance with the applicable rules to determine the facts… [The hearing officer] shall append to his findings of facts conclusions sufficient to inform Congress whether the demand is a legal or equitable claim or a gratuity, and the amount if any legally or equitably due from the United States to the claimant."

*** Judge Robinson's order placed in evidence. In a congressional reference, plaintiff may assert a legal or equitable claim."

**** "Legal Claim; In law and evidence of Judge Robinson, in this case." A legal claim arises from substantive law i.e. the Constitution, a statute, a regulation, or some principle of common law, and is based on the invasion of a legal right.

**** Plaintiffs are barred for any legal claim to relief, by the tort claims statutes of limitations. The term legal claim carries no special meaning in a congressional reference."

**** Equitable Claim. As explained and placed into evidence by Judge Robinson, "in this case. "the principles of right, justice and morality create equitable claims." "The term equity in this context is used in the sense of broad moral responsibility, what the government ought to do as a matter of good conscience." "Equity, thus, contemplates a remedy to a claim in a variety of circumstances where relief would be otherwise unavailable in a traditional jurdicial setting, such as when there is no remedy under existing law or the remedy is time barred by a statute of limitations. While a legal claim founded on the constitution, a statute, a regulation, or a common law principle, such bases are not necessary to warrant relief in equity. [T]o ascertain the proper standard for fixing liability, it is helpful to take in account of the principles governing the particular area of law bearing on the claims. In judging whether there is an equitable claim, however, these legal guidelines need not be rigidly applied. Rather, the particular facts may be examined to determine whether the plaintiffs suffered an invasion of rights for which compensation is due them."

***** Gratuity; In Judge Robinson's order, placed in evidence. "Congress constitutional power to pay debts. In a congressional Reference, an award of monies based upon moral considerations without finding a degree of Government fault is a gratuity.

***** Judge Robinson order early in 1990 Placed into evidence. "That the case would be heard under Colorado Law. This test directs us to the law of Colorado for a determination of what constitutes an unjustified act of omission, fraud, deceit, or ultra hazardous or adherently dangerous activity, trespass or nuisance. It was agreed and stipulated to use circumstantial evidence under Colorado law, which is circumstantial proof because of the probability of the presence of one or more contaminants which may or may not be identified.

***** Colorado Law is based on circumstantial evidence and not based in any way on the theory of a one item basis, such as is the FTCA. Colorado law theory is based on multiple items as a cause of action. Probable cause is somewhere between near submission, but not beyond a shadow of a doubt

***** Judge Robinson ordered and placed into evidence, "the FTCA. or Federal Tort Claims Act. Plays no part in a case before the United States Court of Claims."

***** Judge Robinson ordered and placed into evidence, "This Court shall not bind plaintiffs to a preliminary showing of all elements required to support a strict negligence claim. Also this court's order on September 13, 1990, does not limit plaintiff to a strict showing of negligence. It was concluded in order to recover on an equitable claim, plaintiffs must show that (1) the government committed a wrongful act; (2) this act caused damages to plaintiffs."

**** Judge Robinson ordered and placed into evidence, "Wrongful within the meaning of a Congressional Reference. As stated above the court conceives that there may be an area of non-intentional, non-negligent, "wrongful" action, for which the government may be held liable in equity in a Congressional Reference."

***** Judge Robinson ordered, and placed into evidence, "In regard to whether the activity at the RMA constitutes ultra hazardous or abnormally dangerous activity, defendant has failed to convince the court that Colorado would not consider it such. Defendant has been disposing of toxic chemicals and incendiary munitions since 1942 in a facility located a few miles from the city of Denver. Seven miles from the city of Brighton. and within a few miles of farmland such as plaintiffs'. Defendant was aware as early as 1951 that its chemical disposal techniques posed a risk to neighboring farmers. In the 1950's, the Army began receiving studies and reports finding that contaminants from the RMA were polluting land to the northwest of the Arsenal. The Army also received reports outlining recommendations to stop the polluting. The Army ignored many of the recommendations, however, including one in 1965 to discontinue use of Basin F. The record reveals further that the Army knew, as early as 1969, that Basin F was leaking, yet it took little or no immediate action to resolve the problem. The U. S. Court of Appeals for the 10th Circuit, stated that Colorado could consider such activity ultra hazardous or abnormally dangerous. This court accepts that the manufacture and disposal of toxic munitions is ultra hazardous or abnormally dangerous activity."

***** Judge Robinson, Placed into evidence," The Memorandum from the legislature does not evidence an intent to relieve liability for either the production or destruction of toxic munitions. Moreover, defendant has not shown that its manner of storing and disposing of toxic munitions was authorized, or that any resulting damages were a necessary consequence of that activity.

***** Judge Robinson, placed into evidence, "The referencing statute in the present case contains no concession of liability. However, Congress would not have needed to refer this case to the court for "claims against the United States based upon health problems and other related injuries resulting from the operation and activities at the Rocky Mountain Arsenal, merely to affirm the the FTCA or RMA enabling legislation barred any recovery for liability arising out of the operation of the RMA."

***** Judge Robinson ruling placed into evidence, Defendant has failed to convince the court that operation of an ultra hazardous or abnormally dangerous activity in this instance and under the attendant circumstances, i.e., a few miles from working farmland with knowledge that the basin was leaking, is not a "wrongful" act and that compensation for injuries resulting from such activity. As this court finds that such activity might be considered "Wrongful" "The defendants conduct falls within the scope of a congressional reference.

***** Judge Robinson order placed into evidence. "It is axiomatic that in engaging in a particular activity every person is bound to exercise that reasonable care and caution which would be exercised by a reasonably prudent and cautious person under the same or similar circumstances. as a matter of law, courts will hold every reasonably prudent and careful person to the exercise of the utmost care and diligence in protecting the public from the danger necessarily incident to the carrying on of a hazardous business."

***** Judge Robinson order, and placed into evidence. "There are instances where due to the gravity of the risk created by them, an enhanced measure of care on the part of the defendant is required."

***** Judge Robinson order, and placed into evidence. "Plaintiffs argue that the Army's conduct on the RMA, manufacturing and demilitarization of chemical weapons and disposing of chemical wastes, created a substantial risk. "THIS COURT AGREES." "The point is made that the Army had, from the 1940's been engaged in extremely risky activities that demand the utmost care and diligence to protect the public.

***** Judge Robinson ordered, and placed into evidence. "This Court is satisfied, after resolving all doubts as to the facts, presumptions and inferences in favor of plaintiffs, that genuine issues of material facts exist as to plaintiffs' claims including those sounding in negligence. The Defendant has failed to convince the court that no issue of material fact exist as to defendant's negligence of wrong doing. The defendant has also failed thus far, to convince the court that testimony would be required in this case to establish a standard of care, only when it involves questions beyond the competency of ordinary persons."

CHAPTER NINE

There is no happiness, there is no liberty, there is no enjoyment of life, unless a man can say, when he rises in the morning, I shall be subject to the decision of no unwise judge today.

Daniel Webster, March 10, 1831

***** Shocking Tragedy, an eye opening experience, After three decades of efforts to discover how a pregnant woman's environment can affect the health of her fetus, her baby, her child, researchers are turning their attention to the fathers.

**** The research, much of it in the early stages, suggests that certain substances can cause genetic mutations or other alterations in the sperm, that lead to permanent defects in children.

**** These include familiar birth defects such as heart abnormalities, and retardation as well as the less familiar one like childhood cancer and learning disorders.

**** These few findings may force health officials and occupational safety experts to rethink or expand regulations intended to prevent birth defects that have limited women, but not men, from jobs considered hazardous to the fetus.

**** Each year in the United States, at least 250,000 babies are born with physical birth defects while thousands more develop behavioral and learning defects that appear to have a genetic component.

**** The cause of sixty to eighty percent of the birth defects in the United States is unknown, although many scientists suspect that environmental toxins play a sizable role in a number of them. The male contribution could be substantial, researchers now say. In all honesty I have to assert my opinion and agree with them.

**** Society has focused on the mother and fetus because they are easier to study, and this may not be correct.

**** Since Thalidomide vividly demonstrated that the drugs a woman takes during pregnancy can harm her fetus, scientists have discovered more than thirty drugs, viruses, chemicals and other substances that can cross the placenta and cause structural damage to the fetus. Researchers estimate that another nine-hundred chemicals are toxic to human development.

**** As a result, some American Companies have developed fetal protection policies that banish women of child-bearing age from the factory floor, even if they do not intend to have children.

**** Animal experiments and human epidemiological research previously had linked men's exposures to certain substances with birth defects in their children.

**** Scientists also held to what some refer to as a "mach sperm theory of conception," The idea being that only the fittest sperm were hardly enough to go the distance necessary to fertilize an egg.

**** In fact, research now shows that tiny hairs in the female reproductive tract move sperm along whether they are healthy or not.

**** The tools of molecular biology are now pointing to possible mechanisms in which damage to sperm could lead to birth defects. Scientists can pluck single diseased genes from cells, Examine the hundreds of newly discovered proteins in sperm, place special markers on sperm to follow their development and watch how chemicals interact with sperm proteins and DNA.

**** Researchers have found several childhood cancers that primarily arise from new mutations traced to the sperm, never to the egg.

**** Sperm are particularly vulnerable to mutation because the stem cells that produce them are among the most rapidly dividing cells in the human body. Cells are most vulnerable to genetic damage when they are dividing, they have more opportunities to absorb, metabolize and excrete nasty substances than do quiescent cells.

**** Moreover, researchers now realize that the barrier between blood vessels and tissue in the testes is very thin, allowing many toxic substances to enter testicular structures and seminal fluid.

**** Animal studies have identified more than 100 chemicals that produce spontaneous abortion or birth defects in the offspring fathered by exposed males.

**** Among them are alcohol, opiates, heroin, and methadone, gases used in hospital operating rooms, lead, solvent's pesticides and a variety of industrial chemicals.

**** A very high mortality rate in fetuses, or babies born to women exposed to Mercury Biocides, as in paint, is banned in the United States. If they live through the experience they will most probably suffer Birth defects, severe learning problems, and retardation.

**** In some instances, litter size in greatly reduced or the offspring are deformed. In other cases, the young animals appear healthy but cannot negotiate mazes as well as control animals.

**** In one study, male rats were exposed to low levels of lead, equivalent to amounts encountered by many factory workers.

**** The male rats offspring showed defects in brain development, even if the female rats were not exposed to lead.

**** Other scientists have conducted scores of epidemiological studies looking for links between a father's occupation and birth defects in children, and they have found numerous associations.

**** For example, wives of men exposed to vinyl chloride and waste water treatment chemicals have more miscarriages. Welders who breathe toxic metal fumes develop abnormal sperm, even after exposure stops for three weeks. Firemen who are exposed to toxic smoke have an increased risk of producing children with heart defects.

**** Several studies have found that father's who take two or more alcoholic drinks a day may have smaller than average infants.

**** A British study recently found that men exposed to low levels of radiation at a single nuclear power plant had a higher that expected number of children with leukemia. The greater the exposure, the greater the risk.

**** American investigators have found a link between father's exposure to the defoliant Agent Orange in Vietnam and a variety of birth defects in their children. Agent Orange-Known as 2,4,5-T Their were at least seven variety's of this chemical manufactured during the Vietnam war and used as a defoliant. There were some that were much more dangerous than other's. Much of the variety tested were the lower toxicity agents, and was much of what had been reported.

**** Garage mechanics, auto body workers and other men exposed to hydrocarbons, solvents, metals, oils and paints have a four to eight fold increased risk of having children with Wilm's tumor, a kidney cancer.

**** Larry firmly believes to look at this objectively, he must carry it one step further and look at a growing number of sensitive Americans.

**** A whiff of perfume or a gentle breeze from the neighbor's lawn can leave chemically sensitive people speechless, and they are becoming increasingly vocal about it.

**** They complain that they're getting sick from everyday exposure to common levels of chemicals in the home, office, work, and environment. MULTI CHEMICAL SENSITIVITY SYNDROME, (MSCC).

**** Chemical sensitivity can leave sufferers with headaches, burning in the eye's nose, throat, and lungs. Severe difficulty in breathing, as if the respiratory system is shutting down, nausea, and vomiting, and a paralyzing disorientation that can render one incapable of speaking. It is Known nationally as multi-chemical-sensitivity-syndrome, (MCSS).

***** Larry can tell you first hand since he suffers of MCSS, and has since his 1990 over exposure to Mercury, and to Xylene. It all started out as an allergy from the many chemical poisoning in Colorado at the Rocky Mountain Arsenal. Then in 1985 with the over exposure to a lead based bridge paint at his job, left him suffering a hyper-sensitive allergy to a large number of chemicals even some perfumes. this he was able to tolerate as long as he was careful. But what happened in 1990 The continual over exposure for thirty, Ten hour days to the paint containing Mercury and Xylene took him over the edge with injury to to Central Nervous System, and was diagnosed with MCSS., along with loss of vertigo, motion sickness. This eventually disabled Larry completely from work. Starts as an allergy worsening with time.

**** Skeptics in the insurance, medical and chemical industries say they are not convinced the misery isn't all in the sufferers head. The condition hasn't won recognition in those circles, or the resulting insurance benefits. However Psychologists, and neurologists who deal with the Central Nervous System agree whole heartedly there is a severe chemical problem. Take for instance Mercury enters the body via the nerves, and goes straight to the CNS. (brain), and will cause severe damage very quickly.

**** Yet some scientists insist that society is seeing the first casualties of the proliferation of chemicals. Since World War II, annual United States production of synthetic organic chemicals has exploded from less than One billion pounds, now up to two-hundred seventy-three billion pounds in 1988

**** As of yet we do not understand the mechanisms. It may be neurological or immunological, or it may be biochemical.

**** Then remember all that we know nothing about that as part of wastes that become a bigger monster that is very much invisible to all, from the mixing of wastes, and the inability of taking it apart.

**** Whatever the mechanism, the effects can be dramatic, and those who suffer from chemical sensitivity, also known as environmental illness are, never the less in pain.

**** The illness often begins with a single exposure. Maybe it was new carpet at the work place, or an accidental release of toxic materials fròm a sewage plant, any number of things, but it seems to traumatize the body into a permanent, hypersensitive state.

**** What follows is a bewildering array of symptoms and search for diagnosis and treatment. Then you have Doctor's who aren't aware of the signs and symptoms, or the chemical causing the problem. patients try to avoid the pesticides, paint, plastics, or building materials they fear will trigger an episode.

**** Some sufferers find breathing relief with the use of over the counter spray's, adrenaline, Steroids, but then many sufferers also react to drugs.

**** How many people in the United States are affected is still unknown.

**** According to the National Academy of Sciences, about 15% of the population has heightened sensitivity to chemicals., though fewer people face intense symptoms.

The Medical establishment remains dubious:.

**** "If you believe you couldn't walk outside or walk anywhere where there were exhaust fumes, you would be disabled. If that is the issue, have I seen people like this who are disabled? Certainly, replied one doctor."

**** "Now, whether they have anything wrong with them physically is quite an other matter. If doctors can't find anything physically wrong with them, it must be within the realm of the psyche." Psychologists are now realizing more and more that it is a very real problem, its also a problem that neurologists now understand better than before. It's true most Doctor's don't work with the Central

Nervous system. Not realizing the CNS is very vulnerable to chemicals. Hitler knew it back in the 1930's when it was used to kill people, it knowingly affected the CNS.

**** Some credible organizations have given at least some credence to a physical basis for the disorder.

**** The EPA, in a report to Congress on indoor air pollution, urges financing for research on chemical sensitivity. The Congressional Research Service calls chemical sensitivity one of the consequences of poor indoor air.

**** Some doubters blame it on the employees for malingering. They claim that they are aiming at workman's compensation benefits. They also point the accusing finger at sufferer's subconscious fear of workplace pressures and to fright over frequent media reports about chemicals.

**** Clinical ecologists, doctors who specialize in chemical sensitivity, also are met harshly with skepticism. Main-stream allergists have urged the Health Care Financing Administration to deny reimbursement for their clinical ecologist's work.

***** So, far, scientists have found four groups of people most prone to chemical sensitivity. They are industrial workers, occupants of well insulated, poorly ventilated offices, residents of contaminated communities, and people exposed to chemicals in household air.

***** Larry said "please be careful with all the different spray cans, and bottles, not just pesticides but everything else, you may be contaminating your own home, yourself and your family."

**** Those who believe chemical sensitivity is a physical illness say the skeptics fear an ominous message: If people are getting sick from the chemical-laden world, this may be construed as an omen of things to come.

***** During the winter of 1991 things could not have been worse for Bob and Cindy Allison. They weren't saying that the chemicals in their well 1.5 miles from the Arsenal gave Bob his blood disease, or made a friend's hair turn white, or that the chemicals are potent enough to kill small furry animals.

**** They don't know diseases, but they do know property values. They know that a five-hundred thousand dollar contract on their home and business fell through last summer right after the State Health Department began delivering them bottled water.

**** "It sure didn't help any," Cindy said "when the Health department told us not to drink the water."

**** The health department began delivering bottled water to the Allison's and the owners of eighty-one other wells last summer after it determined that there shouldn't be more than two parts per billion (ppb) of a GB Nerve agent byproduct (DIMP) in the water near the Arsenal.

**** So what is the safe level for Dimp in wells? How much is ok for our children? "No one really knows yet"

**** Pick a number. The United States Army and the EPA. say six-hundred parts per billion, But according to the studies on dogs, rats and mink that did die after drinking (DIMP).

**** Shell Oil Company, which shares the clean up liability with the Army, proposes two-hundred and sixty ppb. The Colorado Department of Health Proposes two ppb, based on a study of mink that died.

***** No one ever thought of the yet invisible killer chemical that is in the same water where you would find DIMP, straight from the Rocky Mountain Arsenal. "On a test done with wheat, DIMP was put in the water at different levels up to 500 ppb, and there were no problems with the wheat

@*@*@ However when they took the water straight from the wells north of the Arsenal, containing various levels of DIMP, there was no wheat growth, the seed would not even germinate.

@*@*@ Larry says. "He completed tests with rabbits, had got his water from the Boxelder ranch wells, not contaminated by the Arsenal, placed small amounts of DIMP in the water for two weeks. there were no problems with the rabbits. However when he took water from wells north of the Arsenal, the rabbits died 2 weeks.

**** DIMP was suppose to be the new tracer for arsenal contamination, even though it was know to be slightly toxic at small levels. However wherever you had DIMP at any level, you had other Arsenal toxins. 5 to 7 ppb DIMP and you 3 to 4 percent Arsenal contamination and again the invisible killer chemicals.

***** It has been discovered that wherever you have DIMP you still have GB-Sarin Nerve Agent at 3% Level which is enough for low level toxicity which can be very punitive and painful.

**** The Water Quality Control Commission came up with sixty ppb by dividing the EPA, formula by ten, and then multiplied another by ten. Then reneged, asking for more time for study.

@*@*@* Larry said, "For the water quality control commission to come up with this, where lives are on the line, they damn well better get some studying in, so they can at least do the job right. by knowing their job. what they done is about the same as throwing numbers in a bag and blindfolded taking them out and say 'Oh well this looks good'."

**** Throughout the Denver metro area, as the noose tightens on pollution, residents are learning the physical and social costs of having a toxin in the neighborhood.

**** While it seems wise to stay on the side of caution, to do so is to sometimes harm people in a very real and immediate way.

**** If the DIMP standard is set at two ppb, for example, dozens of wells in the little community of Henderson will be deemed unsafe and unacceptable, and hundreds of houses will be literally worthless.

**** If it's set at two ppb, the question inevitably will be asked: Is the health department equally ready to spend hundreds of millions of dollars of Colorado taxpayer money to rid the state's waters of every toxin that can kill mink and that shows up in greater quantities than two ppb? "They should be"

**** If it's set at the EPA. six-hundred ppb, everyone's well will pass inspection and maybe they can sell their houses, or maybe they won't because the seller will have to make the buyer aware of the fact DIMP in the well Is a by Product of Sarin or GB nerve agents. Then stop and think, if we were to use the 600 ppb as EPA recommends. Chemical law says for every 100 DIMP, you will have 3% Sarin. for every 100 ppb DIMP you have 3ppb Sarin with 600 ppb you automatically have 18ppb Sarin Agent. EPA standard states that .005 ppb sarin is deadly. Figure (that one out). There will always be low rate poisoning with Sarin agent. Then don't forget even with DIMP, you still have a silent killer in your water because it came from the Rocky Mountain Arsenal.

**** Meanwhile, Bob, Cindy and other Henderson neighbors stare at their glass bottles and their real estate agents and wonder.

**** "The damage to our property values already has been done," said Cindy, whose well measure's two and a half ppb of Dimp, and that damage will stay forever. The Allison's were looking forward to retirement, but now wonder if they can sell their property for half the asking price. "There's no telling when or if we can even sell it," said Bob. But Cindy checked positive for cancer in 1993—died in 1994.

**** So Allison's continue to wonder, will ever end. It will probably like a stigma for life. Things will never be the same.

**** "We've had this house for twenty-five years and for twenty four of those years we've drank that water." Bob said. "I'm not worried about us because we've had a good life, it's almost over. But what about the children, and the grandchildren, what will they have after we are gone?"

**** "You can monitor another twenty-years," replied Cindy, and spend another one-hundred million and be right where we are today. How many holes do they need to monitor? I think the government should start thinking more about the American public, instead of believing they are above the law."

***** "THERE ARE THREE KINDS OF LIES; LIES, DAMNED LIES, AND STATISTICS." "Just ask the Government they have statistics on everything". Larry Land said of the government.

**** The giant Superfund cleanup at the Arsenal was already costing the American taxpayer fifty-two percent more than engineers predicted just five years ago.

**** The partial discovery of unexploded bombs, mortar rounds, polluted prairie dogs and hidden reservoirs of liquid toxin have contributed to $35.3 Million of budget overruns during the cleanup of the Arsenal.

***** "The Engineer's did not take into account all the lies, the damned lies and especially Government statistics. Had they taken it into account, there probably would have been no cost over runs." Larry's thought's for the day, and the simple solution.

**** Records show that since 1986 the Army spent $102.3 million on the forty-seven clean up contracts at the Arsenal. The original estimated cost for those projects was sixty-seven million dollars.

**** These overruns have come early in a cleanup that ultimately is expected to cost $1 billion to $2 billion. The most extensive work is not scheduled to start for another three years.

**** Army officials blame the bloated budgets on lawsuits and extra work required after they found more pollution than was thought to exist.

***** Larry thinks the Army is now talking from experience.

"EXPERIENCE IS THE NAME EVERYONE GIVES TO THEIR MISTAKES."

By Oscar Wilde

**** However, the extra money also paid for major clean up technology research that failed; the construction of a large records center at the Arsenal and a branch office in Maryland; and big fights over obscure issues like how much toxic vapor would be allowed in basements built at the redeveloped Arsenal.

**** The Arsenal and its critics concede that the Arsenal demonstrates the incredible complications that result when anyone tries to clean up a major Superfund site. The truth is that hazardous waste does not just go away. You have to move it, bury it, or put it in barrels and store it. You can even try to incinerate it as they have tried to do with the waste form Basin F, but it does not go away.

**** A February 20, 1994, article in the Denver Post regarding the radioactive waste at the Rocky Flats Nuclear Weapons Plant, clean up executive Pete Swenson states that "decontamination work typically carries a multiplier of 1.6 which means that every 1 pound of detoxified equipment generate another 1.6 pounds of toxic waste." The Energy Department still has no permanent storage facilities for such waste. Many states fight the movement of toxic waste from other areas into theirs for storage. I believe that the phrase "clean up" creates a sense of false security, leading the public to believe that we really can make it all go away.

**** When they first planned this in 1984, nobody else in the country had done a major project like this.

**** They thought at first the total remedial investigation would cost thirty million. Six month later, it was changed to sixty million.

**** The Army has been splitting the cost with Shell Oil Company, under a formula that still hasn't won the needed approval from a federal judge. Under the Agreement, Shell's share of the cost will decrease as the price rises.

**** Here's a look at some of the biggest budget overruns on cleanup contracts.

ARSENAL CLEANUP COSTS.

Contract	Firm	Initial price	Overrun	Final
Records Center	ESE	$209,000	$584,000	$793, 000
Biota Study	ESE	$286,000	$1.3 mill	$1.6 mill
Water/Soil	ESE	$126,000	$500,000	$626, 000
Ground/Water	ESE	$1.6 mill	$2.9 mill	$4.5 mill
Endangerment Tests	Ebasco	$199,000	$756,000	$955, 000
Sewer Tests	Ebasco	$490,000	$555,000	$1 mill
Basin F Cleanup	Ebasco/It	$21.9 mill	$13 mill	$34.9 mill
Well	Weston	$1.1 mill	$2.7 mill	$3.9 mill

Closures

**** Some unexpectedly high costs have resulted from the dangers of excavating earth at the Arsenal, which a top Army official once called the most polluted spot in America.

***** If all the Arsenal's polluted soil were dumped into a ditch 3 feet wide and 3 feet deep, it would stretch from Denver to China.

**** Early in the clean up, one contractor kept hitting metal while drilling wells to collect ground water samples. Further study revealed that the Army earlier had buried dozens of unexploded bombs at the site.

***** All drilling stopped until 2" thick safety mats, made of meshed steel, would be purchased to protect workers from explosions. No bombs blew up, but the delays and extra equipment and work helped swell the contract to Environmental Science and Engineering from $3.1 million to $4 million.

**** Now we ask an Army Official for documented proof that the bombs were removed. They have nothing to share with you.

***** Larry said, "Could this be a Lie, or a Damned Lie, or probably just another Army Statistic".

***** One very reliable source confirmed that this explosive area is only one of many such area's at the Arsenal, many of which the Army remembers nothing about, and have no records.

**** Dozens of potential Arsenal contractors wanted other historical reports before submitting bids. But the Army had more than one million pages of documents to organize. The contract to set up, and operate the library, awarded to Environmental Science and Engineering, rose from $209,000 to $793,,000.

We'll break away for a moment of the Arsenal clean up for a couple updates in Larry's Life complex life, and a few turns it took.

In 1990 after filing the first complaint, the army then filed their answer to the complaints. Then a huge list of questions for everyone involved. These are called interrogatories which are questions about everything in your life, so huge it was almost overwhelming.

Then it was time for Depositions, That is when you appear before their attorneys, with your attorney to answer questions under oath. Their actions are as if they are trying to intimidate you. The first one took over two days literally being barraged with questions.

He kept hammering into Larry Land's financial doings from the time He was able to add and subtract. He felt strongly the reason for this, they were trying to figure out how he seemed to always hold so much financial power in order to do what he was doing as rattling all the doors in Washington D.C., around the World, and even at the Justice Department. He seemed very upset because he was not getting the answers he was evidently looking for. When Larry received copies of my first deposition, all bound up in book form, It was two 8½ by 11 inch books 3 inches thick.

With in two months after the deposition, Larry received a letter in the mail, the IRS. Nice big thick letter. He opened it and said to his wife, dig out your check book, honey we owe the IRS $28,000.00 in back taxes. The poor wife about fell over, as he read to her all the warnings you get if you don't pay up. Larry assured her it would be alright. He had been paying taxes at that time for over thirty (30) years, was never audited. He said well the Justice Department didn't get what they were looking for in the depositions. Now they are targeting me in the audits with the IRS trying to put pressure on.

Well Larry kept good records, all receipts, all canceled checks, and had data sheets for each month. He was upset that it was needed to be done. He got it all together and met with the auditor. She was trying to tell Larry he could not deduct any of the costs for legal fees on the Arsenal incident, Larry was an Entrepreneur, (Inventor) One who does all the research and development of different things, she said them expenses could not be deducted, nor the attorney fees for filing for a patent or copyright. She was literally stripping away everything. Larry even got for her their own IRS rules and regulations, and she would have some excuse as to why it didn't fit the mold. He could see she was rigid and not bending to anything. He told her as he picked up all his papers, That's enough I'll see you in court. Larry was a little worried, this was new territory but he felt confident he had followed all rules and he was an accountant.

He filed for an appeal, filed it in court, was set up for a hearing, he defeated the IRS by APPEAL everything for 1988 and 1989 broke completely even., I owed them nothing, and they owed me nothing. However a couple months later he received another letter in the mail from the IRS. this time they want to audit his taxes for 1990. It all takes time so he prepared for the audit, called them and see up an appointment to do it. It was a different person doing it however in his opinion the guy was a jerk. He wanted Larry to leave all records with him so he could work on it. Larry disagreed and said to the man I have supplied my records to you upon request, the law does not say I have to leave them with you, and I won't. So it took most of the day making copies, he would only give him a month at a time, then of course he had to check to make sure everything was returned. This man was nearly as bad as the last person was not allowing the deductions that by law qualified under IRS regulations. So Larry give him the time he requested to go over all the documents. He was virtually disallowing the same deductions the last auditor had disallowed and lost.

A couple of months went by and Larry received an offer to settle with the IRS, it was like you take half and we'll take half. Larry was very dissatisfied with that, and filed it in appeals Court Washington D.C. The Hearing was almost eight months off so he prepared everything, and within four months he got another letter from the same auditor. He again had Larry owing several thousand in back taxes. He called the guy but he was very chilling not caring whether or not the other case was even heard in Appeals, yet he was disallowing the very same damn deductions. As if they were trying to cut my financial power down to nothing.

Larry called the man and set up an appointment to look at his records. The first thing he as him to sign a paper to extend IRS right to Audit since the time on 1991 was running short, meaning he had to have this completed before the end of three years. Larry signed the paper, after spending several hours there they disagreed on about the same. At that point he also requested the 1992 records since we are running so close to the deadline.

About three months had went by and he had been to appeals a couple of times. He seemed to be the kind of guy you could at least talk to, with out hating him. At this point we were real close to Court when we reached an agreement, and the IRS would have to give me a right to carry forward $3,000.00 that we were actually over for that year. he kept the allowance letter and used in in 1995, since it was not needed in 1992 or 93. Well Larry said by God that's one on'em.

By the time the auditor got back in touch with him, it was already late 1994. He had another deal for me he would allow part of it, and part of it he would not allow. Larry said to him forget it I'll take it to appeals. Then the Auditor has to make out a form showing that he tried and that Larry owed the IRS a certain sum of money, then, that is sent into appeals, along with your request for a hearing., Several months go by, and you are contacted by the appeals office to meet with a person and see if it can be worked out, in the mean time a hearing date is set up.

Larry met with the appeals officer on 1991, and 92, tax years. Larry told the appeals officer, he was willing to just let it stand on it's merits as it was. But since it was the IRS who open this can of worms, he said well they got out, and damned if they didn't multiply. He discussed the Audits, and agreed with me it looks pretty cut and dried. since a lot of it had already had court rulings and approvals. Larry said there are several business deductions I did not take, and at the time I really didn't need them, but now I want paid for them, as he turns over the deduction documents that total nearly $23,000.00 dollars. The appeals officer wrote this all up, agreed to it and said he would have it signed by a judge, which he did.

1993, and 1994 taxes was about the same for each year he would get a letter in the mail stating how many dollars in back taxes you owe, and all the nasty little things they will do to collect, like locking up everything you own including your home and your bank accounts, and or your business. It started out with 1993, and before he was able to talk with the auditor about 1993 he already had the same nasty little letter wanting to audit 1994 taxes. He went in for the third audit and it was pretty much the same many of the deductions they felt could not be allowed even when he showed them in their own schedule it is allowed. Larry in some cases had to go all the way back to the congressional order that the IRS had set their schedules up on to show what was meant for the particular ruling. Any way they made it as hard as hell Its called baiting him, he held his cool, but he fooled them. Just said write out my paper's for Court he said I'm not playing your damn tax games. The auditor would write out the necessary papers to be sent to D.C. for a Court date and appeals.

The papers came in the mail for the appeals court, and a notice that you could talk with an appeals officer and see if this could be resolved before court. Larry again called the appeals office and set up an appointment. Then later went in and seen the appeals officer about three times. Since he filed his 1993 and 94, taxes he did not need any of the deductions that were to be carried forward from 1992 and these tax years were already done. so the $23,000 and the $3,000 were carried forward to 1995. both 1993 and 94, taxes was cleared up as well done. No taxes owed, however received over a thousand dollars back from IRS plus interest. No more than got this cleared up sure enough here come another invitation to a confrontation with IRS. for 1995 with all their dirty little remarks, and if you don't we will.

Every one Larry knew told him that the IRS cannot keep doing this after you win each time. but they are. Even some attorney's said this is harassment. The very next thing Larry done was call his Congressman and let him know what was going on. There were several letters, and an investigation was launched. It sure seemed odd Larry had not been audited for over 30 years, and only two years after he is given permission to sue the United States of America by the United States Congress, something is wrong. We'll talk of another horror story in the work place that caused permanent injury, with life time disabilities that eventually cost Larry his job.

***** It all started when his employer Washington County in Minnesota Sheriff Department assigned Larry and several others to a new unit that was still under construction. No windows, no doors with filtration system not yet completed. In effect the powerful fans were moving and re-circulating inside air The unfiltered air is what you breathed. It was easy to see there was a serious dust problem, dust everywhere, tables books, clothes.

***** The Construction workers were all wearing respirators, Larry ask why are they wearing respirators, and were not? No answer. It was always something that was going to be checked into, it never happened. The consensus of Captain Becker, Captain McGlothlin, and Captain Johnson was that Larry is stirring up problems again, and was ignored completely. Everyone assigned in the building with him was hacking and coughing, by November 23, 1992 Larry was diagnosed with Bronchitis, he reported that problem along with a serious dust problem very, irritating, to the eyes, nose, lungs, and throat. It is believed what caused the bronchitis infection.

***** Larry by the first part of December began reporting it to a supervisor everyday sometimes in writing. It wasn't just him, everyone was coughing, and at least six (6) has Bronchitis, was to their doctor's and given medication and spray inhalants. Larry also was back and forth to the Doctor's sometimes twice a week, each time more severe Bronchitis, medication was not working, and now pneumonia with infiltrate seen in the lungs on the X-ray's. December alone it was reported to supervisor's eleven times. It at least was a good thing for the three holidays Thanksgiving, Christmas, and New Years we had four day weekends off, giving a small amount of extra time to recuperate.

**** During the Month of December Larry collected several air samples unbeknowenced to his supervisor's so if something did go wrong. But again reminded all of them their was something in the air making him and several others sick, in fact along with some of the low echelon supervisors. He asked several times to have the dust in the air were breathing sampled because it was so dusty. Every day the chairs, tables, books in the work areas had to be cleaned. The dust particles could be seen in the air.

**** During the month of December, was nothing but training drills. which would be up and down five flights of stairs, several times a day, along with different types of take downs, boxing drills, control techniques for violent people. Every day a very hard day of workouts, filling our lungs with polluted, dirty air, what ever foreign material there might be. inside the sealed workplace.

***** He also noted that an enormous amount of fiberglass in large rolls were stored on the lower level. He noted this to his supervisors that it appears to be the reason the people working in all the tunnels were wearing respirators. They were doing a lot of insulating, and in the tunnels where ever there were water pipes. They were cutting large amounts of fiberglass in the tunnels at times the fans running without any filtration it was causing like a vortex in the tunnels, you could stand back a little and watch as all the construction dust was being moved to other places. In cutting and working in the tunnels with the fiberglass it became a suspect with the breathing problems. placing the fiberglass around the pipes then taping an sealing each of the pipes. This was going on November, and December, 1992 and still being done in January 1993 became highly suspect for the reason the Captains would not allow the Health Department to monitor the air. At least Larry had his samples of the air just in case.

***** By 7, January 1993 Larry was taken to Hospital by Ambulance unable to breath, and near respiratory shut down. The Hospital diagnosed Bronchitis, Pneumonia, with infiltrates seen in both lower lobes of the lungs. Occupational Asthma at that time was rated at 15%. The Doctor suggested that Larry place a call to the Washington County Health Department, and request air monitoring. He did that as soon as possible because he was in fact injured by the dirty air. It was later found out after Larry had returned to work, he was informed by Captain McGlothlin, Captain Johnson, and Captain Becker, that he could be fired for what he done. He said for what, going to the Health Department. His answer was yes that is our job. They were told it had been reported nearly 30 times, there was too much dust in the air, and nothing was done, except and oh yeah it sure is dusty over here. Since all there officers were back in the Old unit they seen very little of the real problem. Larry was called on the carpet by all three Captains, and told they did not allow the Health Department to test the air as you requested, they had sent them out equipment and all.

**** Larry Called the Health Department to find out what had taken place and what was the air tested for before they were ask to leave by the Captains. He was told they had only tested for Formaldehyde, and Carbon Dioxide in a couple places, he ask if they had checked the air for dust, and what was in the dust. She replied they did not have the time when they were ask to leave and take all equipment, there was nothing we could leave, such as an air sampler for particles.

**** shortly after being called on the carpet about the problem, and Larry again complained of the dust problems, and he was again having a difficult time breathing. He was Immediately reassigned back to the old unit where he remained, until the construction and the entire dust problems was cleared up.

***** Like Larry said, he reported the problem nearly everyday to a supervisor.

"Larry has never in his life had pneumonia. Prior to 1992"

"Larry has never in his life had Bronchitis. Prior to 1992"

"Larry has never in his life had asthma. Prior to 1993"

"Larry had been an Auctioneer for (17) years. Prior to 1984

"Larry's lungs were X-rayed, passed a physical prior to his going to work for Washington County Sheriff Department. 03/07/84"

"Larry's lungs, X-rayed, and a Spirometry test done, for any sign lung problems, and again a physical in September of 1992 at the request of his employer Washington county sheriff Department. all in good health. The only problem ever discussed was the CNS injury from 1990 paint incident".

*** According to work comp. Guidelines this was Occupational Asthma.

"He did not have such condition prior to his employment.

"He also did not have such condition just prior to the incident." It was rated by several Doctor's as such. The Doctor's told him something else about keeping a very close eye on, fiber glass over exposure could be much like asbestos exposure, it Could be progressive, There was something in the construction dust that seemed to have caused the problem, and you have already mentioned fiber glass, and carbon monoxide.

**** By 1996 it was rated by Larry's doctor's at 40% occupational workers comp. As if this weren't bad enough now the employer had to do more x-ray's and testing of the lungs. However their medical doctor's records were requested several times for the rating, and the findings. they would not respond. 1998 they had still not responded, Larry was referred to an occupational specialist for the respiratory system. His condition was getting much worse.

***** Lets talk about Reasonable Accommodations for the disabled. Yes Larry after the toxic paint exposure to Mercury, and Xylene in 1990 has become partially disabled to the Central Nervous System. CNS. He falls now under the American with disabilities Act sponsored by the Justice Department.

***** The Sergeants position at Washington County Sheriff Department was taking applications. Larry put an application in for the position. He also requested assistance should it appear he did not understand what a written or oral question might be.

***** He put in for assistance, but it was not acknowledged in either the written or the Oral board.

**** Larry by this time had nearly 20 years experience in corrections. Had been a riot squad Field Training Officer for the city of Denver Colorado Sheriff Department where he had spent several years as Deputy Sheriff. Before going to work with the Washington County Sheriff's Department as a Corrections Officer.

**** Larry was a Field Training Officer for nearly seven years for Washington County Sheriff's Department.

**** Larry had written and published a Field Training Officer hand book for Washington County, and the State of Minnesota.

He won the suggestion of the Month Award in 1988.

He won the suggestion of the year Award in 1989 for 1988.

***** Out of eight people chosen for the position Larry was not one, and he could not understand why. He ask for the test scores this paper only had partial scores for himself and another older gentleman. By this time he was nearly 50, and so was the other gentleman. The personnel department specifically Judy Honmyhr, and Jeneen Johnson was ask, why are part of our test scores missing, they absolutely refused to give any answers to the problems. I have letters letting them know there will be some legal discussion on this do not destroy any of the material. The next time they were ask for the material or an explanation he was told that it was no longer in the computer, and they could not tell me why part of our scores were missing. they were asked if age had anything to do with it, again no response. He got a hold of the Oral board scores which were fairly high, and satisfactory however the missing scores in two places had it been given would have placed Larry in second place, and the other man in third place for certification. This came from only two people in the personnel department.

***** Then 1994 Larry had asked his employer for reasonable accommodations with a formal letter to the Administrator, Captain McGlothlin. He did get in touch with Larry and discussed it with him.

***

1. accommodations with all new material, written or oral.
2. accommodations on new equipment and existing equipment.
3. until retrieval and retention are complete.
4. extra training and communication on all new material.
5. not to be subjected to carbon monoxide, paint herbicides pesticides, fungicides, or chemical sprays.

***** Reasonable accommodations during the recruitment process.

1. Additional time on any written material for testing.
2. To make sure Larry understands all oral questions.
3. If it seems he is having a problem, ask him.

*** The first set of accommodations was written on papers, was given to all supervisor's, A letter drafted up specifically to Ed Kapler Head of facilities Department.

***** The only problem it was put on paper, looked good, Larry was advised when he received his notification that he would no longer be allowed to do any field training, and as being the notary for the building it would stop also.

***** He got his accommodations on paper only, not one supervisor even understood what it meant nor did they ask, not did any of them offer any accommodation what so ever.

***** As for Ed Kapler he didn't understand either or maybe it was done purposefully. He got his letter that we know. It told Ed Kapler he was to notify Captain McGlothlin if there was to be any pesticide spraying or painting at the LEC. Notification at least 24 hours before the event is to occur. Out of the blue one afternoon his exterminating crew came to the exact location in the entire four story building masks pumps, and all before Larry could get through on the phone for permission to leave the area he was already overcome by the chemical fumes.

He left the area and took his emergency medication, however he has Multi Chemical sensitivity syndrome, becomes deathly ill from exposures.

**** Captain McGlothlin called Kapler about why he did this, he stated he set down Thursday would be the day for that area. But he forgot to notify anyone. According to the letter he was to notify the entire building if he were going to be spraying anywhere.

***** The following week he had sent his exterminating crew over before Larry began his shift and had sprayed his entire work area, and he walked right into it. by now he thinks Kapler is playing games, he informed Captain McGlothlin, Larry had to do his medications quickly and get out of the area. This time it took nearly 3 hours for recovery. He got very nauseated, vomited, difficult breathing, and painful head ache.

***** As if this wasn't enough the following week again he sent his people over to Larry's exact work station. Larry came on shift and caught the people before they began spraying, he ask just who the hell sent them here to my work area, they said orders from the top which is Ed Kapler. The man who is the head of the Washington County Safety Committee. Land sent them on their way and then called Captain McGlothlin to let him know it happened again. He assured Larry he would take care of it.

Well their were 8 times this was done even after he had been given a letter stating it would not happen. It almost seemed as if he were after Larry because of the 1990 Paint problem where OSHA had issued severe citations to the County. It seemed a personal Vendetta.

***** As for the promotion Larry was trying for and he put in for accommodations on the written part and also the Oral part of the process.

**** When Larry took the written test the personnel department done everything they were ask as for the accommodation.

**** The Oral board was different they done nothing that was requested about the Accommodations. Two Sergeants were from Washington County, and one other officer was from another county for the Oral boards. There was not questions as to whether he understood the questions, or was he having problems with the answers. The person from the other County gave a much higher score than those from Washington County, who gave very low scores.

***** Out of all this Larry got no promotions evidently for being a whistle blower during the 1990 paint incident. It seemed as if it were rigged 1992 the part of his written score that was missing and no one could explain why. then the two Washington County Sergeants gave a much lower score than the person from a neighboring County. I wonder why? I do know we had a serious problem of ethics, and possibly a Political problem that affected those decisions.

***** In 1994 Larry won an award of the month, along with another person. She was called into the Administrator's office, given her award and prize. Well Larry waited nothing happened. so he sent a memo to the Administrator, no response, he sent six memo's and finally got a response it had been so long that it got lost, it's not on the shelf anymore.

***** Back with the Engineers at the Arsenal where other predictions also failed to pan out. After reviewing ten years of research on wildlife at the Arsenal, Army engineers didn't think much more work would be required. The prairie dogs, are highly contaminated, and are blamed in the deaths of the bald Eagles.

***** The quality of the Army's existence has led too many failures Larry says, "The price of greatness is responsibility". the Army has taken none

"Winston Churchill"

**** However, biologists soon discovered that while some two, hundred mule deer, were largely free of pollution, others, the hawks, coyotes, and prairie dogs, and fish were not.

**** It was ultimately decided to exterminate the burrows of thousands of prairie dogs to prevent bald eagles from preying on rodents that were permanently contaminated with toxic waste.

**** A more extensive study was required on how dozens of birds and animal species were affected by seven Arsenal contaminants, and Environmental Science and Engineering's contract bloated from $286,000 to $1.6 million.

**** Looking back to 1988, a contractor who had drained four million gallons of pesticide and nerve gas and their byproducts from

Basin F that contained the deadliest and most dangerous Ultra Hazardous toxic pollutants known. Basin F looked empty, the contractor drove into the pond area to get a better look.

***** A crystalline floor cracked beneath the truck tires. Investigation revealed that another ten million gallons of the Ultra Hazardous chemicals was hidden beneath the basins false bottom.

**** That discovery, resulting in additional clean up, to remove the toxic waste. Helped force a contract for Ebasco up from $21.9 million to $34.9 million, and that is not counting the millions when they figure out how to get rid of it.

**** If Army engineers could predict exactly how much contamination would move from polluted soil to clean flowing ground water, cleanup costs would be slashed by millions and the formula could be used at every Superfund site across the nation.

***** Larry says, "This is nothing more than a dream, to make it look as if the Army knows what they are doing. Right at the Arsenal they had polluted by now 100's of thousands of acres. They had polluted the soil they had already went into clean flowing ground water, and by God they polluted that. They sure as hell didn't' learn anything there.

**** Yet after spending and additional $626,000, Environmental Science and Engineering failed to develop a working formula. It could have been of extreme benefit if it had worked, but it unfortunately came to a dead end.

***** Larry Asks? "Why didn't the Army Hygiene, and Environmental Science and Engineering take an interest in the 60's when the they told the Army basin F was leaking, and they knew at that time the Army was moving millions of gallons per day of highly polluted liquid, to clean flowing groundwater to the North and North west of the Arsenal, and they continued doing this for almost an additional 15 years, Knowing it was leaking. Add up today the the 100's of thousands acres are there with the ground water polluted and destroyed. If it were some type experiment the Army was involved in, it didn't work and they got caught deliberately polluting the land, and destroying the lives of thousand's. Larry still says, "Their had to have been payoffs under the table for the Army to allow this kind of thing to go on even the pentagon was involved.

**** Army lawsuits involving Shell Oil, and the Colorado Health Department also piled up millions of dollars in extra cost.

**** From 1975 too 1985, the Army collected 3,600 ground water and surface water samples from the Arsenal in an attempt to find pollution.

**** Army lawyers said those samples might be good enough to help plan a clean up, but not good enough to stand up in court.

**** So the Army collected and additional 12,000 samples, making the 17,000 acre Arsenal one of the most intensively studied tracts in the world. Each sample cost about $3,500.

**** Army officials said extra quality-control lab work required by litigation caused about $4 million in budget overruns.

**** Despite all those tests, Army officials say they probably still haven't found all the pollution.

**** That in itself is not surprising, considering that many of the older chemical tomb's have no records as to what was buried or even where it was buried!

**** One report dug up by Jennifer that lists a 6,000 lb. fork lift being buried, and that's not counting what else was thrown in the same trench, and covered up., Larry says what in hell was so toxic, so potent it could not have been cleaned up.

***** Bare in mind that the Army is only cleaning the area's beneath the basin's and have no intentions, now or ever of doing anything with the countless number of trenches and pits.

**** Army officials say that to be 95% confident that they have detected every contaminated site at least 50 feet in diameter, Army engineers would have to drill 297,000 test bores, or 11,000 per square mile.

**** That in itself would cost more than $1 billion, a price no one is willing to pay. Without those extra bores, many highly contaminated area's will go undiscovered even after a gargantuan clean up effort.

**** The resulting legal liability means the Army probably won't turn over Arsenal ownership to anyone else.

**** Even so, Superfund laws require the Army to consider major redevelopment of the Arsenal before deciding on a final clean up strategy.

**** "The costliest work is yet to come," said Connally Mears who oversees the clean up for the EPA. "I don't recall a case at the Arsenal when cost has been an important factor, but in the future, cost will be a very important issue. At the Arsenal, it's been live, live and learn."

**** And for some of the residents who live or lived, near the Arsenal it's been live, get sick and die. "The United States Army's death sentence, be sick and painful for 40 years and die long before your time," Larry says.

**** Meanwhile, during this same time period, the Army, EPA, and Shell Oil Company had been in serious deliberation concerning one of their earlier ideas, the building of a new, state of the art incinerator.

***** In April Larry and Marie received a letter from a neighbor. It was in response to a question about cancer I had asked of her some weeks earlier.

**** In Part it said: "Our conversation also caused me to do some serious thinking about our cancer rate in this area in the last two years. It was about four years ago that the people in this community of about one-hundred homes began to complain about their wells."

**** "Almost everyone commented that they had noticed changes in their water, Some complained of odd odors. One man said his water smelled the same was Basin "F" did when they supposedly cleaned it up." "He also said he had flammable water, when you would draw a glass of water from the tap, light a match over it, and it would momentarily emit a light blue flame,

**** "Another woman complained that when she went to have a permanent at her hair dresser, her hair turned a light shade of purple. Her beautician was terribly upset and stated that it was like a 'strange chemical reaction, it even foamed up.' Other's said their water tasted as if it were loaded with sulphur and chlorine, other's said it had the smell and taste of kerosene, still other's said their dishes smelled horribly with dirty sulphur smell, Other's also noted an oily film on top of the water."

**** In our case, we noticed the galvanized water tanks, on the metal strips in the aquariums, and in the water closet tank for each bathroom would begin to pick up mineral type deposits in every crack and seam in the tanks, which we had never noticed before. A bright red and white calcium like deposit all soft and mushy. My son has two 25 gallon fish aquariums in his room. they had been run on well water now for three years with no loss of fish. The same as the stock tanks, we always kept large gold fish in them they help keep the tanks free of bugs and algae year round automatically filled and heated. Larry had brought over from the Box Elder Creek farms, 10 large fish he had for three years in the stock tanks, the fish would first turn mostly white, and lost all scales, within just days they would be belly up.

**** We had taken sample of the white and reddish material to a few different laboratories, who would check for materials within their expertise.

***** One commercial laboratory found all the metals used by both the Army, and Shell Oil Company in their labs, and for production they found Lead, Barium, Praseodumium, Cerium, Lanthanum, Cesium, Iodine, Tin, Molybdenum, Niobium, Zirconium, Yttrium, Strontium, Rubidium, Selenium, Gallium, Bromine, Arsenic, Gallium, Zinc, Copper, Nickel, Cobalt, Iron, Manganese, Chromium, Vandaium, Titanium, Scandium, Calcium, Potassium, Chlorine, Sulphur, Phosphorus, Silicon, Aluminum, Magnesium, Sodium, Fluorine, Boron, Beryllium, Lithium, a few of these items one may find as a natural occurrence, some are very rare and used only in the manufacturing of other Industrial chemicals. A great deal of them were used directly at the Arsenal, as imported heavy metals used in the Laboratories, and for Manufacturing.

***** Still another Laboratory who was previously connected to the Army immediately recognized the bright Red colored material, we were shown looking at it through a powerful microscope they appeared to be groups of tiny bright red spheres, we were told it was used by both the Army and Shell Oil Company. It was called Ion Exhange Resin. It was used to separate one chemical from another. Put in a tank with two chemicals, it was able to latch on to one, then removing the red from the tank it will have collected one of the chemicals only and removed it from the tank. These sphere's were so tiny they could not be seen by the naked eye. Only under a microscope they were tiny bright red spheres of the 300 to 400 hundred mesh size. They could travel anywhere water went.

***** Larry was told it was used by the Army, and Shell Oil Company to trace any leaks of chemicals from the Arsenal. When they were faced with the fact chemicals were leaking from the Arsenal, they both denied using it at all in a continued and botched cover up. It was not natural, someone had to have put it there. This one person, an ex-Army Scientist stated the Army used it for tracing any leaks, from the Arsenal, they know exactly what the hell it is.

***** What ever was in the material in the stock tanks, out of all 12 huge 300 to 500 gallon stock tanks on Larry's ranch, the material ate all the seams totally out as if it were caustic, or acid, in the water tests the PH levels would run from 7.5 to 8.9. The stock tanks had all been purchased new, and with in 6 months they would not hold water. Larry also had several teeth tested by this same method, teeth were deformed, and turned black, and notably the contained large amounts of the chemicals named above.

***** Yet another laboratory tested eight water samples, and found in all samples, Hydrazine rocket fuel, Aerozine Rocket fuel, along with small yet detectable amounts of cyanide. He explained to Larry that is why were seeing the young heifers in death the mammary glands are blue. He said that is one of the first signs found when looking at cyanide poisoning.

***** Spectran Laboratory, in Denver with his new sophisticated Infra-red Spectrometry, came up with two different water samples states he has a rather large curve saying phosphonate but what it is related to he had no charts or grafts. Like he said only the Army would know. It was later found out to be DIMP, IMPA, and other related pesticides. DIMP by product of GB, and Sarin Nerve agents, IMPA by product of BZ Nerve agent. At the time of the testing the Army was the only person able to Identify it. In 1972, 1973, and 1974, they would not cooperate in it's identification. However it was the United States Army, and Shell Oil Company that did Identify DIMP, IMPA, Aldrin, Endrin, Deildrin and other pesticides in the Land wells. As well as the State Health Department finding also Nemagon, and high levels of Chloride in the Land Wells, Noting that the Land wells contained between four to seven percent Arsenal pollution in their wells.

***** "Larry, claims within the last two years, seven of his close neighbors have been diagnosed with varying types of cancer. We've had liver, brain, uterine, bone marrow, and leukemia, resulting in six deaths. Several were extremely fast acting. Some died within two months of being diagnosed. Besides cancer, there have been unexplained cases of seizures, eye problems, and general health decline in several other residents, their children's children being born with terrible learning defects, still some being born suffering mental retardation. Some of this has also been seen in those moved away from the Arsenal.

**** Still there are many of them who have accepted the theory, that we are unaffected and even a recent widower can see no possible connection to the water, even knowing his water was polluted from the Arsenal. People are too often unwilling to accept the unpleasant possibilities.

***** "Yet they silently hope, and pray people, like Larry who laid his life on the line for everyone as their watch dog does much more than just watch. Larry "who's cry" was heard around the World and back to Washington D.C.. As doors rattled, They say he is the most tenacious of all, when he gets hold of them he never lets go until justice is served. He has had them in his clutches since 1972, he has been the leader in closing the 27 square mile Army base, and he calls it Death Factory U. S. A. is in our midst. The Army and Industry Giant Shell Oil Company, a large part of the Death Factory's amongst us. At a trial in Denver Colorado Larry rattled his saber as Shell was trying to get permission to force pump under hundreds of even thousands of pounds per square inch PSI, toxic chemicals under the ground into the Dakota, and Lakota Sands in Morgan County close to the City of Brush. Larry lashed out with proof to the Courts that if Shell were allowed that as a cause, the effect would be disastrous polluting the water for everyone in Brush, and Fort Morgan etc, clear into Nebraska. He was able to point out to the Court the exact route the "vehicle" pollutants would take to get to every ones water wells. He was able to overwhelm some of Shell Oil Company experts as they testified, he would demand they answer his question, and when they could not answer the questions, and name the vehicle route of the pollutants into all the wells in that area and beyond. Larry would demand they answer the question, then he asked the Courts permission to enter into evidence the very reason why Shell Oil Co. would be polluting the water if they were allowed to do this. He noted what he was saying was a published fact going back to the 1800's, and where it could be found. There were some pretty happy people there, and there were some very unhappy people there, because Larry just stalled out all possibility for Shell Oil Company to remain in Colorado. When the Court adjourned to make a decision. One of the unhappy people kinda short with balding head walked up to Larry face to face his head turned red as he stated "you're the most radical little son of a Bitch we have ever had to deal with". News camera's running tight in our faces, as Larry states back to the man "Well, thank you that makes me feel that I am doing my job". Just outside the Court room those with happy faces lined up clear out the corridor to the street just to line up and shake Larry's hand, and say radical or not you did one hell of a job here today. It was on the news that night beeping out the nasty wording. Still today thirty years later he struggles on for the safety of our environment not just for him self but everyone Worldwide. What affects other parts of the world, must not be dealt with lightly, for what affects their environment will someday have an effect on our environment, Larry says it has become a multi National effort to clean up the environment world wide, stopping both Industry, and Government, willful-wanton, reckless negligent disasters, like those at the Rocky Mountain Arsenal affecting millions of lives now and in the future. "the most polluted piece of land on earth.

***** Larry says, "There is a difference between right and wrong, and what the Army, and Shell Oil Company done was totally wrong and they knew it. Where you have a cause there is an effect. There are right causes that obtain the right effect, and when one deliberately makes the wrong choices, then you will have the wrong effect. you cannot say whoops, as the Army, and Shell did, what they caused was done deliberately, and knowingly caused the wrong effect. Shell Oil's 258 insurance Companies in the largest trial in the United States history, proved Shell Oil Co. was wrong and they knew it. The Army, and Shell deliberately had a planned strategy to cover it up for over twenty years. This information was withheld from Congress, and was known by the Pentagon when they admitted Basin F was in fact leaking as part if not all the protective membrane was absent. Yet they allowed the Army to continue polluting the area for an additional fifteen years. When they were caught red handed, in 1972 by Larry, they outright lied, admitted and then denied, what they had said, and then classified the information they did not want made public for an additional ten years, until their permanent closure in 1983, by cease and desist State of Colorado.

***** Larry says, "what they caused to happen, in Minnesota to him in 1984, "The hit and run incident." as far as he is concerned it was no accident, and was shown to be purposefully done. Then a short time thereafter memo's were found referring to Larry Land in Shell Oil Company files, stating "let the sleeping dog lay."

***** "Larry hopes that this information will help everyone in some way to help each other." As stated earlier, the subject of incineration once again came before the public scrutiny. Many heated debates had arisen concerning the use of an incinerator but the bottom line was, did the public want more pollutants in the air? If not, what was the best alternative to incineration to eliminate the millions of tons of contaminated soil, along with the millions of gallons of liquid waste contamination. In the end, it made little difference what the public felt or wanted. The public was brought into the picture by the Army, Merely out of necessity to make the community believe they were truly a part of the input on the clean up efforts, and how it was to be done. Many people, along with Larry and Marie, feel that this was a cut and dried decision, already made, long before this meeting.

**** On Thursday, June 27, 1991, the Rocky Mountain Arsenal participated in a meeting sponsored by Citizens Against Contamination. This is the information given to the public.

**** The Army was invited to discuss submerged Quench Incineration (SQI), a process that will be used to dispose of literally millions of gallons of liquid removed from Basin F. This also includes projections of additional leachate generated from the waste pile. Leachete is the liquid that gradually drains to the bottom of the waste pile. Additional design work will have to be done to determine how the storage tanks will be cleaned out, a process that will generate some additional waste that will be treated by the SQI process.

**** During the meeting, citizens were told how the SQI will operate and were apprised of emergency response planning measures that would take place in the unlikely event of a problem. An update was provided regarding the current status of the SQI, and a time frame for construction completion and operation were discussed.

**** The site of the facility, located near Basin F, has been cleared and a construction trailer brought in. Approved design packages for the SQI equipment have been released to the contractor for fabrication.

**** Upon completion, the SQI will treat approximately 10,000 pounds of waste per hour, or 15-to-20 gallons each minute. The Army is looking for all incineration to be completed by the summer of 1995.

**** The actual incineration process takes place in several steps. Liquid is first transported via an enclosed double pipe system from the pond of storage tanks to the facility. There, it is placed in special day tanks, which are capable of holding one day's supply of liquid to be burned. From the day tanks, the liquid is injected at the top of the incinerator through five injection nozzles that spray it into a natural gas flame.

**** After being injected, the organics and ammonia contained in the liquid are burned, while heavier elements, such as heavy metals and salts, flow down the incinerator walls to the quench tank, located at the bottom of the tank, where they are dissolved and retained in quench liquid. AS gasses bubble through the quench liquid, they are cooled and partially scrubbed. After leaving the quench tank, the gasses pass through two additional scrubbing steps before being exhausted out of the stack. Quench liquid, which is saltwater brine, is then sent to one of several off-site facilities for additional treatment.

**** Saltwater, which is all that remains following the additional treatment is discharged to the Ocean. This statement is disturbing for a number of reasons. One, the American public does not know for sure that saltwater is truly all that remains. Again we are asked to trust the very organization responsible for the devastation in the first place. Now, considering the impact one federal agency, like the Army, has over the other this seems foolish. For example, Congress funds both the Environmental Protection Agency, and the Army, with the Army being the more dominant and powerful of the two. Why not have civilians, plain, ordinary everyday working people, who are not connected to the Federal Government in any way, go in and verify the facts and report back to the public. Then compare apples to apples, oranges to oranges, they should match. If you are comparing avocados to apples one can immediately ascertain there is a problem.

**** Number two, the Army is saying that this so-called saltwater project is then discharged into the ocean, but as of yet there is no plan how this is going to happen. It's quite obvious that the Army did not construct a 1,400 mile pipe line from Denver to the Pacific Ocean, and if they are not using tanker trucks or the railroad, they can't ship it by boat, so this saltwater must be discharged into one of the fresh water streams that abound near the Arsenal. What are they thinking?, that it will eventually end up in the ocean. This deserves and answer.

**** It is painfully obvious we can not fully trust the Federal Government, or more to the point, the Army, Industry giants, and the Pentagon. How can we be sure this amount of so-called salt water isn't creating yet another set of problems. possibly a few years down the line when everyone's well water begins to make the skin pucker, and is salty to the taste, maybe then, and only then will we really know what they had done with the salt water left over after incinerating the toxic waste.

**** If Larry is sounding too cynical regarding the Federal Government the Army, the giant Industry or the Pentagon, "He does not apologize." He says they have gone unchecked for too long for us to roll over with our eyes shut, all the while hoping it goes away, because it won't. There is too much greed, pay offs, and too much power, and more than anything else, there is too much money involved for it to just go away. "Someone needs to guard the hen house, not them they have lost all credibility, and that would be like letting the fox continue guarding the hen house.

**** The SQI treatment process has built-in automatic control that will initiate a safe shutdown in case of problems. Shutdowns will occur in one of two ways:

**** 1- They system will automatically shut off the liquid waste feed, leaving the natural gas burner on so that the incinerator remains at the same temperature and only natural gas is burned. Variances in temperature or water levels in the air pollution control system, or continuous monitoring of the stack, could all result in shutting off the liquid feed.

**** 2- The other type of shutdown would be a total shutdown of the entire facility. This would occur in the event of a power failure or pressure drop in the natural gas line. The liquid feed and the natural gas would shut off and water in the quench tank would seal the incinerator outlet.

**** The SQI process is a 30-year old commercially available technology. SQI is being used at more than 20 sites in the United States and more than 116 sites worldwide. The Army has only added a few special improvements with regards to the injection nozzles and quench tank to handle the unique properties of Basin "F" liquids, containing hundreds of heavy metals.

**** Two other pieces of information came to Larry in July of 1991. they are as follows.

**** Construction is slated to begin this month on the CERCLA Waste water System Interim Response Action (IRA). This facility will be located just north of Seventh Avenue on D. Street, outside of Commerce City, and will take approximately 12-to-14 months to build.

**** The system will treat waste water generated from lab operations, drill purgings, decontamination wastes and waste water as appropriate from other sources such as equipment washing.

**** Prior to treatment, waste water will be stored in tanks, one of which is an 87,000-gallon tank outside. Four smaller tanks will be located inside. Waste water will be transported to the tanks via truck or underground pipeline. Most of the water to be treated is decontamination water. This plant is designed to treat 15 gallons of water per minute.

**** Several different types of technology will be used to treat the water. Treatment will vary depending upon the individual water samples. The following are descriptions of the treatment processes:

> *Activated Carbon, Ultraviolet/ Oxidation, Air Stripping-These three technologies are designed to treat organic contaminants. The bulk of the water will be treated using these processes.
>
> *Chemical Precipitation—-This treatment process precipitates metals out of the water. Water

is pumped into the system,
Chemicals are added, and metals
separate and settle to the bottom
of the tank.

*Activated Alumina—This process
is designed to treat other inor-
ganic's within the water

**** In addition to the treatment system, two enclosed bays will be constructed to wash and decontaminate vehicles used during Arsenal cleanup projects. Treated waste water is sent to the decontamination pad for initial vehicle washing. One of the bays is fully automated like a typical car wash, with electric eyes that activate the high pressure wash system. Once a vehicle travels through the automated wash system, a final rinse is completed using potable water.

**** The other decontamination bay will utilize water spray wands like those found in a manual car wash.

**** Water is treated in batches rather than in a continuous flow. The discharged water is tested for compliance with Applicable or Relevant and Appropriate Requirements (ARAR's), which are outlined in the Decision Document. If treated water is not diverted to the decontamination bays, it is discharged to the Basin A, Neck recharge trenches. If the water fails to meet the standards, it will be recycled again through the treatment system until the criteria are met.

**** The Army Corps of Engineers designed the treatment system and Shell Oil Company is responsible for its construction. "Once again the Wolves have been allowed to watch over the sheep."

**** This last piece of information totally escapes all reasoning and is particularly offensive. Both Larry and Marie think that after everything that has happened in the last twenty or thirty years, the Government would stop thinking of and for the American public as if they were a bunch of foolish imbeciles who are incapable of realizing their own destiny, yet they keep trying.

**** "Remediation of Other Contamination Sources: Interim Response Action-Shell Section 36 Trenches" is well underway.

**** This IRA was designed to stop the migration of contamination from the soils into the ground water from a site that was used, To dispose of liquid and solid waste generated during the production of pesticides. The trenches were used from 1952 through 1965 and are a known source of ground water contamination. The Shell Trenches are located between the North and South Plants in section 36 of the Arsenal.

**** Trenches covering a 7-to-10 acre site were excavated, partially filled with laboratory and plant wastes, and covered with excavated soils.

**** A source containment system was selected to curtail the migration of contaminants from the soils to the ground water through the source of contamination. Upon completion, a 360-degree slurry wall would surround the trenches and a vegetative cap would cover the entire site to reduce the flow of ground water through the source of contamination.

**** The cap was not approximately 60 percent completed, and installation of the slurry wall would begin in June. The 2,200 linear-foot slurry wall would be completed by mid July. The wall is between approximately 14-to-30 feet deep, and keyed into bedrock.

**** Once the vegetative cover was completed, monitoring wells were installed to determine the effectiveness of the containment system. Construction was scheduled for completion in September 1991.

**** The selected slurry wall installation method was to eliminate contaminant releases into the air. The selected technique, known as the vibratory beam method, involve vibrating a steel beam in the soil. As the beam was removed, the slurry mix was injected into opened space, resulting in a slurry wall approximately four inches thick. The more common method of slurry installation involves excavation of the trenches using a two to three foot back hoe, and then filling it up with the soil/clay slurry mix.

**** As part of this IRA, the Final Implementation Document has been issued to the Organizations and State, and field design data have been collected. The next phase of activity will involve construction of a haul road, slurry wall working platform and slurry wall; capping of the trench's site; installation of monitoring wells and eventual site revegitation.

**** Construction was scheduled for completion in September 1991.

Here's the part that smells of another cover up. By their own admission the Army and Shell Oil Company do not remember nor do they have records for all the buried trenches made during this time period of operation, and before. So what becomes of all the thousands of barrels left unchecked?

**** Here's another point. Nothing was said about removing the barrels or the contaminated soil from the trenches they are working on.

**** What is the point of wasting millions of our tax dollars on a job that is only partially done? What's more who gave the OK for this to go ahead knowing full well that all of the trench's had not yet been rediscovered.

**** Larry says, "Can you imagine being an archeologist one thousand years from now, and you discover the very spot the Rocky Mountain Arsenal was. The Death Factory in our Midst. Be careful for whenced you dig, it will be more venomous than a deadly snake.

CHAPTER TEN

PERSIAN GULF WAR SYNDROME

FOUND DEEP IN THE ARCHIVES OF THE UNITED STATES CONGRESS AT THE VERY BOTTOM OF EVERYTHING ELSE, WE FIND SOME TRUTH AS TO WHY SO MANY OF THE VETS WHO FOUGHT IN THE GULF WAR, CAME BACK HOME EXPERIENCING SIGNS, AND SYMPTOMS, OF EXPOSURE TO BOTH CHEMICAL AND BIOLOGICAL WARARE.

***** Over ten years now, there has been a committee examining the U. S. export policy to Iraq prior to the Gulf War. They have learned that materials licensed for export by the United States may have aided Iraqi chemical, biological, nuclear, and missile system development programs. There is now reason to believe some of these materials may have contributed to the illnesses being suffered by the Gulf War veterans and their families.

***** Larry says, "It's been ten years here, and they are still discussing the possibility, of actual responsibility." "It sounds to me as if it is one of three kinds of lies, 'Lies' 'Damned Lies' or is it just another statistic'." everything becomes just another statistic. The responsibility lies with the United States, and the Pentagon, The hub of the military.

*** "Both Chemical and Biological war fare in what started out to be Desert Shield, for the liberation of Kuwait, in less than 24 hours it would be called Desert Storm, a war with Iraq."

**** "Obviously once Desert storm went into action, there was a lot of shelling, and bombing all around you. The air was heavy with smoke, and you could hear air raid sirens everywhere. The brass was spreading the word to everyone, everywhere there is nothing to worry about, every ones eyes were red, inflamed as if someone had just thrown a handful of sand right in your eyes, irritated throats, chests were tight, difficult breathing, everyone just about as irritable as hell."

**** "When they began blowing the Bunkers, soldiers going through them would find bunkers filled with ammunition, and it was not Iraqi ammunition either. There were a lot of American munitions stored in each and every bunker they went into, and there were hundreds. They were wired up to C4 explosives, and blown up, totally destroyed them and all the while were three miles away. It was a lot like throwing a lit match into the fireworks factory. The explosions began sending hardware our way right over our heads, and into the area."

**** "I also learned that they were involved in chemical scares, when the alarms would sound. Everyone had to dawn the chemical mask and suits. These were supposedly false alarms, they said don't worry, there is only trace amounts of chemicals in the air. There thought here was it was because the plants have been blown away and the air if filtering this way, they said to everyone, "Don't' Worry There Is Not Enough To Hurt You."

**** It was very difficult to even know whether or not you were smelling something, the constant smell in the air was just completely overwhelming. They were giving everyone the antidote in bottles as if a prescription pill. Perignostigmine-Bromine, little white chemical pills. Everyone seems to think they got sick, and feeling the influence, from the antidote, that was to make sure they didn't get sick from potential chemical warfare. There was nausea, shakes, severe headaches, and vomiting everywhere.

***** "There is a real sad story behind all this now, for when the soldiers came home they brought part of the Nerve agents, or Biological agents home to their families. What the veterans were exposed to in the Gulf war, they have brought home, and the transmissibility is very real." Obviously said, "Dr. Victor Gordon A Veterans Affairs Medical Center Physician in Manchester, New Hampshire." Then Quoting Dr. Gordon, "There is definitely transmissibility., Either there are some chemicals or a virus or God knows what that is causing this."

***** "Incidentally what storage areas were bombed, and releasing chemicals toxins, Bio-war-fare-agents, and Viro-war-fare-agents. These horrible and horrendous agents were supplied to Iraq during the 1980's, during the Reagan Administration. "It is our own Chemicals, Toxins, Bio-warfare, and Viro-warfare The United States sold to Iraq, that has caused the illness to these brave men and women in uniform, during the Desert Storm, in Iraq."

***** "There were at least a quarter of a million Americans who left their families and friends, and were quickly taken by plane, by ship, and by truck, to the Persian Gulf, Saudi Arabia, Kuwait, and

Iraq. Active duty personnel, reservists, and the National guard were all called out to the inevitable; Desert Storm, war with Iraq. They learned something very quick that Saddam Hussein, would be using Chemical, and Biological weapons, the same as was used against the Kurds in Northern Iraq, Southern Iraq, The Kurds in Northern Iran, and on the Iranian soldiers themselves in the Iran, Iraq war of the 80's He threatened openly against the United States Forces and promised to use this type warfare on them. This was a new threat from Saddam Hussein, but old news to the United States who had supplied him, and was aware of his suppliers from France, Germany, and Russia. During the 1980's The United States not only encouraged, but also supplied him with the finished product, and the equipment, technology, and in some cases technicians to assist getting himself up in manufacturing of deadly chemicals and deadly germ warfare."

**** As we watched the war while it was actually happening, it seemed we were a part of a big book, as we became engrossed in every page, and almost knew when a chapter was ending, and a new one would begin. Everyone writing home telling family and friends how well things were going, and not to worry, we'll be home soon.

**** "Incidentally, the storage areas we were bombing, releasing what chemicals? And how about those oil well fires?" Word is out that each well was wired to nerve agents and various Bio-war-fare agents, that were released into the thick oily smoke" that would be carried well into the air, and drift right into Kuwait city and the area's where the American soldiers were. Suddam Hussein had it planned and knew exactly what would take place, and they used the direction of the wind in carrying whatever chemicals they had directly to the Americans, and to Kuwait." It was just like glue anywhere, anyplace or any person his/her clothes became contaminated with toxins, Viro-toxin, Bio-Toxin or chem-Toxins. Suddam Hussein actually thought he was winning the war at that point as he kept announcing the Armageddon is upon us, this is the mother of all wars."

**** From the time President Ronald Reagan, signed the Proliferation agreement with many other nations in 1983, and at the same time signing agreements to limit nuclear weapons, and we all thought this is great. I personally was astonished, We actually had a President, and Government that cared." All until I began learning that the Pentagon was looking for a newer and more effective Nerve agents., to be manufactured right here in the United States. In 1987, they had secrets to a deadlier toxin which as stated is being manufactured in the United States today already large stock piles exist.

***** "All the time the United States, Russia, Germany France, were selling the Neuro-Toxins, Bio-Toxins, and viro-toxins to third world Countries, The few that may have received the actual product, Technology, Technical assistance, and Technicians. Iraq, Turkey, Iran, Libya, Afghanistan, India, Pakistan, North Korea, China, North Vietnam, to name a few that are highly suspect. Iraq, there is no doubt, most of it came from the United States, the technical support, from all those involved, shipping and handling was free with the purchase." This was probably all done without the knowledge of President Reagan." "The exact Idiology, and separation of power that allowed Hitler, and his Henchmen to organize a reign of Terror, on a free World creating a monster that terrorized the civilized world, causing World War II, mass executions by the millions, committee by those in power that produced a wide spread terror every where.

**** "A very painful death, for the Kurds in Iraq, Iran, and the Iranians was just a "spray" away for Suddam Hussein. He had towns and villages sprayed with Lewisite, and Mustard Agents, then behind that came the GB Nerve Agent (Sarin Gas), and other's believed to be Biological." This left people dying in their homes, dying in the streets as they tired carrying their babies, and children in an effort to find safety, They would die clutching their child as they take their last breath or have the final convulsions. With these materials there is no safety, Their lungs, and Brains destroyed, while the Lewisite and Mustard Agents destroy the flesh. I know this sounds absolutely Horrible, but it was the truth in Northern, Southern, Iraq and Iran.

***** "This evil man Suddam Hussein, and his disciples rule with a heavy fist, you either do what they say or you are dead. I wanted to make this very clear, there are "Death Factories in our Midst." The United States, Russia, France, and Germany are all involved in Suddam Hussein's conspiracy to rule the World. I'm sure the United States didn't' know when they joined his conspiracy, that those weapons, and technology would be used against the United States armed forces, but they still try to deny their own responsibility to the Veterans, their loved ones, their families who were poisoned with, and injured for life, for which there is nothing to stop it now, or the injuries that will be passed on down from generation to generation These were Neuro-Toxins, made in the United States, who is, and was very much aware of the Catastrophic consequences. As we move on we'll have a chance to share what I have in my possession, and again like everything else the Government does, it was put aside and it got brushed under the rug, what you don't see, or ask will not be answered. remember the three types of lies, "Lies", "Damned Lies", then they just become "Statistics".

***** "Never tell a lie...Unless lying is one of your strong points." The credibility of the United States is shrouded in lies, and covered in statistics.

GEORGE WASHINGTON PLUNKETT

***** "It is said from those who were there in Kuwait City, that the first eight weeks, the sun looked like the Moon, and there was no moon or very little. The smoke was like soot, it was in your clothes, your eyes, nose throat, when you spit, it looked like gear grease." Once you had it on you it stayed there was no water for a shower for nearly three weeks, and then after the shower you had to wear the same dirty sooty clothes, because there wasn't enough water to do the laundry, for nearly five weeks.

**** "There was some concern over many dead animals, and it was thought that the Americans may be shooting them., Well first of all the Zoo animals were all shot, and left dead by the fleeing Iraqi's. There was something figured in the area of 20,000 dead animals. As a veterinarian We was asked to check into the situation, and get a report back as soon as possible. As you probably are aware in the Desert there are a lot of insects, lotta flies. Well We found the first dead animal, it was closely examined for the cause of death. There were no bullet holes in those outside the Zoo, but we also examined the Zoo animals we did find gun shot wounds. Then we begin to put two and two together because we noticed all around the dead animals, and even with the rotting blood on those in the Zoo, there were some dead insects around the animals, There were no insects or flies on the animals. It was only then we realized that oily substance on the animals was very toxic, and acting as a deterrent to the insects. Part of the training in chemical and biological, warfare defense is to be on the lookout for anything like that. You are not to touch it, and required to report it to the higher Authority.

***** The pentagon is unwilling to help the sick, and dying Gulf war vets. Yet they were the one's in the 1980's who was doing big business with Iraq, helping them win the war over the Kurds, and Iran. During the Desert Storm he turned the business end of all the Germ warfare, Chemical warfare, and Nuclear warfare at the Americans. The United States created the Monster, and really don't know what to do with it. How do they kill it? At least there is a certain part of our Government feels a real responsibility to the Vets, and there are others who don't care. "Larry Says, If your Senators, Governors, or Representatives don't jump in and help anyway they can, Get rid of them." The Department of Defense has totally ignored all the facts, and are not available to discuss it with any one."

***** It was much the same with the Agent Orange incident in Viet Nam. They have ignored it, lied about and now it has become a statistic they can file away. There were seven types of Agent Orange, and it was the Pentagon that kept saying we need something more potent, and it was made much more potent, Vets were sprayed, or in the area after being sprayed, and will suffer the consequences through out a lifetime. It is a material that penetrates the human body, it not immediately washed off, Clothes, and anything you touch by your body It will penetrate, and may take months even years before the real effects become apparent, much like the Bio chemistry, the Chemicals, it has usually a light showing of signs, and symptoms, that get worse as time passes, Months, and even Years later the handicaps, and disabilities become apparent before your time, you will live a life of pain and suffering.

***** "1985 to 1989 Biological Cultures had been shipped directly to Iraq with approval from the United States Commerce Department the Executive Branch of Government. The Approval for delivery was given, specifically for Saddam Hussein. It is a known fact by our Government that when they sent the cultures to Saddam Hussein, they also sent him all the technology, equipment, and technical support for him to construct, and build the kind of Biological, and Chemical weapons that would create the exact kind of symptoms were seeing here now. If troops were out in the field and exposed to those very items we just talked about fingerprinted directly to Saddam Hussein."

***** "This is a horror story, you think could not have happened, and yet it did, from the highest levels of the United States. The firms that sold this broke no law, they needed full Government approval, and they got that approval without question. To send directly to Saddam Hussein, live cultures Bacillus-Anthracis, (Anthrax), Clostridium-Botulinum (Botulinum Toxin), and much much more later as I will describe the Symptoms, and Health Maladies. The wives, and families are sickened with the same types symptoms after the Vets came home."

***** "The Department of Defense, doesn't even want to talk about it, they've told enough damn lies, and they have filed it as a statistic. Basically the bottom line, the veterans who's live's and their families lives are basically destroyed, are ignored and thought to be disposable when the war was over. The Department of Defense are looking at the next war, and figuring out all the lies they can produce for the next one until it will become a statistic." They work one way the Department of Defense, and the Pentagon, there are three types of lies, Lies, Damned Lies, and then you have a statistic. It's a lot like taking the Monkey off their back and placing in on someone else's back.

***** The United States has shamefully turned their back on the wounded vets and their families, of the Persian Gulf War. This is something that happened during a time of war, and the United States will not face up to what they were very much a part of. The Senate approved and gave the Pentagon several million dollars to do some medical studies on chemical, biological, and radiation type effects to the human body. They don't have to spend a dime but they are still sitting on that money, they have not hired any medical or scientific people to do these studies. The truth is they don't want to admit the mistakes that were made because they don't have to have to pay the bills. Who the hell was the idiots who sent biological warfare to Saddam Hussein in the first place? This entire Iraq problem, the Government created the just want to go away. Like the Ostrich burying his head in the sand. They are very anxious to call it another statistic.

***** "Several hours in the air war, chemical alarms started going off, and this was daily three times or more per day, it was like somebody blew the whistle again. The soldiers would actually peak out of the tents to see if it was real or not. To tell the truth What the hell were they looking for, If the Biological or chemical agents were in the air you could not see it. It could have been real and they didn't even know it. Most did not even dawn the mask. This went on for weeks on end. They were always called dry runs, and now we pay the price.

***** "Soon we'll get into all the names of Chemicals, and Germ Warfare that was made available directly, "To Saddam Hussein, Iraqi Chemical and Biological Warfare Program"." From the United States of America." GB, Sarin, VX, and BZ, are all Nerve agents that were sent by the United States to Iraq. These are known to block the neuro transmitters in the brain, and depending on which ones can effect any organ in the body. The Liver, Heart, and kidneys which it enlarges, and leaves it in a weakened condition or until it's to weak to go on."

***** This was seen, and heard first hand from specialists. Histoplasma-Capsulatum, Causes a disease superficially resembling "Tuberculosis" Pneumonia, and enlargement of internal organs. Cancer seems to be also on the increase amongst veterans of the Gulf war, There are several chemical agents that are known carcinogens, and Saddam had this material, shipped to one of many addresses he used, by the United States Commerce Department, this was done as late as 1989 just ahead of the war.

***** "GULF WAR SYNDROME: THE CASE FOR MULTIPLE ORIGIN MIXED CHEMICAL AND BIOTOXIN WARFARE RELATED DISORDERS"

***** "The following information was made available to all Senators and all Congressmen, by U. S. Senator Donald W. Riegle, Jr. Shortly after the Persian Gulf War took place and Vets began to report their Warfare Related Disorders by Chemical and Biotoxic Warfare Munitions, in which a large percent were American Made, with American and Russian Technology."

***** "The first group of Over 4,000 U. S. Veterans of the Gulf War suffering from a myriad of illnesses collectively labeled "Gulf War Syndrome" are reporting symptoms of Muscle and Joint Pain, Memory Loss, Intestinal and Heart problems, Fatigue, Running noses,

Urinary Urgency, Diarrhea, Twitching, Rashes, and Sores". "This is just the beginning, many are afraid to talk about it because of their jobs, and the jobs they will never again be able to do." Some of these material have long term effects, and won't get any better possibly worse over time, and some will never be pronounced until years later." (Some are Cumlitive, and Progressive). "Many of the Veterans showed symptoms consistent with exposure to a mixed agent attack."

***** "Czech Chemical Decontamination Unit had detected traces of a chemical warfare agent, (SARIN) in areas of Northern Saudi Arabia. In 1983 there were documents of the sale of Sarin and all related compounds were shipped to Iraq. We Were shocked by Photos released from Iraq showing the same 1,000 and 2,000 pound containers of GB, Sarin, VX, and BZ at a storage depot, and one appeared to have painted on it's face U. S. Army, were the very same containers used by the United States Army for storage. The Department of Defense maintains that there is no evidence that U. S. Forces were exposed to chemical warfare agents. Yet all the Iraqi Manufacturing Complex's that were bombed at Samarra, Al Falluja, Muthanna, were in the business of manufacturing Nerve agents, Sarin 1,000 tons per month, with the assistance of foreign firms, and probably supplies. VX, BZ, Hydrogen Cyanide, Cyanogen Chloride, Lewisite, and mustard agents."

***** "The Key Essential Element of this entire investigation on Gulf War Syndrome would have been careful, (Medical Monitoring). This is the first step necessary in order to have a hands on experience for what happened. More importantly, the result of Non-Disclosure is a continuing failure by the United States to provide adequate medical care to thousands of veterans of the Gulf War who have been wounded in action on the Chemical and Biological Battlefield."

***** "Iraqi Chemical and Biological Warfare Capability: Over the last ten years, Iraq, a signatory to the Geneva Protocols of 1925 prohibiting the use of poisoned gas and to the Biological Warfare Convention of 1972 banning biological weapons, has expended an enormous amount of research and energy in the development of these and other prohibited weapons of mass destruction."

***** "By the start of the Gulf War, it was learned the Iraqi forces had developed Chemical delivery capabilities for rifle grenades, 81-mm mortars, 152l-mm, 130-mm, and 122-mm, artillery rounds; Bombs; 90-mm Air-to-ground rockets; 216 kilogram FROG and 555 kilogram SCUD warheads; and possible land mines, and cruise missiles, with cluster bomb capability."

***** "Chemical Warfare weapons which survived allied bombing were inventoried at Muthanna Manufacturing Facility for destruction; 13,000 155-mm artillery shells loaded with mustard gas.

6,200 Rockets loaded with nerve agents.

800….Nerve Agent aerial bombs.

28…..Scud Warheads loaded with Sarin.

75….. Tons of Sarin Nerve Agents.

60-70..Tons of Tabun Nerve Agents.

150….Tons of Mustard gas, and stocks of thiodiglycol, mixture for mustard gas."

***** "Iraqi Military adhered to Soviet Military Doctrine; which is a mixing of toxic agents; (often referred to as cocktails). This enhances the capabilities of the Nerve agents, and defeats the precautions taken by the enemy. "Use of mixed agents could account for the wide variety of symptoms displayed by the Gulf War Veterans. It is done by combining a variety of Biotoxins, Nerve agents, Vesicants, and Blister agents.

***** Also discovered at various locations, was evidence of research into certain biological agents, including Botulinus Toxin, Anthrax, and an organism responsible for gangrene.

IRAQ GOVERNMENT HAD THE CAPABILITIES OF COMBINING THE FOLLOWING.

BIOTOXINS:

**** Biotoxins are natural poisons, chiefly of cellular structure A distinction is made between exotoxins, given off by an organism while it is alive, and endotoxins, given off after a cell's death. The former cause the injurious effects of biological weapons, but the latter guarantee those of chemical weapons and do not cause the widespread disease outbreaks associated with biological warfare. Some examples of biotoxins include botulinus toxin and Staphylococcic enterotoxin. Iraq was also known to be experimenting with the use of anthrax and an organism responsible for gangrene.

CHEMICAL NERVE AGENTS:

**** Nerve Agents kill by disrupting the metabolic processes, causing a buildup of a chemical messenger (Acetylcholine) by inhibiting the production of Acetylcholinesterase, a key regulator of neurotransmission. Lethal exposure to chemical nerve agents is generally characterized by drooling, sweating, cramping, vomiting, confusion, irregular heart beat, convulsions,, loss of consciousness and Coma. Little is known, however, about the long term effects of non lethal exposure.

Sarin – A colorless and practically odorless liquid, Sarin dissolves well in water and organic solvents. The basic military use of Sarin is that of a gas and a persistent aerosol. A highly toxic agent with a clearly defined myopic effect, symptoms of intoxication appear quickly without any period of latent effect. *Sarin has a cumulative effect*, independent of its method of entry into the body. The progressive signs of initial Sarin intoxication include myosis (contraction of the pupil), photophobia, difficulty breathing and chest pain.[19]

Soman – A neuro-paralytic toxic agent, it is a transparent, colorless, involitile liquid smelling of camphor. Soluble in water to a limited degree, Soman is absorbed into porous and painted surfaces. Soman is similar to Sarin in its injurious effects, but more toxic. When it acts on the skin in either droplet or vapor form, it causes a general poisoning of the organism.[20]

Tabun – A neuro-paralytic toxic agent, it is a transparent, colorless liquid. The industrial product is a brown liquid with a weak sweetish smell; in small concentrations it smells of fruit, but in large concentrations, it smells of fish. Tabun dissolves poorly in water but well in organic solvents; it is easily absorbed into rubber products and painted surfaces. Injury occurs upon skin contact with Tabun vapor and droplets. The symptoms of injury appear almost immediately. Marked myosis occurs.[21]

VX – This colorless, odorless, liquid has a low volatility, is poorly soluble in water, but dissolves well in organic solvents. The danger of pulmonary VX intoxication is determined by meteorological conditions and the delivery method used. VX is thought to be very

19

20

21

effective against respiratory organs when in the form of a thinly dispersed aerosol. The symptoms of VX intoxication are analogous to those of other nerve agents, but their development is markedly slower. *As with other nerve agents, VX has a cumulative effect.*[22]

Vesicants and Blood Agents:

Lewisite – A vesicant toxic agent, industrial lewisite is a dark-brown liquid with a strong smell. Lewisite is a contact poison with practically no period of latent effect. Lewisite vapors cause irritation to the eyes and upper respiratory tract.[23]

According to the Center for Disease Control, lewisite would cause stinging and burning. Its smell, generally characterized as the strong smell of geraniums, could be confused with the smell of ammonia (the reaction to which is regulated by pain fibers rather than smell).[24]

Cyanogen Chloride – The French first suggested the use of cyanogen chloride as a toxic agent. *U.S. analysts have reported that it is capable of penetrating gas mask filters*. Partially soluble in water, it dissolves well in organic solvents. It is absorbed easily into porous materials; its military state is a gas. Cyanogen chloride is a quick acting toxic agent. Upon contact with the eyes or respiratory organs, it injures immediately. Lethal exposures result in loss of consciousness, convulsions and paralysis.[25]

Hydrogen Cyanide – A colorless liquid smelling of bitter almonds, hydrogen cyanide is a very strong, quick acting poison. Hydrogen cyanide affects unprotected humans through the respiratory organs and during the ingestion of contaminated food and water. It inhibits the enzymes which regulate the intra-cell oxidant-restorative process. As a result, the cells of the nervous system, especially those affecting breathing, are injured, which in turn leads to quick death. An important feature of hydrogen cyanide is the absence of a period of latent effect. The military state of hydrogen cyanide is as a gas. The toxic and physiologic properties of hydrogen cyanide permit it to be used effectively in munitions – predominantly in rocket-launched

22

23

24

25

artillery. Death occurs after intoxication due to paralysis of the heart. Non-lethal doses do not cause intoxication.[26]

Blister Agents:

Mustard Gas – A colorless, oily liquid which dissolves poorly in water, but relatively well in organic solvents, petroleum, lubricant products, and other toxic agents. The injurious effect of mustard gas is associated with its ability to inhibit many enzyme systems of the body. This, in turn, *prevents the intra-cell exchange of chemicals and leads to necrosis of the tissue. Death is associated mainly with necrosis of the tissue of the central nervous system.* Mustard gas has a period of latent effect (the first signs of injury appear after 2-12 hours), but does not act cumulatively. *It does not have any known antidotes.* In military use, it can come in gas, aerosol, and droplet form. It therefore acts through inhalation, cutaneously, perorally and directly through the blood stream. The toxic and physico-chemical properties of mustard gas allow it to be used in all types of munitions.[27]

***** WHAT I'M ABOUT TO GIVE YOU SECOND HANDED, AND STRAIGHT OUT OF CONGRESS, THE SHIPPING ORDERS FOR SADDAM HUSSEIN, FROM THE UNITED STATES OF AMERICA, FREE TRADE POLICY, WAS APPROVED BY THE UNITED STATES COMMERCE DEPARTMENT. ALL SENATOR'S, AND CONGRESSMEN HAD PERSONAL KNOWLEDGE TO ALL THE MATERIAL, INCLUDING THE PRESIDENT OF THE UNITED STATES GEORGE W. BUSH, AND WILLIAM CLINTON, AS WELL AS THE VICE PRESIDENTS, AND THE SECRETARY OF STATE. THANKS TO SENATOR DONALD RIEGLE, JR. WHO RELEASED THIS TO THEM ALL FOR THEIR BENEFIT, AND THE BENEFIT OF THE VETERANS AND THE FAMILIES WHO HAVE PAID THE ULTIMATE PRICE TO THE UNITED STATES.

***** "EVIDENCE THE UNITED STATES ARMED IRAQ; WITH BIOLOGICAL EXPORTS, PRIOR TO THE GULF WAR." "In reports issued by Senator Riegle, he ask today February 9, 1994 that the Department of Defense and Department of Veterans Affairs

26

27

establish disability compensation for these Veterans and their families, consistent with their degree of disability regardless of their ability to arrive at a definitive medical diagnosis."

***** "Senator Riegle announced also on February 9, 1994 he called for hearings too investigate the export of these materials and the possible link to Gulf War Syndrome. Mr. Riegle states, "I think we need to change our policy on the export of these materials to assure that such deadly agents don't fall into the hands of countries with known biological warfare programs."

***** "The United States and Iraq are signatories of the 1972 Biological Warfare Convention which restricts the use and proliferation of biological Warfare agents. Following the Persian Gulf War, U. N. inspectors confirmed that Iraq was conducting biological warfare research. According to the Pentagon's official report to Congress on the Conduct of the Persian Gulf war, Written in 1992: "By the time of the invasion of Kuwait, Iraq had developed biological weapons. Its advanced and aggressive biological warfare program was the most advanced in the Arab World. Large scale production of these agents began in 1989 at four facilities near Baghdad. Delivery means for biological agents ranged from simple aerial bombs and artillery rockets to surface to surface missiles."

***** "President Ronald Reagan, The United States of America I believe signed a Treaty with Russia banning the proliferation of both Chemical and Biological Weapons. The Corporations, the United States Army, who sold or traded Biological and Chemical warfare materials, and Technology to Iraq are guilty of treason for the proliferation of Chemical and Biological warfare material. The United States Department of Commerce who approved them sales and delivery are as guilty as Sin itself."

***** Following you will see all the various types of Biological agents, names, dates sold and shipped to in Iraq. Like the Ministry of higher Education, Iraq Atomic Energy agency. Some of the various signs and symptoms suffered, by Vets, and families.

The following is detailed listing of biological materials, provided by the American Type Culture Collection, which were exported to agencies of the government of Iraq pursuant to the issuance of an export licensed by the U.S. Commerce Department:

Date: February 8, 1985

Sent to: Iraq Atomic Energy Agency
Materials Shipped:
Ustilago nuda (Jensen) Rostrup
Date: February 22, 1985
Sent to: Ministry of Higher Education
Materials Shipped:
Histoplasma capsulatum var. farciminosum (ATCC 32136)
Class III pathogen
Date: July 11, 1985
Sent to: Middle and Near East Regional A
Materials Shipped:
Histoplasma capsulatum var. farciminosum (ATCC 32136)
Class III pathogen
Date: May 2, 1986
Sent to: Ministry of Higher Education
Materials Shipped:

1. Bacillus Anthracis Cohn (ATCC 10)
   Batch # 08-20-82 (2 each)
   Class III pathogen
2. Bacillus Subtilis (Ehrenberg) Cohn (ATCC 82)
   Batch # 06-20-84 (2 each)
3. Clostridium botulinum Type A (ATCC 3502)
   Batch# 07-07-81 (3 each)
   Class III Pathogen
4. Clostridium perfringens (Weillon and Zuber) Hauduroy, et al
   (ATCC 36324) Batch# 10-85SV (2 each)
5. Bacillus subtilis (ATCC 6051)
   Batch# 12-06-84 (2 each)
6. Francisella tularensis var. tularensis Olsufiev (ATCC 6223)
   Batch# 05-14-79 (2 each)
   Avirulent, suitable for preparations of diagnostic antigens.
7. Clostridium tetani (ATCC 9441)
   Batch# 03-84 (3 each)
   Highly toxigenic.
8. Clostridium botulinum Type E (ATCC 9564)

Batch# 03-02-79 (2 each)
Class III pathogen

9. Clostridium tetani (ATCC 10779)
Batch# 04-24-84S (3 each)
10. Clostridium perfringens (ATCC 12916)
Batch# 08-14-80 (2 each)
Agglutinating type 2.
11. Clostridium perfringens (ATTC 13124)
Batch# 07-84SV (3 each)
Type A, alpha-toxigenic, produces lecithinase
C.J. Appl.
12. Bacillus Anthracis (ATCC 14185)
Batch# 01-14-80 (3 each)
G.G. Wright (Fort Detrick) V770-NP1-R.
Bovine anthrax,
Class III pathogen
13. Bacillus Anthracis (ATCC 14578)
Batch# 01-06-78 (2 each)
Class III pathogen.
14. Bacillus megaterium (ATCC 14581)
Batch# 04-18-85 (2 each)
15. Bacillus megaterium (ATCC 14945)
Batch# 06-21-81 (2 each)
16. Clostridium botulinum type E (ATCC 17855)
Batch # 06-21-71
Class III pathogen.
17. Bacillus megaterium (ATCC 19213)
Batch# 3-84 (2 each)
18. Clostridium botulinum Type A (ATCC 19397)
Batch# 08-18-81 (2 each)
Class III pathogen
19. Brucella abortus Biotype 3 (ATCC 23450)
Batch# 08-02-84 (3 each)
Class III pathogen
20. Brucella abortus Biotype 9 (ATCC 23455)
Batch# 02-05-68 (3 each)
Class III pathogen
21. Brucella melitensis Biotype 1 (ATCC 23456)

Batch# 03-08-78 (2 each)
Class III pathogen

22. Brucella melitensis Biotype 3 (ATCC 23458)
Batch# 01-29-68 (2 each)
Class III pathogen
23. Clostridium botulinum type A (ATCC 25763)
Batch# 8-83 (2 each)
Class III pathogen
24. Clostridium botulinum type F (ATCC 35415)
Batch# 02-02-84 (2 each)
Class III pathogen

Date: August 31, 1987
Sent to: State Company for Drug Industries
Materials Shipped:

1. Saccharomyces cerevesiae (ATCC 2601)
Batch# 08-28-08 (1 each)
2. Salmonella choleraesuis subsp. Choleraesuis
Serotype typhi
(ATCC 6539) Batch# 06-86S (1 each)
3. Bacillus subtillus (ATCC 6633)
Batch# 10-85 (2 each)
4. Klebsiella pneumoniae subsp. Pneumoniae
(ATCC 10031)
Batch# 08-13-80 (1 each)
5. Escherichia coli (ATCC 10536)
Batch# 04-09-80 (1 each)
6. Bacillus cereus (11778)
Batch# 05-85SV (2 each)
7. Staphylococcus epidermidis (ATCC 12228)
Batch# 11-86s (1 each)
8. Bacillus pumilus (ATCC 14884)
Batch# 09-08-80 (2 each)

Date: July 11, 1988
Sent to: Iraq Atomic Energy Commission
Materials Shipped:

1. Escherichia coli (ATCC 11303)
Batch# 04-87S
Phage host

2. Cauliflower Mosaic Caulimovirus (ATCC45031)
   Batch# 06-14-85
   Plant virus
3. Plasmid in Agrobacterium Tumefaciens
   (ATCC37349)
   (Ti plasmid for co-cultivation with plant integration vectors in E.
   Coli) Batch # 05-28-85

Date: April 26, 1988
Sent to: Iraq Atomic Energy Commission
Materials Shipped:

1. Hulambda4x-8, clone: human hypoxanthine
   Phosphoribosyltransferase (HPRT) Chromosome(s) X q26.1
   (ATCC 57236) Phage vector; Suggested host: E.coli
2. Hulambdal4-8, clone: human hypoxanthine
   Phosphoribosyltransferase (HPRT) Chromosome(s): X q26.1
   (ATCC 57240) Phage vector; Suggested host: E.coli
3. Hulambdal5, clone: human hypoxanthine
   Phosphoribosyltransferase (HPRT) Chromosome(s) X q26.1
   (ATCC 57242) Phage vector; Suggested host: E.coli

Date: August 31, 1987
Sent to: Iraq Atomic Energy Commission
Materials Shipped:

1. Escherichia coli (ATCC 23846)
   Batch# 07-29-83 (1 each)
2. Escherichia coli (ATCC 33694)
   Batch# 05-87 (1 each)

Date: September 29, 1988
Sent to: Ministry of Trade
Materials Shipped:

1. Bacillus anthracis (ATCC 240)
   Batch#05-14-63 (3 each)
   Class III pathogen
2. Bacillus anthracis (ATCC 938)
   Batch#1963 (3 each)

Class III pathogen

3. Clostridium perfringens (ATCC 3629)
Batch#10-23-85 (3 each)
4. Clostridium perfingens (ATCC 8009)
Batch#03-30-84 (3 each)
5. Bacillus anthracis (ATCC 8705)
Batch# 06-27-62 (3 each)
Class III pathogen
6. Brucella abortus (ATCC 9014)
Batch# 05-11-66 (3 each)
Class III pathogen
7. Clostridium perfringens (ATCC 10388)
Batch# 06-01-73 (3 each)
8. Bacillus anthracis (ATCC 11966)
Batch# 05-05-70 (3 each)
Class III pathogen
9. Clostridium botulinum Type A
Batch# 07-86 (3 each)
Class III pathogen
10. Bacillus cereus (ATTC 33018)
Batch# 04-83 (3 each)
11. Bacillus ceres (ATCC 33019)
Batch# 03-88 (3 each)

Date: January 31, 1989
Sent to: Iraq Atomic Energy Commission
Materials Shipped:

1. PHPT31, clone: human hypoxanthine phosphoribosyltransferase (HPRT) Chromosome(s) X q26.1 (ATCC 57057)
2. plambda500, clone: human hypoxanthine phosphoribosyltransferase pseudogene (HPRT) Chromosome(s): 5 p14-p13 (ATCC 57212)

Date: January 17, 1989
Sent to: Iraq Atomic Energy Commission
Materials Shipped:

1. Hulambda4x-8, clone: human hypoxanthine Phosphoribosyltransferase (HPRT) Chromosome(s)

X q26.1
(ATCC 57237) Phage vector; Suggested host: E.coli

2. Hulambda14, clone: human hypoxanthine Phosphoribosyltransferase (HPRT) Chromosome(s): X q26.1
(ATCC 57240) Cloned from human lymphoblast Phage vector; Suggested host: E.coli
3. Hulambda15, clone: human hypoxanthine Phosphoribosyltransferase (HPRT) Chromosome(s) X q26.1
(ATCC 57241) Phage vector; Suggested host: E.coli

F-15C Eagle Gulf War
• Air superiority • Highest scoring fighter • Unbeaten in combat

Arming Iraq: U.S. Exports of Biological Warfare-related Materials

"The [Iraqi] program probably began in the late 1970s and concentrated on the development of two agents – botulinum toxin and anthrax bacteria." - The Conduct of the Persian Gulf War: Final Report to Congress, DoD, 1992

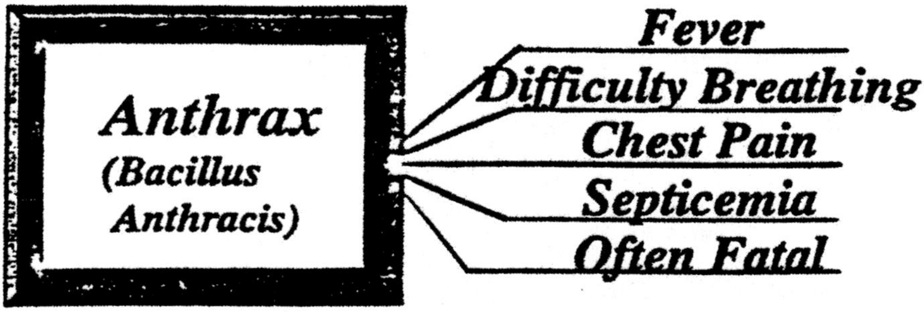

***Shipped: May 2, 1986, Ministry of Higher Education
September 29, 1988, Ministry of Trade***

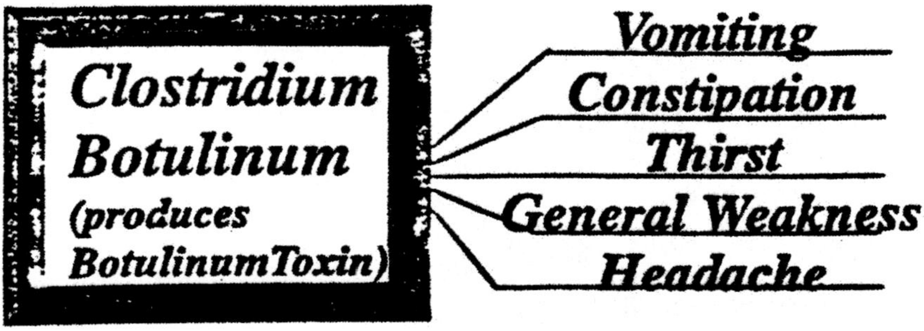

***Shipped: May 2, 1986, Ministry of Education
September 29, 1988, Ministry of Trade***

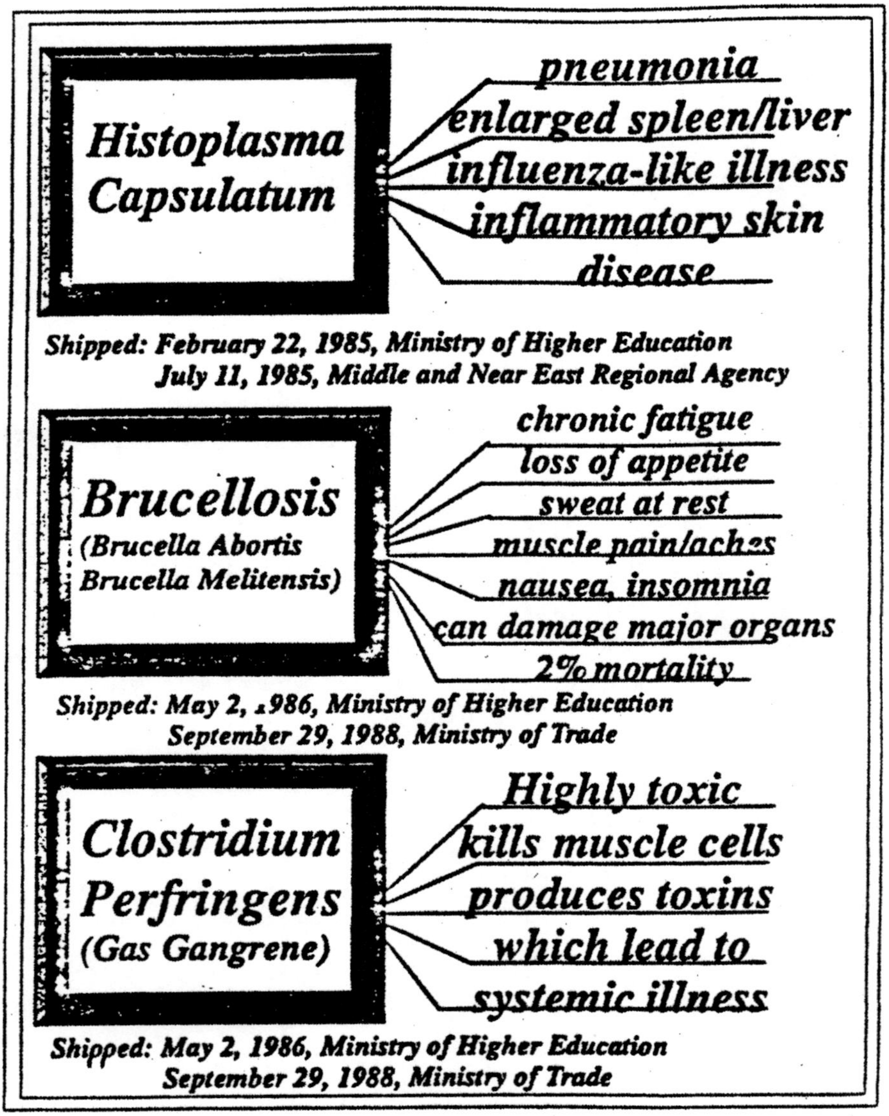
Histoplasma Capsulatum
pneumonia
enlarged spleen/liver
influenza-like illness
inflammatory skin disease
Shipped: February 22, 1985, Ministry of Higher Education
July 11, 1985, Middle and Near East Regional Agency
Brucellosis
(Brucella Abortis Brucella Melitensis)
chronic fatigue
loss of appetite
sweat at rest
muscle pain/aches
nausea, insomnia
can damage major organs
2% mortality
Shipped: May 2, 1986, Ministry of Higher Education
September 29, 1988, Ministry of Trade
Clostridium Perfringens
(Gas Gangrene)
Highly toxic
kills muscle cells
produces toxins
which lead to
systemic illness
Shipped: May 2, 1986, Ministry of Higher Education
September 29, 1988, Ministry of Trade

Several shipments of Escherichia Coli (E.Coli) and genetic materials (human and bacterial DNA) were sent to the Iraq Atomic Energy Commission

Shipped: August 31, 1987, April 26, 1988, January 17, 1989, January 31, 1989

Other Materials Shipped to Iraq:

- *Saccharomyces cerevesiae*
- *Bacillus subtillus*
- *Staphylococcus epidermis*
- *Salmonella cholerasuis*
- *Klebsiella pneumoniae*
- *Bacillus pumilus*
- *Bacillus megaterium*
- *Clostridium tetani*

Part A. Iraqi Chemical and Biological Warfare Program
1. Iraqi Chemical and Biological Warfare Capability
2. Soviet Military Doctrine and the Use of Combined Agent Warfare

Part B. Gulf War Syndrome
1. The Reported "Chemical Weapons Attacks" on the 644th Ordinance Company and the Naval Reserve Construction Battalion 24
2. The Relationship Between These Attacks and Gulf War Syndrome (Group I Disorders)
3. Saddam Hussein's History in the Use of Chemical Warfare Agents
4. Department of Defense Inconsistencies in Statements About the Chemical Warfare Capabilities of the FROG-7 Rocket

Part C. Gulf War Syndrome – Group II Disorders
1. The Coalition Bombing of Iraqi Chemical, Biological, and Nuclear Facilities and the Detection of Chemical Agents by Both Czech and American Forces
2. Soviet Estimates on the Use of Chemical Weapons and the Dispersion of Chemical Agents.
3. The Relationship Between The Bombings and Gulf War Syndrome (Group II Disorders

Part D. Gulf War Syndrome – Group III Disorders
1. Administration of Chemical and Biological Warfare Agent Pre-treatment Drugs (Group III Disorders)

Part E. Conclusions
1. Why Wasn't Everyone Affected – The Need for Immediate Advanced Medical Research
2. Conclusions and Recommendations

Map of Iraq-Northern Saudi Arabia - Indicating area of Group I Exposures:
SYRIA
IRAQ
SYRIAN DESERT
SAUDI ARABIA
An Nafūd
BAGHDĀD
Karbalā'
An Najaf
Al Ḥillah
Kirkūk
Irbil
Ar Raqqah
Ḥalab (Aleppo)
Ḥims (Homs)
Dimashq (Damascus)
KUWAIT
Al Kuwayt
Al Basrah
Ar Riyāḍ (Riyadh)
Al Madīnah (Medina)
Makkah (Mecca)
Aṭ Ṭā'if
Al Hufūf
Hamadān
Qazvīn

Map of Iraq-Norhtern Saudi Arabia - Indicating area of Group II Exposures:

treat affected Veterans and their families."

***** "We must ensure that those Men and Woman who served this Country during the Gulf War, on active duty, in the reserves, and those who have since left the military services, receive proper medical attention. The National Archives has retained many letters, the unheard pleas and appeals of the veterans who returned home after World War I complaining of illnesses as a result of their exposure to mustard gas. Surely, we cannot tolerate turning a deaf ear on the thousands of veterans who served in the Gulf War. Without proper testing and treatment, their conditions will worsen. They cannot wait, Many are now destitute-their savings spent on medical care not being provided by the Government. Others, unable to work, receive no pension or compensation because the Department of Veterans Affairs is unable to diagnose their illnesses."

***** Senator Donald W. Riegle Jr.: "I believe that this issue needs to be resolved, in order to ensure that our Armed Services are properly prepared for future conflicts that might involve the use of these weapons. I know that you share my concerns, both about the well-being of those who wear the uniforms of the United States Armed Forces, and about the preparedness of this nation to protect its forces in future conflicts."

***** Larry said, "The research done to date has uncovered a great deal of evidence that U.S. Forces, and their families at home were in fact exposed to Chemical, Bio-Chemical, and Biological warfare agents as a result of the Bombings of 18 Chemical, 12 Biological, and 4 Nuclear facilities in Iraq during the Persian Gulf War."

***** "There have been compelling accounts from eyewitnesses, including Chemical Officers, of many events which appear to be best explained as direct Chemical agent attacks.

***** January 27, 1991 82nd Airborne Div. Chemical alert alarms, that actually registered traces of chemicals in the air.

***** "They kept telling everyone, we know the alarms are sounding but there isn't enough of anything to hurt you. They told us they seen a lot of Bunkers that were filled with ammunition, It wasn't Iraqi ammunition, it was American munitions. We blew about a 100 bunkers, wired them with C4, and got about 3 miles away. It was like throwing a lit match into a fireworks factory. It literally blew them to hell. There were Chemical alarm scares, and trace amounts was in the air, we were told there is nothing to worry about. It's probably the manufacturing plants that have been blown up, and the air is filtering this way."

***** There is something I cannot understand for the life of me, I know we supposedly have chemical alarms. But what kind of alarm is available for Biological Warfare. If they were hit with Biological warfare, there was no alarm for it, and no one would have even known if they had been hit with any biological warfare used or were blown up in Iraq, that may have contaminated them.

Statement of Senator Donald W. Riegle, Jr.

Arming Iraq: Biological Agent Exports Prior to the Gulf War

On September 9, 1993, I released a report which suggested that "Gulf War Syndrome," that disabling and sometimes fatal collection of illnesses afflicting thousands of veterans with debilitating muscle and joint pain, memory loss, intestinal and heart problems, fatigue, running noses, urinary and intestinal problems, twitching, rashes, and sores, could have resulted from exposure to chemical and biological warfare agents, either from direct exposure or from the downwind fallout of the coalition bombings of Iraq.

My initial inquiry focused on exposure to chemical agents, due to the many reports of chemical alarms sounding before and during the war and the compelling accounts of eyewitnesses to events which appear to be best explained as chemical agent attacks. Since that time, a number of researchers have contacted my office with a more disturbing proposal.

These researchers believe that the symptoms experienced by these veterans may be the result *not only* of exposure to chemical warfare agents and other environmental hazards, but possibly also as a result of exposure to biological warfare agents.

This is an extremely serious issue with serious consequences, but it may explain the alarming and growing evidence that the illness appears to be spreading to the spouses and children of the affected veterans.

All government agencies and institutions, including the U.S. Congress, have a responsibility to uncover every available lead which might assist medical researchers in discovering the nature and scope of these illnesses. This Administration must defend the health and well-being of its people, especially those who have been willing to lay down their lives for the United States. It has been nearly three years since these young men and woman began suffering, and too many have died.

The Senate Committee on Banking, Housing, and Urban Affairs, which I chair, has oversight responsibility for the Export Administration Act. Pursuant to this Act, Committee staff contacted the U.S. Department of Commerce and requested information on the export of biological materials to Iraq during the years prior to the Gulf War.

After receiving that information, we contacted a principal supplier of these materials to determine what, if any, materials were exported to Iraq which might have contributed to a offensive or defensive biological warfare program.

Records available from the supplier for the period from 1985 until the present show that during this period, pathogenic, meaning "disease producing", toxigenic, meaning "poisonous," and other materials, were exported to Iraq pursuant to application and licensing by the U.S. Department of Commerce. Records prior to 1985 were not available, according to the supplier. These exported biological materials were not attenuated or weakened and were capable of reproduction. Thus, from at least 1985 through 1989, the United States government approved the sale of quantities of potentially lethal biological agents that could have been cultured or grown in large quantities in an Iraqi biological warfare program.

I find it especially troubling that, according to the supplier's records, these materials were requested by and sent to Iraqi government agencies, including the Iraq Atomic Energy Commission, the Iraq Ministry of Higher Education, the State Company for Drug Industries, and the Ministry of Trade.

While there may be legitimate needs for pathogens in medical research, closer scrutiny should be exercised in approving exports of materials to countries known or suspected of having active and aggressive biological warfare programs.

Iraq has long been suspected of conducting biological warfare research, in addition to its chemical and nuclear warfare research programs.

Indeed, according to the Department of Defense's own report to Congress on the Conduct of the Persian Gulf War, written in 1992: "by the time of the invasion of Kuwait, Iraq had developed biological weapons. Its advanced and aggressive biological warfare program was the most advanced in the Arab world… [The] program probably began in the late 1970s and concentrated on the development of two agents – botulinum toxin and anthrax bacteris … Large scale production of these agents began in 1989 at four facilities near

Baghdad. Delivery means for biological agents ranged from simple aerial bombs and artillery rockets to surface to surface missiles."

U.N. inspectors after the war found four facilities involved in biological warfare-related research. While no evidence of production was noted, at least one of those facilities could produce up to 50 gallons of biological agents each week. The United States government approved the export of materials which could have been used to support such a program.

Included in these approved sales are the following biological materials which have been considered by various nations for use in war, with their associated disease symptoms:

Bacillus Anthracis

[BAS-SILL-US ANN-THRA-SIS] or *ANTHRAX* is disease-producing bacteria which was identified by the Department of Defense in *The Conduct of the Persian Gulf War: Final Report to Congress* as being a major component in the Iraqi biological warfare program.

Anthrax is an often-fatal infectious disease due to ingestion of spores; it begins abruptly with high fever, difficulty in breathing, and chest pain. The disease eventually results in septicemia, or blood poisoning, and the mortality rate is high. Once Septicemia is advanced, antibiotic therapy may prove useless, probably because the exotoxins remain, despite the death of the bacteria.

Clostridium Botulinum

[CLAWS-TRI-DEE-UM BOTCH-YOU-LIN-UM], a bacterial source of BOTULINUM TOXIN, causes vomiting, constipation, thirst, general weakness, headache, fever, dizziness, double vision, dilation of the pupils, paralysis of the muscles involving swallowing and is often fatal.

Histoplasma capsulatum

[HIS-TOE-PLAZ-MA CAP-SUL-A-TUM] causes a disease superficially resembling tuberculosis that may cause pneumonia, enlargement of the liver and spleen, anemia or an influenza-like illness and an acute inflammatory skin disease marked by tender red nodules, usually on the shins. Reactivated infection usually involves the lungs, brain and spinal membranes, heart, peritoneum, and adrenals.

Brucella Melitensis

[BREW-CELLA-A MEL-IT-TEN-SIS], a bacteria which can cause chronic fatigue, loss of appetite, profuse sweating when at rest, pain in joints and muscles, insomnia, nausea, and can result in damage to major organs.

Clostridium Perfringens

[CLAWS-TRI-DEE-UM PER-FRIN-JENS] is a highly toxic bacteria which causes GAS GANGRENE. The bacteria produce toxins that move along muscle bundles, killing cells and producing necrotic tissue that is favorable for further growth. Eventually, these toxins and bacteria enter the bloodstream and cause a systemic illness.

In addition, several shipments of E. Coli [COLE-EYE] and genetic materials, human and bacterial DNA, were shipped directly to the *Iraq Atomic Energy Commission.*

I offer this and other specific information on the nature of the materials exported for the use of medical researchers seeking to diagnose and treat the affected veterans and their families.

Today, I am asking the Department of Defense and Department of Veterans Affairs to establish a disability compensation rating for Gulf War veterans consistent with the true extent of their disability and regardless of the ability to arrive at a definitive diagnosis.

I am also asking the Department of Health and Human Services, the Department of Veterans Affairs, and the Department of Defense, and their newly formed task force addressing this issue, to study the reported transmission of these illnesses to the spouses and children of these veterans, and to assess what, if any, public health hazard might exist.

I am asking that the Secretaries of each of these Departments respond to these concerns not later than March 31, 1994.

Over the next several months, the Committee on Banking, Housing, and Urban Affairs will be reviewing the Export Administration Act, which is due for reauthorization. As chairmen of this Committee, I will call hearings to examine the policies that led to the export of these materials as well as the consequences of these policies.

I assure the veterans, their families, and the people of the United States that the policy under which these licenses were granted will be examined and strengthened. The defense of the United States should not be undermined by export policies that allow this government to assist any pariah nation, such as Iraq, in the furtherance of nuclear, chemical, or biological weapons programs.

I ask that the remainder of my statement be inserted into the record as if read and that supporting attachments be inserted into the record in the appropriate place.

"WATCH FOR" DEATH FACTORY U.S.A. PART II will follow soon.

ABOUT THE AUTHOR

It was Larry's dream to one day own the very place he was born and raised. He was born one mile north of the Rocky Mountain Arsenal that had begun manufacturing death gas at the very same time. Little did Larry realize he was on a path that would one day again cross that with the Rocky Mountain Arsenal, but in 1972 it did. Larry felt he had fulfilled his dream when he purchased the home place. It didn't dawn on him that his parents were really run down and had no reason why. Until he drilled a couple of new wells and within just days he had nearly 700 cattle lying dead or near dead, and they suffered from toxic poisons in their water. Larry's parents died long before their time. Larry was broke, he had 1.5 million dollars net worth in livestock, and they were all dead or destroyed by GB and Sarin nerve agents that which had found it's way to the land wells. Larry began an investigation that would last over thirty years. During the investigation Larry discovered there were over two hundred burial dumps for the toxic waste.

Over the thirty-year time that Larry worked to investigate the events surrounding his place of birth, he would encounter many trials and difficulties. Following Larry's legal attempts to 'cease and desist' the state of Colorado files similar actions against the United States Army and Shell Oil Company. After years of effort, teams of lawyers, chemists, geologists, and hydrologists, in 1983 the Shell Oil Company was removed from Colorado and the US Army ordered to clean up the mess. Estimated time to complete the clean up project range as far into the future as 2060.

In 1983 Larry moved to Minnesota to begin anew. Toxic poisons would continue to haunt his future. In 1984, just days before his chance to testify before congress, Larry was a victim of a hit and run by a car swerving over from the opposite side of the road and struck him on his bicycle throwing him to the ground, and then, after coming to a grinding stop. The driver first tried to strangle him, and was over powered by Larry, the man went to his car cursing, Larry thought OH" OH" it's over now. He could invision the man coming back with a gun to end it. Instead he got into his car hit reverse with the tires smoking tried backing over top of him, he was just barely able to clear the wheels as he rolled over and over. As the driver sped out to the street his tires howling, I believe he thought he had ran over Larry. To put an ending to this story, four years later in 1988 when Shell oil was trying to make their insurance companies pay the price for clean up. One insurance company had came across some documents that could shed some light on the hit and run may have been orchestrated by Shell Oil Company, then another memo dated a few days after the hit and run stating "Let the sleeping dog lay." To say they had accomplished all that was necessary for now.

The Washington County Sheriff Department, in Stillwater Minnesota Mr. Land's Employer used a very toxic and hazardous industrial Lead based Bridge Paint that required an oxygenated airline respirators at all times. This was used to paint the interior of our work area without ventilation, using high pressure sprayers. Then in 1990 they completed painting his work area again with a Mercury based paint. The Sheriff's department in which Larry worked poisoned Larry twice more, enacted steps to reduce his physical abilities to perform his job, were affected. Larry over exposed with high levels of fiberglass particulate and then compounded the problem by refusing to work with the now disabled Larry. The two toxic poisonings involved using deadly lead based paint and an alkyd, alkyd, epoxy paint containing high levels Mercury Biocide banned for use in the United States by the EPA. The paint was so toxic it continued to affect Larry for an additional few years. Larry was totally disabled by his employer Washington County, by negligent use of Toxic and Hazardous Materials used in his work place. Then for 2 ½ months while cutting, wrapping and installing fiberglass on all plumbing pipes in the new building, without ventilation. Air samples taken for examination by a laboratory specializing in indoor pollution in the work areas discovered Larry had been over exposed and a Toxicologist from the University of Minnesota testified there was a minimum of 12 to 15% fiberglass for every breath taken.

Printed in the United States
1068100001B/181

9 781410 710857